AF167656

Communications
in Computer and Information Science **2240**

Series Editors

Gang Li, *School of Information Technology, Deakin University, Burwood, VIC, Australia*

Joaquim Filipe, *Polytechnic Institute of Setúbal, Setúbal, Portugal*

Zhiwei Xu, *Chinese Academy of Sciences, Beijing, China*

Rationale

The CCIS series is devoted to the publication of proceedings of computer science conferences. Its aim is to efficiently disseminate original research results in informatics in printed and electronic form. While the focus is on publication of peer-reviewed full papers presenting mature work, inclusion of reviewed short papers reporting on work in progress is welcome, too. Besides globally relevant meetings with internationally representative program committees guaranteeing a strict peer-reviewing and paper selection process, conferences run by societies or of high regional or national relevance are also considered for publication.

Topics

The topical scope of CCIS spans the entire spectrum of informatics ranging from foundational topics in the theory of computing to information and communications science and technology and a broad variety of interdisciplinary application fields.

Information for Volume Editors and Authors

Publication in CCIS is free of charge. No royalties are paid, however, we offer registered conference participants temporary free access to the online version of the conference proceedings on SpringerLink (http://link.springer.com) by means of an http referrer from the conference website and/or a number of complimentary printed copies, as specified in the official acceptance email of the event.

CCIS proceedings can be published in time for distribution at conferences or as post-proceedings, and delivered in the form of printed books and/or electronically as USBs and/or e-content licenses for accessing proceedings at SpringerLink. Furthermore, CCIS proceedings are included in the CCIS electronic book series hosted in the SpringerLink digital library at http://link.springer.com/bookseries/7899. Conferences publishing in CCIS are allowed to use Online Conference Service (OCS) for managing the whole proceedings lifecycle (from submission and reviewing to preparing for publication) free of charge.

Publication process

The language of publication is exclusively English. Authors publishing in CCIS have to sign the Springer CCIS copyright transfer form, however, they are free to use their material published in CCIS for substantially changed, more elaborate subsequent publications elsewhere. For the preparation of the camera-ready papers/files, authors have to strictly adhere to the Springer CCIS Authors' Instructions and are strongly encouraged to use the CCIS LaTeX style files or templates.

Abstracting/Indexing

CCIS is abstracted/indexed in DBLP, Google Scholar, EI-Compendex, Mathematical Reviews, SCImago, Scopus. CCIS volumes are also submitted for the inclusion in ISI Proceedings.

How to start

To start the evaluation of your proposal for inclusion in the CCIS series, please send an e-mail to ccis@springer.com.

Udunna Anazodo · Naren Akash · Moritz Fuchs ·
Celia Cintas · Alessandro Crimi ·
Tinahse Mutsvangwa · Farouk Dako ·
Willam Ogallo
Editors

Medical Information Computing

First MICCAI Meets Africa Workshop, MImA 2024, and First MICCAI
Student Board Workshop on Empowering Medical Information
Computing and Research through Early-Career Expertise,
EMERGE 2024, Held in Conjunction with MICCAI 2024, Marrakesh,
Morocco, October 6, 2024, Revised Selected Papers

 Springer

Editors
Udunna Anazodo 🆔
McGill University
Montreal, QC, Canada

Moritz Fuchs 🆔
Technical University of Darmstadt
Darmstadt, Germany

Alessandro Crimi 🆔
AGH University of Krakow
Kraków, Poland

Farouk Dako 🆔
University of Pennsylvania
Philadelphia, PA, USA

Naren Akash 🆔
International Institute of Information
Technology
Hyderabad, Telangana, India

Celia Cintas 🆔
IBM Research Africa
Nairobi, Kenya

Tinahse Mutsvangwa 🆔
IMT Atlantique Brest
Plouzané, France

Willam Ogallo 🆔
Google Kenya
Nairobi, Kenya

ISSN 1865-0929   ISSN 1865-0937 (electronic)
Communications in Computer and Information Science
ISBN 978-3-031-79102-4   ISBN 978-3-031-79103-1 (eBook)
https://doi.org/10.1007/978-3-031-79103-1

This Springer imprint is published by the registered company Springer Nature Switzerland AG
The registered company address is: Gewerbestrasse 11, 6330 Cham, Switzerland

If disposing of this product, please recycle the paper.

Preface

This volume contains papers presented at the MICCAI Meets Africa and MICCAI EMERGE workshops, held in conjunction with the Medical Image Computing and Computer Assisted Intervention (MICCAI) conference. The events took place during October 6th, 2024, in Marrakech, Morocco.

The papers presented describe cutting-edge research from computational scientists and clinical researchers working on a variety of medical image computing challenges relevant to the African and broader global contexts, as well as emerging techniques for image computing methods tailored to low-resource settings. This compilation offers insights into the latest advances in machine learning, medical imaging analysis, image segmentation, disease detection, diagnosis, and prognosis, with a focus on global health applications and innovations for real-world clinical impact.

The volume is divided into two main part: The first part comprises the paper submissions to the MICCAI Meets Africa workshop, which focused on unique challenges, datasets, and applications in African healthcare. The second part contains a selection of papers from the MICCAI EMERGE workshop, which emphasized novel methods and tools that support medical image computing in low-resource environments, including artificial intelligence applications designed for broader accessibility and deployment.

The aim of the first part was to showcase the latest contributions from the medical imaging community in Africa. It emphasizes research addressing disease detection, diagnosis, and treatment in the African context, along with machine learning models tailored for the region's data. By bringing together researchers, clinicians, and industry stakeholders, this session fostered collaboration to tackle region-specific health challenges through the use of advanced image analysis and AI techniques.

The second part highlighted innovations in medical image computing suited for low-resource settings. These papers explore strategies for enhancing the robustness, efficiency, and usability of machine learning and image analysis tools, thereby making them more accessible and applicable to healthcare systems with limited resources. The authors showcase advancements in algorithmic approaches, annotation strategies, and software tools, contributing to the ongoing efforts to democratize medical imaging technology.

We sincerely hope that this volume will inspire continued research and collaboration in the field of medical imaging, particularly in addressing the unique challenges faced by healthcare systems in diverse environments.

December 2024

Udunna Anazodo
Naren Akash
Celias Cintas
Alessandro Crimi
Tinashe Mutsvangwa

Organization

Organizing Committee

Udunna Anazodo	McGill University, Canada
Harry Anthony	University of Oxford, UK
Celia Cintas	IBM Research Africa, Kenya
Alessandro Crimi	AGH University of Krakow, Kraków, Poland
Farouk Dako	University of Pennsylvania, USA
Camila Gonzalez	Stanford University, USA
Anees Kazi	Massachusetts General Hospital, Harvard Medical School, USA
Benjamin Killeen	Johns Hopkins University, USA
Amar Kumar	McGill University, Canada
Moritz Fuchs	TU Darmstadt, Germany
Yanis Najy Miracoui	Stanford University, USA
Akash Naren	International Institute of Information Technology, India
Ahmed Nebli	Forschungszentrum Juelich, Germany
Tinashe Mutsvangwa	IMT-Atlantique, France
William Ogallo	Google, Kenya
Antonio R. Porras	University of Colorado Anschutz Medical Campus, USA
Amin Ranem	TU Darmstadt, Germany
Constantin Ulrich	German Cancer Research Center (DKFZ), Germany
Advaith Veturi	University of Colorado, Anschutz Medical Campus, USA
Paul Wilson	Queen's University, Canada
Weina Jin	Simon Fraser University, Canada
Anna Zapaishchykova	Brigham and Women's Hospital, USA

Program Committee

Kumar Abhishek	Simon Fraser University, Canada
Roa'a Al-Emaryeen	University of Jordan, Jordan
Mbangula Lameck Amugongo	Namibia University of Science and Technology, Namibia

Contents

First MICCAI Meets Africa Workshop, MImA 2024

Early Detection of Liver Fibrosis

Cheikh Yakhoub Maas[1]([✉]) [iD], Mamadou Bousso[2], Mouhamad Allaya[1,2],
Ousmane Sall[1,2], and Papa Ba Gaye[1,2]

[1] Iba Der Thiam University, Thies, Senegal
cyakhoub.maas@univ-thies.sn

[2] Gastro-Enterology Department, Iba Der Thiam University, Dakar, Senegal
https://www.univ-thies.sn/

Abstract. Liver cirrhosis, a widely prevalent pathology, is characterized
by a insidious progression, often asymptomatic until an advanced stage.
Despite medical advances, this condition remains potentially fatal. Early
detection of hepatic fibrosis is crucial to improve survival rates. However, current screening methods have significant constraints: ultrasound
images, often with insufficient resolution, can complicate interpretation
and lead to misdiagnoses; moreover, non-invasive alternatives such as
FibroTest or FibroScan, while accurate, are expensive and less accessible. Furthermore, the delays associated with shipping samples abroad
for detailed analysis can be detrimental, with speed being a key factor in
managing this disease.

In this context, the application of artificial intelligence in the diagnosis and detection of liver fibrosis has become indispensable. With this
in mind, three new models have been developed: an EfficientNet model
designed for precise identification of ultrasound images, a model based
on Vision Transformer technology for classifying livers into 'healthy' or
'fibrotic' categories, and an occlusion model to identify important areas
on which our vision model relies to classify our images as healthy or
fibrotic livers.

Keywords: Deep Learning · Computer Vision · Models · Images ·
Liver Ultrasound · Cirrhosis · Early Detection

1 Introduction

In the Senegalese context, nearly 85% of the population exhibits at least one
marker of hepatitis B virus (HBV), with a chronic carriage rate of 11%. Chronic
liver diseases (CLD) pose an increasing challenge in global public health. Hepatic
fibrosis is the primary risk factor for mortality associated with liver diseases.
Unfortunately, CLD is often diagnosed at an advanced stage, reducing survival
chances. Therefore, early diagnosis of hepatic fibrosis is crucial to improving
CLD prognosis. It is recommended to assess hepatic fibrosis in patients with risk
factors such as excessive alcohol consumption, metabolic syndrome, obesity, and
diabetes.

U. Anazodo et al. (Eds.): MIImA 2024/EMERGE 2024, CCIS 2240, pp. 3–13, 2025.
https://doi.org/10.1007/978-3-031-79103-1_1

Liver condition assessment methods are divided into two distinct categories: invasive procedures, which require the introduction of instruments or needles into the body to obtain liver tissue samples or accurate information, such as biopsy; and non-invasive methods, such as Fibrotest and FibroScan, which avoid direct intervention inside the body. However, the costs of these exams, especially Fibrotest, are not always affordable for everyone, and the turnaround times for exams conducted abroad can be considerable. Thus, in 2019 in Senegal, these assessments were primarily available through the Health Sciences Faculty of Iba Der Thiam University in Thiès and the Infectious Diseases Department of the National University Teaching Hospital of Fann.

Given the high prevalence of Hepatitis B among patients, abdominal ultra-sounds are regularly prescribed to monitor liver status and detect potential cir-rhosis. However, the resolution of these ultrasounds does not allow for the early detection of liver fibrosis or cirrhosis. It is in this context that a model based on Transformer technology proves promising for improving the diagnosis of liver fibrosis from ultrasound images.

The main objective of this study is therefore to develop a deep learning model based on transformers, capable of accurately classifying liver images based on the degree of liver fibrosis. This model will thus allow a significant advance in the use of ultrasound as a tool for early detection of liver fibrosis.

2   Related Work

The adoption of ViT architectures is currently experiencing rapid growth in the field of medical diagnostics. Numerous studies have significantly demonstrated the exceptional accuracy of these models. This paper provides an in-depth anal-ysis of the deep learning methods employed by researchers for the detection and diagnosis of liver fibrosis. In recent years, interest in the integration of artificial intelligence (AI) in computer-assisted medical decision-making, particularly for liver fibrosis, has significantly increased. This trend is primarily driven by the exponential benefits associated with these technologies.

Nd.P. Diagne et al. [1] applied a transfer learning-based model to identify the presence or potential absence of liver cirrhosis. Their dataset, consisting of 118 images, was split into a training set (60% of the data), a validation set (20%), and a test set (20%). Their transfer learning-based classification model displayed an accuracy of 84%. In comparison, our model achieved an average accuracy of 91.67% and an F1 Score of 93%.

Yasaka et al. [2] demonstrated that a neural network could predict liver fibrosis, achieving an AUC of 0.76 for significant fibrosis (F3) in 286 CT patients, and an AUC of 0.85 using Primovist in 634 patients for F3 fibrosis.

Dandan et al. [3] utilized CNN to extract image features and lightGBM for liver classification, achieving respective accuracies of 82.1%, 85%, and 80.9% for normal, steatotic, and fibrotic liver image classification.

R. Anteby et al. [4] explored the use of deep learning technology for non-invasive assessment of liver fibrosis. The researchers analyzed 40, 405 radiological images from $15,853$ patients, using ultrasounds, CT scans, and MRIs. Most studies reported an accuracy above 85% compared to histopathology.

Ilias Gatos et al. [5] (2017) employed a Support Vector Machine (SVM) on Ultrasound shear wave elastography (SWE) imaging and achieved an accuracy of 87.3% with a dataset of 126 images.

Previous studies on the application of deep learning to the early detection of liver fibrosis are listed in Table 1.

Table 1. Summary of Studies on Liver Fibrosis Detection

Authors	Year	Data Type	Model Name	Accuracy	Number Images
Nd.P. Diagne et al.	2022	Ultrasound image	Transfer learning in CNN	84%	118
Yasaka et al.	2018	CT images	Neural network	AUC 0.76 (F3)	286 (CT), 634 (Primovist)
Dandan et al.	2019	Ultrasound images	CNN + LightGBM	82.1% 85%	79
R. Anteby et al.	2021	Ultrasound images	Neural network	85%	$15,853$
Ilias Gatos et al.	2017	Ultrasound images	SVM	87.3%	126

3   Methodology

3.1   Data Presentation

Our dataset consists of a collection of liver images from both clinical data (Fibroscan) and radiological data (ultrasound images). These images are categorised into five degrees of fibrosis: F_0 (no fibrosis), F_1 (minimal fibrosis), F_2 (moderate fibrosis), F_3 (severe fibrosis) and F_4 (cirrhosis). Initially, these images varied in size, and in order to standardise them, we resized them to 329×375, isolating the part relating to the liver.

Figure 1 shows ultrasound images of a healthy and fibrotic liver, and Fig. 2 depicts the structure of our dataset. We have a total of 118 liver ultrasound images, divided into different categories of fibrosis, with 38 images for F_0, 9 for F_1, 2 for F_2, and 69 for F_4. However, there were no reference images for weaning fibrosis (F_3) and a high prevalence of cirrhosis (F4) among the patients whose images were evaluated.

3.2   Data Preprocessing

The data were classified into two distinct groups: class 0 represents healthy individuals, while class 1 includes patients with significant fibrosis. Given that there are fewer images for stages F_1 and F_2 compared to stages F_0 and F_4, we merged the images from stages F_0 and F_1 into a single class (class 0), while the images from stages F_2 to F_4 were combined into class 1. Class 1 contains 82 images, while the remaining images are assigned to class 0. The training dataset, representing 80% of the data, is used to train the model. The validation dataset,

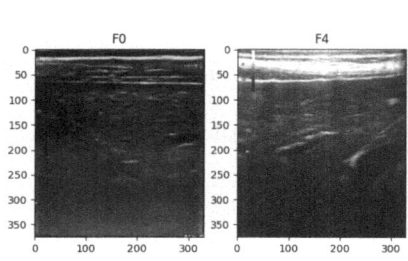

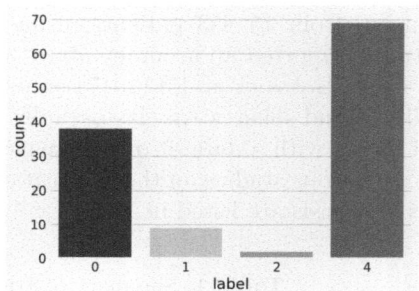

Fig. 1. Ultrasound image of the liver   **Fig. 2.** Data structure

accounting for 20% of the data, is used to evaluate the model's performance during training.

A set of geometric augmentations was applied, including rotation, horizontal and vertical flips, cropping, and resizing to dimensions of 224 × 224. Additionally, random adjustments to brightness and contrast were made, along with the application of ColorJitter and Gaussian blur. Perceptual analytic transformations, affine transformations, and a custom lambda function with a 50% probability were also applied randomly. A binary value of "1" was assigned to images labeled as fibrosis, while a value of "0" was given to normal liver images. Data augmentation was performed exclusively on the training set.

3.3   Vision Transformer Model

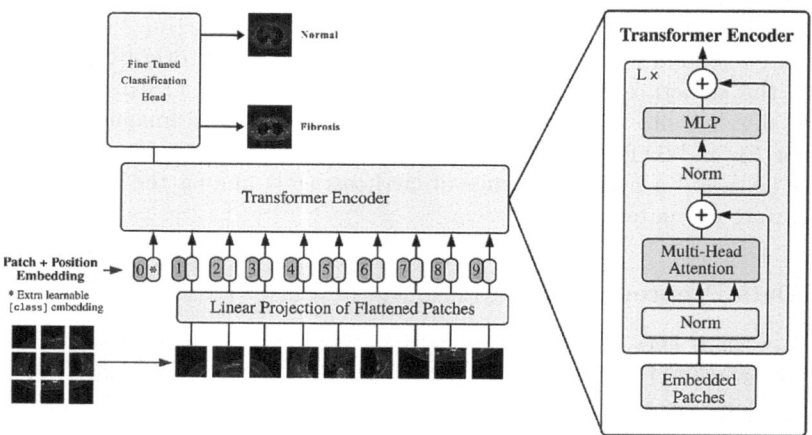

Fig. 3. Vision Transformer architecture

ViT Fig. 3 is designed to process images using attention mechanisms and linear transformations. The Vision Transformer (ViT) architecture divides the image into small square patches. These patches are then transformed into feature vectors using linearisation. Position embeddings are added to the feature vectors to encode the spatial information of the patches. Using a Transformer encoder, the model applies attention calculations to the feature vectors along with the position embeddings. The Transformer encoder is composed of multi-headed self-attention modules and fully connected neural networks (MLPs). Finally, the output of the encoder is sent to a linear classifier to predict the class labels of the image.

In this section, the objective of the ViT model is to determine whether the image provided represents a healthy or fibrotic liver.

3.4   Efficientnet Model

EfficientNet is a neural network architecture Fig. 4 that seeks to maximise performance while maintaining resource efficiency, making it a popular choice for tasks such as image classification. This is achieved using a technique called 'scaling'. Instead of simply increasing the size of all network components, EfficientNet intelligently adjusts the width, depth and resolution of the layers. This allows the network to become more powerful without becoming excessively complex.

In this section, the objective of the EfficientNet model is to determine whether the image provided represents an ultrasound scan of the liver or not. The data for the binary classification has been divided into two classes. Class 0 contains 118 ultrasound images of the liver, while class 1 contains 118 different images that are not related to the liver. For the experiment, we used a training set representing 70% of the data and a test set to evaluate performance, representing the remaining 30%. For image pre-processing, we resized them to an input size of (224, 224) pixels.

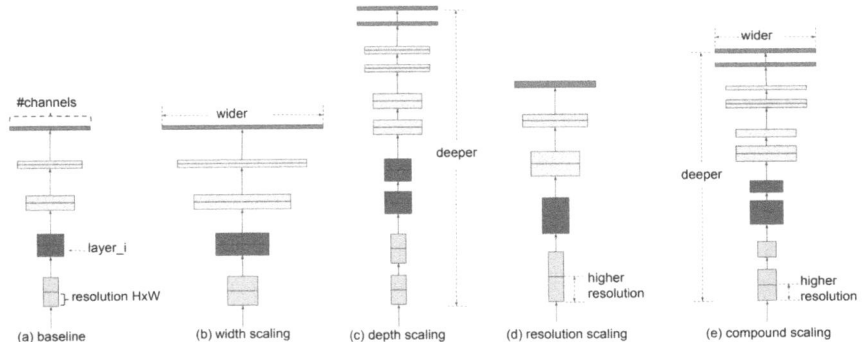

Fig. 4. EfficientNet architecture

3.5   Visualization and Interpretability

Deep learning networks are often seen as "black boxes" because it's not always clear why a network makes a certain decision. To make the network's behavior more understandable, we use interpretability techniques to translate its outputs into information that humans can interpret. These techniques allow us to answer questions about the network's predictions by providing visual explanations of what the network "sees". In our study, we will seek to identify important areas on which our vision model relies to classify our images as healthy or fibrotic liver.

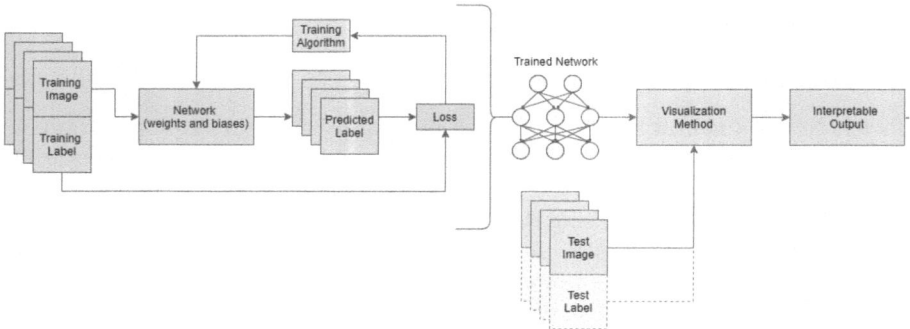

Fig. 5. Occlusion model architecture.

Occlusion Sensitivity. Figure 5 is a straightforward technique to measure the sensitivity of the network to small perturbations in the input data. This method disrupts small areas of the input by replacing them with an occlusion mask, typically a gray square. The mask moves across the image, and the change in the probability score for a given class is measured.

4   Results

In Fig. 6, the graph depicts the evolution of the error on the training and test sets as a function of the number of epochs. There is a trend of error reduction until reaching a nearly zero level. This decrease in error is closely linked to a precision rate of 98% on the test data, highlighting the model's effectiveness in distinguishing between liver ultrasound images and others.

These results are achieved using the following optimal hyper parameters: a batch size of 32 images per batch, for a total of 30 epochs, with a learning rate of 0.0001.

Figure 7 illustrates the progression of the F1 score on the training and test sets as a function of the number of epochs.

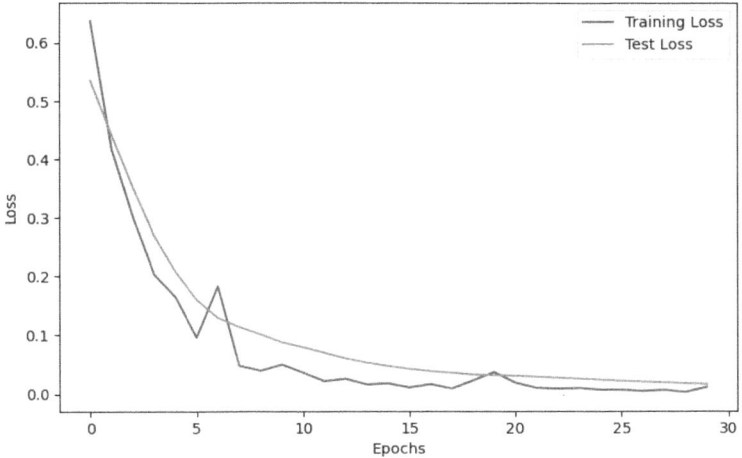

Accuracy - Train: 0.99, Test: 0.98
F1 Score - Train: 0.99, Test: 0.98

Fig. 6. Results for liver ultrasound image detection.

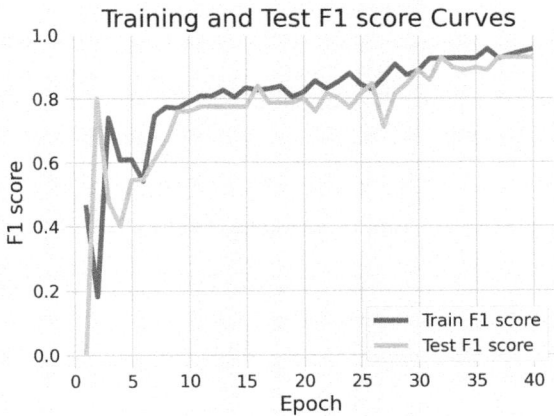

Test loss: 0.37
Test accuracy: 91.67%
Test AUROC: 0.91
Test F1 score: 0.93

Fig. 7. Results for classification as healthy or fibrosed liver.

EARLY DETECTION OF LIVER FIBROSIS

The classification model based on the Vision Transformer is used to determine
whether an image represents a healthy or fibrotic liver.

Insert Image

Drag and drop file here
Limit 200MB per file • JPG, JPEG, PNG, JPG, BMP, PNG   Browse files

foirel.jpg 14.4KB   ×

Input image   Invalid image:

                                     The provided image does not correspond
                                     to a liver ultrasound.

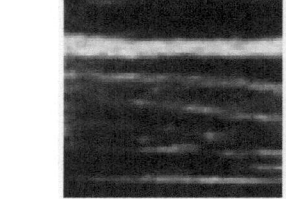

Fig. 8. Invalid Image   **Fig. 9.** Example of occlusion of a healthy
                                   liver.

Following the application of the skorch functions and Grid Search CV on our
Vision Transformer model, we identified the following optimal hyper parameters:
batch size of 2, learning rate of 0.0004, Hidden dimension of 8, 2 blocks in the
model, 4 attention heads and 3 patches per side of the image. The search process
required 46 minutes.

By retraining our model with these optimal hyper parameters for a period
of 40 epochs, we observed improved performance, as evidenced by the following
metrics: a test loss of 0.37, a test accuracy of 91.67%, a test AUROC of 91%,
and a test F1 score of 93%.

Due to the imbalance of classes in our dataset, we opted for the F1 score
curve as the optimal metric to evaluate the model's performance. It is observed
that the training and test F1 score curves converge towards the value of 1.

Figure 9 illustrates the result of applying the occlusion technique to an image
of a healthy liver, thereby highlighting the most important parts of the image
for classification.

We deployed our two models in a Streamlit application. The application
works as follows: If the user inputs an image that isn't a liver ultrasound, EfficientNet predicts it as invalid Fig. 8. However, if the image is valid, EfficientNet
confirms its validity, and the Vision Transformer model then predicts whether
the liver is healthy Fig. 10 or fibrotic Fig. 11.

We have also tested our models using publicly available data to ensure their
robustness and accuracy.

5 Discussion

Table 2 summarizes the different models tested for liver image detection. The
EfficientNet stands out for its flexibility and lightweight nature. Its trained model
has a size of 15 MB, with a training time of 5 min and a deployment time of
2 min. In contrast, the ResNet model is more massive, making its deployment
more challenging. Its trained model has a size of 91 MB, a training time of 5 min,
and a deployment time of 45 min. The number of parameters is approximately

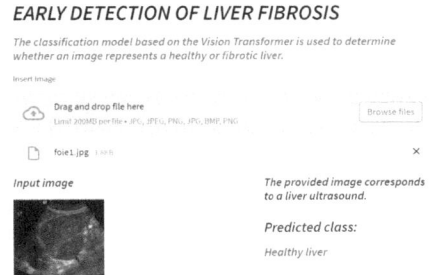

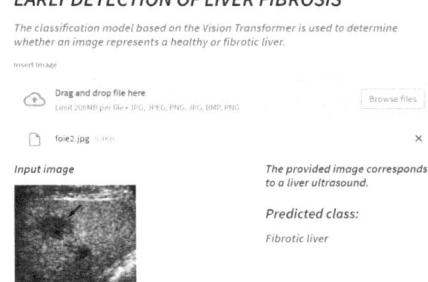

Fig. 10. Liver predicted healthy

Fig. 11. Liver predicted fibrotic

5, 288, 548 for the EfficientNet and 25, 557, 032 for the ResNet-50. Table 3 summarizes the different models tested for classification as healthy liver or fibrosis. The Vision Transformer outperformed the other models with the highest accuracy of 91.67% and an F1 score of 93%, indicating strong overall performance. ResNet-50 also performed well, achieving an 84% accuracy and matching F1 score, making it a solid alternative but less effective than the Vision Transformer. EfficientNet, while still competent, lagged behind with a 78% accuracy and an 82% F1 score, suggesting it may not be as suitable for this task compared to the other models. The Vision Transformer model proves to be more efficient and flexible than other models. Its trained model has a size of 960 KB and a training time of 5 min.

Table 2. Performance of different models for detecting liver ultrasound images.

MODELS	ACCURACY	F1 SCORE
EfficientNet	98%	98%
ResNet-50	98%	98%
Vision Transformer	86%	85%

Table 3. Performance of different models for classification as healthy liver or fibrosis.

MODELS	ACCURACY	F1 SCORE
Vision Transformer	91.67%	93%
ResNet-50	84%	84%
EfficientNet	78%	82%

6   Conclusion

In conclusion, our paper presents a solution for the early detection of liver cirrhosis, a serious disease often diagnosed too late. We utilized artificial intelligence, specifically an EfficientNet model to detect ultrasound images and a Vision Transformer model to identify whether livers are healthy or affected by fibrosis from these images.

The results demonstrate a success rate of 98% for the first model and an F1 score of 93% for the second model on test data. This study highlights the superior performance of the Vision Transformer in accurately identify and distinguish liver fibrosis from ultrasound images, potentially facilitating early and effective detection of liver cirrhosis.

Future goals include expanding our dataset with more ultrasound images and identifying different levels of fibrosis (no fibrosis, minimal fibrosis, moderate fibrosis, severe fibrosis, cirrhosis), while enhancing the accuracy of our ViT model. We also aim to enhance the interpretability of our model by using occlusion-based techniques to better identify and highlight critical areas within the images, a work that is currently in progress.

References

1. Diagne, Nd.P., et al.: Help in the early diagnosis of liver cirrhosis using a learning transfer method. In: 2nd Pan-African Artificial Intelligence and Smart Systems Conference (PA-AISS 2022) (2022)
2. Yasaka, K., et al.: Deep learning with convolutional neural network for differentiation of liver masses at dynamic contrast-enhanced CT: a preliminary study. Radiology **287**(1), 146–155 (2017). https://doi.org/10.1148/radiol.2017170706
3. Dandan, L., et al.: Classification of diffuse liver diseases based on ultrasound images with multimodal features. In: 2019 IEEE International Instrumentation and Measurement Technology Conference (I2MTC), pp. 1–5 (2019). https://doi.org/10.1109/I2MTC.2019.8827174
4. Anteby, R., et al.: Deep learning for noninvasive liver fibrosis classification: a systematic review. Liver Int. (2021). https://doi.org/10.1111/liv.14966
5. Gatos, I., et al.: A machine-learning algorithm toward color analysis for chronic liver disease classification, employing ultrasound shear wave elastography. Ultrasound Med. Biol. **43**(9), 1797–1810 (2017). https://doi.org/10.1016/j.ultrasmedbio.2017.05.002
6. Shamrat, F.J.M., et al.: High-precision multiclass classification of lung disease through customized MobileNetV2 from chest X-ray images. Comput. Biol. Med. **155**, 106646 (2023). https://doi.org/10.1016/j.compbiomed.2023.106646
7. Teramoto, A., et al.: Automated classification of idiopathic pulmonary fibrosis in pathological images using convolutional neural network and generative adversarial networks. Diagnostics **12**, 3195 (2022). https://doi.org/10.3390/diagnostics12123195
8. Syed, A.H., et al.: Deep transfer learning techniques-based automated classification and detection of pulmonary fibrosis from chest CT images. Processes **11**, 443 (2023). https://doi.org/10.3390/pr11020443

9. Dosovitskiy, A., et al.: An image is worth 16×16 words: transformers for image recognition at scale. arXiv preprint arXiv:2010.11929 (2020). https://doi.org/10.48550/arXiv.2010.11929

10. Christe, A., et al.: Computer-aided diagnosis of pulmonary fibrosis using deep learning and CT images. Investig. Radiol. **54**, 627 (2019). https://doi.org/10.1097/RLI.0000000000000574

11. Ambita, A.A.E., Boquio, E.N.V., Naval, P.C.: COViT-GAN: vision transformer for COVID-19 detection in CT Scan images with self-attention GAN for data augmentation. In: Farkaš, I., Masulli, P., Otte, S., Wermter, S. (eds.) ICANN 2021. LNCS, vol. 12892, pp. 587–598. Springer, Cham (2021). https://doi.org/10.1007/978-3-030-86340-1_47

12. Kothadiya, D., et al.: Attention based deep learning framework to recognize diabetes disease from cellular retinal images. Biochem. Cell Biol. 1–12 (2023). https://doi.org/10.1139/bcb-2023-0151

13. Al Rahhal, M.M., et al.: COVID-19 detection in CT/X-ray imagery using vision transformers. J. Pers. Med. **12**(2), 310 (2022). https://doi.org/10.3390/jpm12020310

14. Decharatanachart, P., et al.: Application of artificial intelligence in chronic liver diseases: a systematic review and meta-analysis. BMC Gastroenterol. **21**(1), 10 (2021). https://doi.org/10.1186/s12876-020-01585-5

15. Wang, J., Wei, L., Wang, L., Zhou, Q., Zhu, L., Qin, J.: Boundary-aware transformers for skin lesion segmentation. In: de Bruijne, M., et al. (eds.) MICCAI 2021. LNCS, vol. 12901, pp. 206–216. Springer, Cham (2021). https://doi.org/10.1007/978-3-030-87193-2_20

Optimized Brain Tumor Segmentation for Resource Constrained Settings: VGG-Infused U-Net Approach

Mizanu Zelalem Degu[1]([✉]) [iD], Confidence Raymond[2,3] [iD], Dong Zhang[4] [iD],
Amal Saleh[5], Udunna C. Anazodo[2,3,6,7] [iD], and Gizeaddis Lamesgin Simegn[8] [iD]

[1] Faculty of Computing and Informatics, Jimma University, Jimma, Ethiopia
mizanu143@gmail.com
[2] Medical Artificial Intelligence Laboratory (MAI Lab), Lagos, Nigeria
[3] Lawson Health Research Institute, London, ON, Canada
[4] Department of Electrical and Computer Engineering, University of British Columbia,
Vancouver, Canada
[5] School of Medicine, College of Health Sciences, Addis Ababa University, Addis Ababa,
Ethiopia
[6] Montreal Neurological Institute, McGill University, Montréal, Canada
[7] Department of Clinical and Radiation Oncology, University of Cape Town, Cape Town,
South Africa
[8] School of Biomedical Engineering, Jimma University, Jimma, Ethiopia

Abstract. Semantic segmentation of brain tumors plays a crucial role in assisting treatment by precisely delineating tumor boundaries in brain images. This aids clinicians in formulating surgical plans and targeted therapies, ultimately enhancing patient outcomes. Automatic brain tumor segmentation poses a significant challenge due to its complexity and resource-intensive nature, particularly in low-resource settings where access to high quality data and expertise is limited, hindering effective medical image analysis and treatment planning. In this study, the complex multi-label 3D segmentation problem was simplified into more manageable 2D single-label segmentation tasks and a U-Net architecture was adapted, with the integration of a pre-trained VGG19 model to reduce computational demands while effectively extracting features. Six models were trained in total, three using the conventional U-Net and three employing the VGG-infused U-Net (VIU-Net). The Experiment was performed using a total of 1424 magnetic resonance images. The study shows that the VIU-Net model achieved lower dice losses of 0.188, 0.068, and 0.106 for the segmentation of Necrotic Core, SNFH, and Enhancing Tumor regions respectively, showcasing the effectiveness of our method and it's suitability for implementation in low resource settings.

Keywords: Automatic Segmentation · Brain Tumor · Optimized U-Net · Low Resource Setting

U. Anazodo et al. (Eds.): MImA 2024/EMERGE 2024, CCIS 2240, pp. 14–23, 2025.
https://doi.org/10.1007/978-3-031-79103-1_2

1 Introduction

A brain tumor is an anomalous mass of tissue characterized by the uncontrolled and unrestrained proliferation of cells. In this condition, cellular growth and multiplication occur without apparent restraint from the regulatory mechanisms that typically control normal cells. This unregulated growth can lead to the formation of a mass within the brain, potentially impacting its normal functions and necessitating specialized medical attention and intervention [1].

Brain tumors, particularly gliomas, are among the deadliest cancers, with 80% of individuals diagnosed with glioma dying within two years. In contrast, over 90% of those diagnosed with prostate or breast cancer survive for at least five years. In Sub-Saharan Africa, glioma death rates have risen by 25%, sharply contrasting with the 30% declining rates in the Global North [2]. While earlier studies indicated a relatively lower number of cases in developing countries, recent investigations focusing on low- and middle-income countries, particularly in sub-Saharan Africa (SSA), reveal the substantial prevalence of brain tumors in the region [3, 4]. According to recent studies [2] the death rate from glioma in SSA are among the highest in the world and continuing to rise. Delayed presentation, incident of other infectious disease such as HIV, shortage of health facilities, low level of income is among the contributors [5]. In response to this intricate pathology, an accurate diagnosis technique is essential for successful treatment planning, and magnetic resonance imaging (MRI) has become the principal imaging modality for diagnosing brain tumors and their extent.

Brain Tumor segmentation from MR images is essential in standard of care, enabling the clinicians to identify tumor location, extent, and subtypes. This not only helps with initial diagnosis but also aids with administering and monitoring treatment progress. Given the critical nature of this task, the precise demarcation of the tumor and its sub-regions is traditionally carried out manually by experienced neuro-radiologists. This manual process is inherently challenging, not only due to its tedious nature but also because it demands the expertise of highly experienced professionals. The complexity arises from multiparametric image contrast, the presence of heterogeneous tumors, and the reliance on labeling that is subject to both inter and intra-rater variability [6–8].

Developing computer-aided decision support systems, particularly those involving automatic semantic segmentation and classification tailored for low resource settings such as Sub-Saharan Africa countries (SSA), is challenging due to the scarcity of human resources, infrastructure, insufficient funding for cancer research and an insufficient number of trained personnel (neuroradiologists, oncologists, and physicists) [3, 9, 10]. To this end, the aim of this study is to develop a brain tumor semantic segmentation model with minimal resource by breaking down the demanding 3D multilabel segmentation problem into more manageable 2D tasks, where tumor features of the image can be effectively extracted. This will enable accessible brain tumor segmentation approach suitable for resource-constrained environments.

2   Related Works

Before the availability of sophisticated computational devices and large dataset, attempts on automatic brain tumor segmentation were limited to handcrafted feature extraction and shallow machine learning algorithms. Since its establishment, the brain tumor segmentation (BraTS) [11] Challenge has served as a significant motivator for the development of state-of-art algorithms for tumor segmentation. The challenge has been providing a community benchmark dataset and environment for adult glioma over the past 11 years [12]. Almost all the proposed models for the challenge used deep learning models as a backbone [13–16]. In recent challenges, a specific deep learning model, U-Net and its modified versions, are being used in addressing the brain tumor segmentation problem. U-Net is a deep learning model based on convolutional neural network specifically designed to better address segmentation problems in biomedical computer vision [16].

In recent studies, an extension or modification of the U-Net architecture has been applied, involving the integration of a 3D U-Net with an additional variational decoder branch. This enhancement is designed to provide extra supervision and regularization to the encoder branch [15]. A two-stage cascaded U-Net was also proposed for segmentation [13]: the first stage was trained to produce coarse segmentation mask and the second stage was trained to refine the output of the first stage. The nnU-Net [17], a self-configuring framework that automatically adapt U-Net to diverse datasets was proposed to streamline the process of medical image segmentation. The winner of BraTS-2021 [6] suggested utilization of large nnU-Net model with group normalization, and axial attention in the expanding path of the network.

Although previous methods on brain tumor segmentation produced promising performance, these models were trained using high quality images captured in resource-rich settings and higher resolution multiparametric (mpMRI) techniques, resulting in unreliable and decreased performance in low-resource settings [18]. In the context low-resource settings, medical image acquisition presents distinct challenges characterized by limited expertise, unoptimized imaging protocols, and the shortage of imaging infrastructure, leading to low-quality images with limited contrast and resolution. Moreover, fine-tuning large models equipped with multiple automated augmentation modules is computationally expensive and nearly impractical in these settings.

3   Methods

3.1   Dataset

The dataset used for training the models was acquired from the BraTS-Africa Challenge [2]. A total of 60 adult glioma cases collected retrospectively from clinical studies in SSA were used in this study. Details of the data collection and image pre-processing are described in the BraTS-Africa Challenge [5]. Briefly, each case consisted of four pre-processed three-dimensional (3D) mpMRI volumes (T1-weighted (T1w), post gadolinium contrast T1 weighted (T1ce), T2-weighted (T2w) and T2-FLAIR (FLAIR)) with one multi-labelled tumor binary mask, comprising of three sub-regions (enhancing tumor, surrounding non-enhancing FLAIR hyperintensity (SNFH), and non-enhancing tumor core) [5].

3.2 Dataset Processing

All data were further processed before model training. Firstly, all image and mask pairs were cropped to remove redundant background voxels limiting noise and computational cost of training the model. To clip voxel values and improve ease of network convergence, we performed the min-max normalization. Then, generated two-dimensional (2D) image slices from the 3D isotropic ($1mm^3$) image volumes by taking one slice at a time iteratively along the sagittal view where the quality of the image is better relative to trans axial or coronal views. Each iteration of the slicing produced 9,856 number of $128 \times 128 \times 1$ images for each T1CE, T2w, and FLAIR volumes. The T1w volume was not resliced to 2D since the T1CE with the same dimensions, was already included. Individual slices from the three volumes were then added together to produce a 2D merged image. During the slicing process, we intentionally excluded certain 2D frames to ensure a balanced representation of segmented and non-segmented regions. Specifically, we maintained a ratio of at least 5% segmented regions to non-segmented regions. Finally, a total of 1,424 2D images were obtained from slicing the 60 3D volumetric datasets and used in training and evaluating the model. The generation of 2D images from the 3D label data (Mask image) followed a similar process. The multilabel mask images were then split into three single-class masks, facilitating the training of a single-class segmentation model. Figure 1 illustrates the 3D to 2D data decomposition workflow.

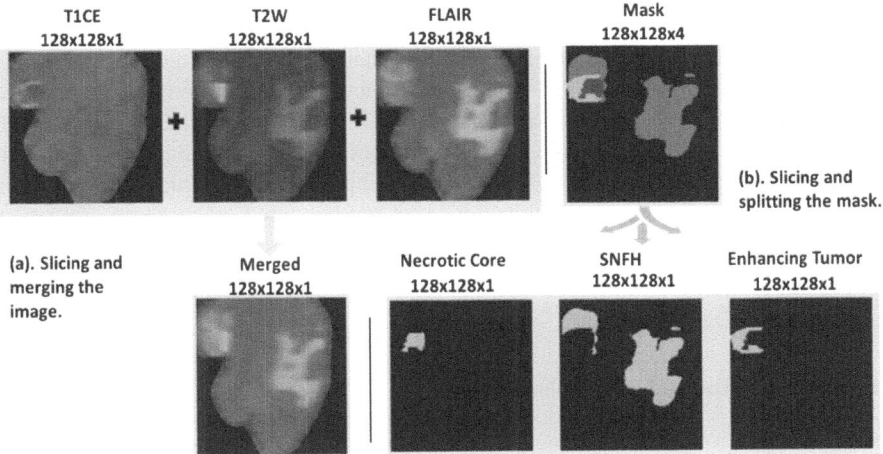

Fig. 1. The 3D to 2D data decomposition workflow illustrated using one case.

3.3 The Segmentation Models

With the objective of designing an efficient brain tumor segmentation model for a low-resource setting scenario, characterized by a small dataset, and limited computational.

resources, the strengths of the U-Net model, which was initially designed for semantic segmentation tasks, was combined with the capabilities of a pretrained image classification model - VGG19, in extracting features. Consequently, our proposed model,

VGG-infused U-Net model (VIU-Net)[1], takes the form of a 2D U-Net with a VGG component serving as the feature extraction enhancement module. Beyond feature extraction another silent novelty and strength of this work is in the deployment of VGG19, which is deployed at a very vital position in the U-Net model (Skip connection) where it helps denoise high dimension features to further improve the performance of the model [19, 20]. The VGG module is employed on the last skip features, which represent the output of the final down-sampling layer. The choice of utilizing VGG over other available pretrained models, such as ResNet [21] and EfficientNet [22], is due to its structural simplicity that allows the extraction of specific parts to be integrated with the U-Net architecture. Following the VGG19 block, a 2D up-sampling layer was added to ensure the output tensor shape aligns with the shape of the first up sampling layer for the concatenation. The architectural detail of the VIU-Net model is summarized in Fig. 2.

3.4 Model Training

In this step, separate 2D segmentation models were trained for each of the three masks (Necrotic Core, SNFH, and Enhancing Tumor). A comparative analysis was conducted between the proposed model and the default U-Net model, resulting in a total of six trained models: three default U-Net models and three VIU-Net models. To maintain consistency for ease of comparison, identical hyperparameters were used for all models. The batch size across all models was set at 64, constrained by memory limitations. Training extended over 200 epochs, incorporating an early stopping technique with a patience value of 3. Gradient optimization employed the Adam optimization technique, initializing with a learning rate of 0.0001. The average Dice Loss and Dice Coefficient values were calculated for the training loss function and for model evaluation, respectively. The calculations are based on the standard computation in segmentation tasks of the similarity between the multi-label segmentation output from the models and the binary tumor masks [23]. The model architecture was implemented using Keras framework with TensorFlow 2.11.0, with CUDA 11.2 on an NVIDIA Tesla T4 GPU.

4 Results

4.1 Results of the Default U-Net Model

The default U-Net model was trained as the baseline for our segmentation problem. Three U-Net models, for each segmentation labels (Necrotic Core, SNFH, and Enhancing Tumor) were trained on 997 of the dataset (70%; training data) and evaluated on the remaining 427 (30%; validation data) dataset.

A dice coefficient of 77.64%, 90.17%, and 83.54 and dice losses of 0.22, 0.098, and 0.16 were obtained for segmenting the Necrotic Core, SNFH, and Enhancing Tumor regions, respectively (Table 1). Figure 3 shows the training and validation curves of the model.

---

[1] The VIU-Net code can be found at [https://github.com/CAMERA-MRI/SPARK2023/VIU-Net].

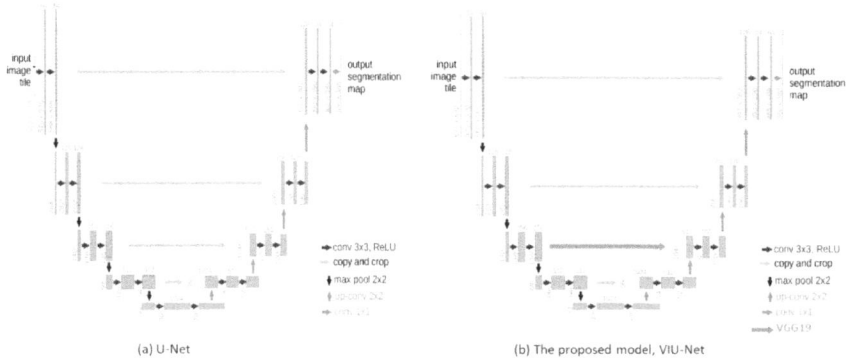

(a) U-Net   (b) The proposed model, VIU-Net

Fig. 2. (a) Architecture of the original U-Net model used with default settings [17] and (b) the architecture of the proposed VGG-Infused U-Net (VIU-Net) model introduced here.

Table 1. Dice Coefficient (Coeff) and Dice loss of the default U-Net and VGG-Infused U-Net (VIU-Net)

Model	Necrotic Core		SNFH		Enhancing Tumor	
	Dice Coeff	Dice Loss	Dice Coeff	Dice Loss	Dice Coeff	Dice Loss
U-Net	77.64%	0.22	90.17%	0.09	83.54%	0.16
VIU-Net	81.31%	0.19	93.16%	0.06	89.39%	0.11

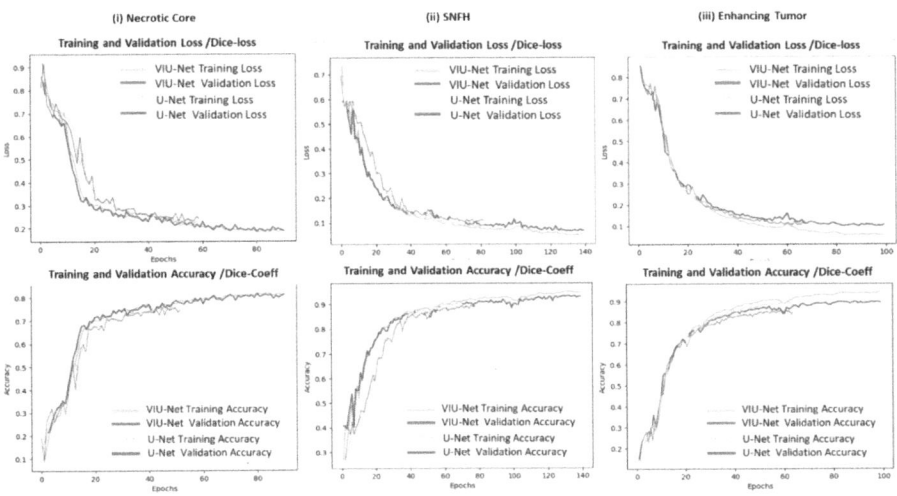

Fig. 3. Training history plot of the U-Net and VIU-Net model.

4.2   Results of the Proposed Model (VGG-Infused U-Net (VIU-Net))

The three VIU-Net models for each segmentation labels were trained and validated on the same dataset at the default U-Net models. The dice coefficients of 81.31%, 93.16%, and 89.39% and dice losses of 0.188, 0.068, and 0.106 were obtained for Necrotic Core, SNFH, and Enhancing Tumor regions, respectively (Table 1). Figure 3 shows the training and validation curves of the proposed model. As shown in Table 1, the proposed VIU-Net model in general, produced higher Dice coefficient and lower Dice losses compared to the default U-Net model. An illustration on a single case (Fig. 4), shows better representation of the predicted tumor sub-region masks using the VIU-Net model compared to the default U-Net model.

Table 2. Comparison of the performance the proposed approach with other reported models for two-dimensional (2D) Brain Tumor segmentation.

Study	Dataset	Data size	Method	Dice Coefficient		
				ET	NC	SNFH
Tseng et al. [24]	BraTs 2015	274	Fused based attention	0.69	0.68	0.85
Li et al. ([25]	BraTs 2017	285	Pair + Fusion	0.75	0.71	0.88
Wang et al. ([26]	BraTS 2018	285	Feature fusion	0.89	-	-
Liu et al. [27]	BraTS 2020	285	Attention based feature fusion	0.76	0.8	0.88
The proposed Model, VIU-Net	BraTS- Africa 2023	60	VGG infused U-Net	0.89	0.81	0.93

Data size is the number of 2D images in the training data used by each study. Enhancing Tumor (ET), Necrotic Core (NC), surrounding non-enhancing FLAIR hyperintensity (SNFH).

5   Discussion

Image segmentation plays a pivotal role in understanding the extent and nature of brain tumors for further treatment decisions. In low-resource settings, where data and computational resources are constrained, the challenge for accurate brain tumor segmentation increases. This study explores the importance of incorporating pre-trained models into the architecture of segmentation models, with a specific emphasis on their effectiveness in improving the performance of models for accurate brain tumor segmentation in environments with limited resources.

The inherent ability of pre-trained models to capture features in images can enhance the performance of underlying segmentation models. A VGG19 pretrained model was integrated with a U-Net architecture for this purpose. The selection of VGG19 was motivated by its simplicity, enabling the extraction and seamless integration of internal components into the U-Net. This approach is particularly advantageous compared to

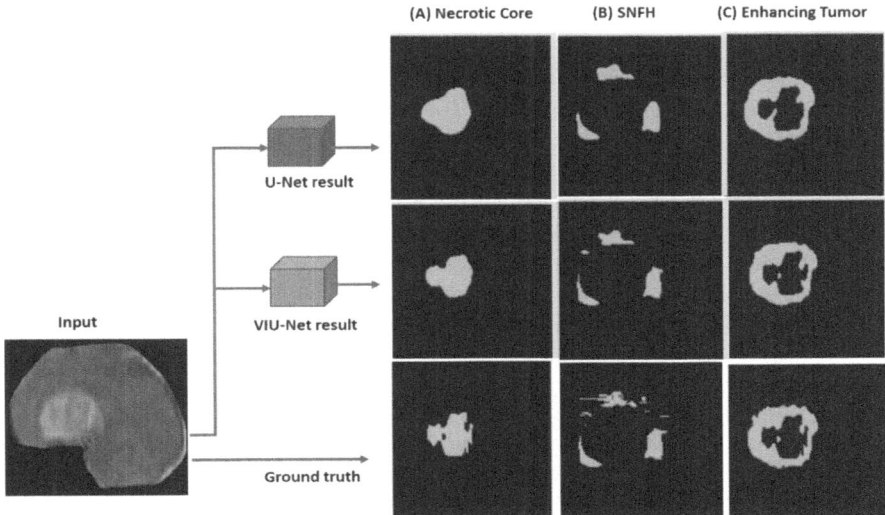

Fig. 4. Comparison the U-Net and the VIU-Net with respect to the ground truth on one slice in a single case.

more complex models like ResNet [21] or EfficientNet [22], where interconnections between layers can pose challenges for selective integration. The data pre-training processing step of decomposing the 3D images into 2D and the transformation of multilabel segment masks into its single-label component masks, enabled focused feature extraction of the tumor sub-regions. Consistent hyperparameters were maintained across all models to facilitate a fair and meaningful comparisons.

While the U-Net model exhibited commendable performance, the proposed model outperformed the baseline model in producing more detailed segmented regions for all three sub-regions, at greater accuracy. This demonstrates how insertion of the pretrained image classification model, VGG19, helps the U-Net to capture complex features and return better accuracy in limited number of datasets, although at nearly twice the training speed (number of epochs). The VGG infused U-Net also produced comparable results to others reported methods were 2D brain tumor segmentation was implemented with much larger datasets (Table 2), even outperforming some methods, suggesting that the proposed model may have promising use for automated brain tumor segmentation in lower resource settings.

We acknowledge that the study's focus on low-resource settings may fully address challenges in these settings, particularly relating to data quality or computational resources. However, these challenges are unmet needs, affecting up to 80% of the world's population where medical imaging capacity is still severely limited and for which further research to explore techniques that can effectively mitigate these constraints is necessary. In addition, the proposed model's generalizability to diverse datasets and its robustness under varying conditions or tumor types requires further investigation.

6 Conclusion

This study aimed to address the challenges of brain tumor segmentation, specifically in low-resourced settings, by proposing a U-Net architecture enhanced with a VGG19 pretrained model. The integration of a VGG19 pretrained model with a 2D U-Net architecture presents a viable solution for improving brain tumor segmentation in resource-constrained environments. The decomposition of higher dimensional data to smaller subsets made segmentation of complex data more feasible in constrained settings. In general, the proposed VGG infused U-Net outperformed the baseline U-Net as well as other previously reported methods developed on much larger dataset. These results contribute valuable insights into the development of efficient segmentation models for medical imaging that have potential real-world applications across many clinical settings.

Acknowledgement. . The author would like to thank the Lacuna Fund for Health and Equity (PI: Udunna Anazodo, 0508-S-001), the National Science and Engineering Research Council of Canada (NSERC) Discovery Launch Supplement (PI: Udunna Anazo-do, DGECR-2022–00136), and the Digital Research Alliance of Canada (the Alliance) for their support. The McGill University Doctoral Internship Program and McMedHacks, Linshan Liu, Lab Manager at the MiND Lab, Montreal Neurological Institute, McGill University, are also thanked for their invaluable contributions. Finally, sincere appreciation is expressed to the SPARK Academy 2023 Instructors and assistants for their dedication and hard work in providing us with valuable training.

References

1. AANS. Brain Tumors (2023). https://www.aans.org/en/Patients/Neurosurgical-Conditions-and-Treatments/Brain-Tumors. Accessed 21 Dec 2023
2. Adewole, M., et al.: The Brain Tumor Segmentation (BraTS) Challenge 2023: Glioma Segmentation in Sub-Saharan Africa Patient Population (BraTS-Africa). PubMed (2023)
3. Kanmounye, U.S., et al.: Adult brain tumors in Sub-Saharan Africa: a scoping review. Neuro Oncol. **24**(10), 1799–1806 (2022)
4. Mbi Feh, M.K.N., Lyon, K.A., Brahmaroutu, A.V., Tadipatri, R., Fonkem, E.: The need for a central brain tumor registry in Africa: a review of central nervous system tumors in Africa from 1960 to 2017 (2021)
5. Oprea, C., et al.: Brain opportunistic infections and tumors in people living with HIV—still a challenge in efficient antiretroviral therapy era. J. NeuroVirol. **29**(3), 297–307 (2023)
6. Luu, H.M., Park, S.H.: Extending nn-UNet for brain tumor segmentation (2021)
7. Rios Piedra, E.A., Taira, R.K., El-Saden, S., Ellingson, B.M., Bui, A.A.T., Hsu, W.: Assessing variability in brain tumor segmentation to improve volumetric accuracy and characterization of change. In: IEEE-EMBS International Conference on Biomedical and Health Informatics (2016)
8. Visser, M., et al.: Inter-rater agreement in glioma segmentations on longitudinal MRI. NeuroImage: Clin. **22**, 101727 (2019)
9. Njei, B., et al.: Artificial intelligence for healthcare in Africa: a scientometric analysis. Heal. Technol. **13**, 947–955 (2023)
10. Omotoso, O., et al.: Addressing cancer care inequities in sub-Saharan Africa: current challenges and proposed solutions. Int. J. Equity Health **22**(1), 189 (2023)
11. Menze, B.H., et al.: The multimodal brain tumor image segmentation benchmark (BRATS). IEEE Trans. Med. Imaging **34**(10), 1993–2004 (2014)

12. Kazerooni, A.F., et al.: The brain tumor segmentation (BraTS) challenge 2023: Focus on Pediatrics (CBTN-CONNECT-DIPGR-ASNR-MICCAI BraTS-PEDs). PubMed (2023)
13. Jiang, Z., Ding, C., Liu, M., Tao, D.: Two-Stage cascaded U-Net: 1st place solution to BraTS challenge 2019 Segmentation Task. In: Brainlesion: Glioma, Multiple Sclerosis, Stroke and Traumatic Brain Injuries. Springer, Cham (2020)
14. Kamnitsas, K., et al.: Ensembles of multiple models and architectures for robust brain tumour segmentation. In: International MICCAI Brainlesion Workshop. Springer, Cham (2018)
15. Myronenko, A.: 3D MRI Brain Tumor Segmentation Using Autoencoder Regularization, in Brainlesion: Glioma, Multiple Sclerosis, Stroke and Traumatic Brain Injuries. Springer, Cham (2019)
16. Ronneberger, O., Fischer, P., Brox, T.: U-Net: convolutional networks for biomedical image segmentation. In: Medical Image Computing and Computer-Assisted Intervention – MICCAI 2015. Springer, Cham (2015)
17. Isensee, F., et al.: nnU-Net: a self-configuring method for deep learning-based biomedical image segmentation. Nat. Methods (2021)
18. Zhang, D., Confidence, R., Anazodo, U.: Stroke lesion segmentation from low-quality and few-shot MRIs via similarity-weighted self-ensembling framework. In: Medical Image Computing and Computer Assisted Intervention – MICCAI 2022. Springer, Cham (2022)
19. Trung, N.T., et al.: Low-dose CT image denoising using deep convolutional neural networks with extended receptive fields. SIViP 16(7), 1963–1971 (2022)
20. Bodavarapu, P., Srinivas, P.V.V.S.: Facial expression recognition for low resolution images using convolutional neural networks and denoising techniques. Indian J. Sci. Technol. 14, 971–983 (2021)
21. He, K., et al.: Deep residual learning for image recognition. In: 2016 IEEE Conference on Computer Vision and Pattern Recognition (CVPR) (2016)
22. Tan, M., Le, Q.V.: EfficientNet: rethinking model scaling for convolutional neural networks. ArXiv, 2019. abs/1905.11946
23. Sudre, C.H., Li, W., Vercauteren, T., Ourselin, S., Jorge Cardoso, M.: Generalised dice overlap as a deep learning loss function for highly unbalanced segmentations. In: Lecture Notes in Computer Science, pp. 240–248. Springer, Cham (2017)
24. Zhou, T., Canu, S., Ruan, S.: Fusion based on attention mechanism and context constraint for multi-modal brain tumor segmentation. Comput. Med. Imaging Graph. 86, 101811 (2020)
25. Li, Y., Shen, L.: Deep learning based multimodal brain tumor diagnosis. In: Brainlesion: Glioma, Multiple Sclerosis, Stroke and Traumatic Brain Injuries: Third International Workshop, BrainLes 2017, Held in Conjunction with MICCAI 2017, Quebec City, QC, Canada, September 14, 2017, Revised Selected Papers 3. Springer (2018)
26. Wang, H., et al.: Global and local multi-scale feature fusion enhancement for brain tumor segmentation and pancreas segmentation. In: Brainlesion: Glioma, Multiple Sclerosis, Stroke and Traumatic Brain Injuries. Springer, Cham (2020)
27. Liu, C., et al.: Brain tumor segmentation network using attention-based fusion and spatial relationship constraint. In: Brainlesion: Glioma, Multiple Sclerosis, Stroke and Traumatic Brain Injuries. Springer, Cham (2021)

Optimizing Classification of Congestive Heart Failure Using Feature Weight Importance Correlation

Iyabosola B. Oronti[1]([✉]) [iD], Muhammad S. Haleem[2] [iD], Leandro Pecchia[1,3] [iD],
and Thomas Popham[1]

[1] University of Warwick, Coventry CV4 7AL, UK
{Iyabosola.B.Oronti,l.pecchia,t.popham}@warwick.ac.uk
[2] Queen Mary University of London, 327 Mile End Rd, Bethnal Green, London E1 4NS, UK
m.haleem@qmul.ac.uk
[3] Università Campus Bio-Medico di Roma, Via Álvaro del Portillo 21, 00128 Rome, Italy

Abstract. In this work, a novel method for selecting the optimal set of input features for classifying the presence of congestive heart failure (CHF) using a supervised machine learning approach is presented. A random forest classifier (RFC) was utilized to carry out the binary classification task and two different models were explored. We employed the embedded RFC feature importance attribute for the first model, and a multi-classifier technique which integrates the feature weight importance correlation (F-WIC) method was adopted for the second model. Our results show that the second model using the F-WIC method offers superior performance (100% accuracy) and provides a generalized approach to feature engineering for machine learning models irrespective of the algorithm used. This work offers a novel method for selecting the optimal set of input features for classifying the presence of congestive heart failure (CHF) using a machine learning approach.

Keywords: machine learning · feature engineering · congestive heart failure

1 Introduction

Our nerve and muscle cells communicate with each other using electrical and chemical signals. Regular electrical signals also control our heartbeat [1]. The electrocardiogram (ECG) is the measurement and graphical rendering of the electrical signals generated within the heart muscles and has become a staple clinical tool in depicting a plethora of cardiac clinical states and disease conditions, which has significantly improved diagnosis and patient care. The ECG embodies a physiological phenomenon – heart rate variability (HRV), which is a measure of the differences in interval between succeeding beats of the heart [2–4]. Over time, HRV measures have been developed as a quantitative marker of autonomic activity. Usually, HRV is depressed in patients with congestive heart failure (CHF) compared with healthy subjects. CHF can be a long-lasting or acute progressive condition that affects the ability of the heart muscles to pump blood to other parts of the body because of fluid accumulation around the heart. In this paper, the discriminatory power of HRV measures is used to optimize the binary categorization of CHF.

© The Author(s), under exclusive license to Springer Nature Switzerland AG 2025
U. Anazodo et al. (Eds.): MImA 2024/EMERGE 2024, CCIS 2240, pp. 24–31, 2025.
https://doi.org/10.1007/978-3-031-79103-1_3

In previous studies, Pecchia et al. investigated how short-term HRV features vary in CHF patients according to disease severity, as well as in healthy people using the Normal Sinus Rhythm and the CHF RR Interval (i.e., peak-to-peak ECG interval) online databases [5]. They progressed this work by employing the power of these features to discriminate between normal subjects and CHF patients using the classification and regression tree (CART) method [6], achieving sensitivity and specificity values of 79.3% and 100%, respectively. These metrics were further improved by incorporating two nonstandard HRV features [7], which further pushed the sensitivity and specificity values to 89.7% and 100%, respectively. Subsequently, in [8], Pecchia et al. proposed a disease management platform integrating the CART classifier in [7] to enhance the effectiveness and efficiency of home monitoring of CHF using data mining for early detection of any worsening in patient's condition. The system achieved an accuracy and precision of 96.39% and 100.00% in detecting CHF, and 79.31% and 82.35% in classifying severe versus mild HF, respectively.

Similarly, Melillo et al. [9] studied the potential of standard long-term HRV measures for CHF diagnosis, also using the CART method and four public Holter databases. Their best performing classifier achieved a specificity and sensitivity rate of 100.00 and 89.74%, respectively. In [10], an automatic CART risk classifier was later developed by the authors for risk assessment in CHF patients using standard long-term HRV measures. Results obtained from this study (i.e., sensitivity rate of 93.3% and specificity rate of 63.6%) showed that HRV could be a useful tool for risk assessment in identifying higher risk CHF patients. Another study which focused on the risk assessment of CHF was carried out by Chen et al. [11]. The authors proposed the use of dynamic indices of HRV, as opposed to long- and short-term indices, to capture the dynamics of 5-min short term HRV measurements for quantifying autonomic activity changes of CHF. The multistage risk assessment model achieved CHF detection and quantification analysis at an accuracy of 96.61%

Besides the studies mentioned above, other robust machine learning approaches and input feature types have been used in the literature to detect and classify CHF. For instance, in [12–15], a combination of decision tree (DT), naïve bayes (NB), random forest (RF), logistic regression (LR), artificial neural network (ANN), Fuzzy analytic hierarchy process (Fuzzy_AHP), k-nearest neighbor (KNN), and support vector machine (SVM) were employed with total model accuracies varying between 87.27% to 93.19% for CHF detection and classification. More recently, studies combining different types of data such as electronic health records (EHR), claims data, clinical trials data, inpatient registry, and data obtained from research cohorts with different deep learning (DL) methods to detect [16, 17], classify [18] or predict the risk of CHF [19, 20] abound in the literature with good to excellent performance metrics. Notably, Porumb et al. achieved 100% CHF detection accuracy using a convolutional neural network (CNN) model that accurately identifies CHF on the basis of one raw ECG heartbeat only. However, most of these DL models suffer from lack of interpretability and transparency, which are core issues to consider if these models are to be deployed in low- and middle-income countries due to trust issues and limited expertise and knowledge of AI systems. This work proposes a simple but effective method of selecting the optimal set of input features for detecting and classifying the presence of CHF in patients with the disease using a machine learning approach.

2   Methods

2.1   Datasets and Data Preprocessing

Secondary data in the form of long-term (20 h) ECG Holter monitor data for healthy subjects (Massachusetts Institute of Technology-Beth Israel Hospital (MIT-BIH) Normal Sinus Rhythm Database) [21] and CHF patients (Beth Israel Deaconess Medical Centre (BIDMC) Congestive Heart Failure Database) [22] were obtained from PhysioNet [21]. The BIDMC dataset had a mix of patients with New York Heart Association (NYHA) III and IV functional classes [23] (i.e., severe CHF). Figure 1 shows both the compacted and expanded views of the raw annotated signal for Record 4 in the BIDMC database. HRV features [2, 3] were extracted from the ECG signals for both datasets using the pyHRV Python Library [24, 25] across the time, frequency and non-linear domains, leaving 73 features in all after the removal of irrelevant features. Additionally, the class of the respective datasets were mapped to the ground truth (i.e., 0 and 1).

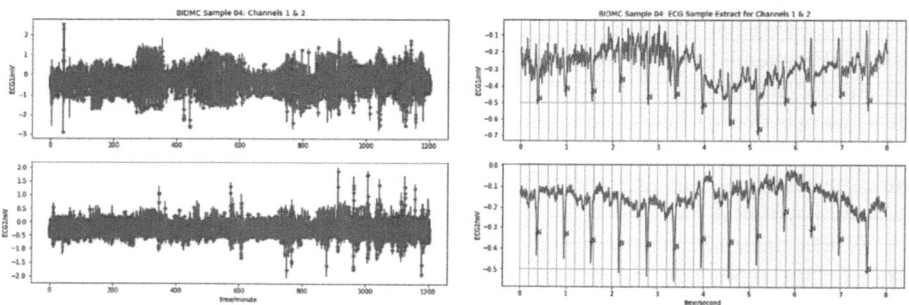

Fig. 1. *Left* - 20 h raw annotated signal; *right* - 8 s extract of sampled record.

2.2   Feature Engineering and Model Development

Two main models were developed based on the feature engineering approach adopted. For Model 1, the 10 most important HRV features were selected from time, frequency and non-linear domain using the embedded random forest classifier (RFC) feature importance attribute. The *mean decrease in impurity* (MDI) method was adopted for the selection. This method computes feature weight importance as the mean and standard deviation of the accumulation of impurity decrease within each tree in the classifier, implemented through the Gini Log Loss or mean squared error (MSE), and consistently returned the same features at every code run. The main disadvantage of this method is that impurity-based feature importances can sometimes be misleading for high cardinality features (i.e., features with many unique values), and can rank numerical features to be the most important. This was, however, not an issue in this work because all input features were continuous values. The original dataset (containing 73 input features) was split into training and test sets in the ratio 80:20. The training set was used to train the model while the test set served as a collection of data points to evaluate the generalizability of the model to new, unseen data. The shuffle function was applied to randomly change the order of the sample rows, while the random state function was used to initialize the seed used to split the dataset.

Table 1. Input Features Selected for Models 1, 2a, and 2b.

*Input Feature	Domain	Variable Name/Unit	Model 1	Model 2a	Model 2b
hr_mean	time	mean heart rate [bpm]	✓	✓	✓
hr_max	time	Maximum heart rate [bpm]	✗	✓	✗
nni_mean	time	mean normal-to-normal interval (NNI) [ms]	✓	✗	✗
dfa_alpha1	non-linear	short-term detrended fluctuation analysis slope	✓	✗	✗
nni_counter	time	number of NNIs [-]	✓	✓	✓
nni_max	time		✗	✓	✗
sd1	non-linear	standard deviation of data series along major axis of Poincare plot	✓	✗	✗
fft_ratio	frequency	fft LF to HF ratio [-]	✓	✗	✗
hr_max	time	maximum heart rate [bpm]	✓	✗	✗
sdsd	time	standard deviation of NNI differences [ms]	✓	✗	✗
fft_peak 1	frequency	peak frequency of fft band 1 [Hz]	✓	✓	✗
fft_rel 0	frequency	relative power of fft band 0 [%]	✓	✗	✗
fft_rel 2	frequency	relative power of fft band 2 [%]	✗	✗	✓

*Input features are arranged in no particular order.

fft = fast Fourier transform; VLF = very low frequency; LF = low frequency; HF = high frequency; Hz = hertz; bpm = beats per minute; ms = milliseconds; band 0 = VLF: [0.00Hz - 0.04Hz]; band 1 = LF: [0.04Hz - 0.15Hz]; band 2 = HF: [0.15Hz - 0.40Hz]

The second model was designed to adopt a more generalized approach to feature engineering for machine learning models. Fourteen commonly used classifiers were aggregated to select the 10 best features each, using the rated order of importance and score (1- highest; 10 - lowest). The idea was to explore how likely it was for a feature to be picked by any of the classifiers and to document the frequency of occurrence of each feature across all classifiers as well as the associated weighted feature importance score. A frequency table was created where features occurring more frequently were rated higher in the table. Where features reported the same frequency across classifiers, a higher value of the mean weighted feature importance score was used to decide their hierarchy in the frequency table. Features occurring more than once were further subjected to iterative correlation analysis based on the order of importance of the feature in the frequency table.

Features with correlation values higher than the set *correlation iteration threshold* (CIT) ($\leq$0.45 & 0.40 respectively) were removed from the analysis. These two thresholds were set to evaluate the terminal performance of Model 2 in terms of the number of features selected and their relevance to the classification task, resulting in Models 2a and 2b. Input features for Modes 1, 2a, and 2b are presented in Table 1 while Fig. 2 summarizes the overall feature engineering pipeline for Model 2. In total, 16 features were identified with a frequency of occurrence greater than one across all classifiers.

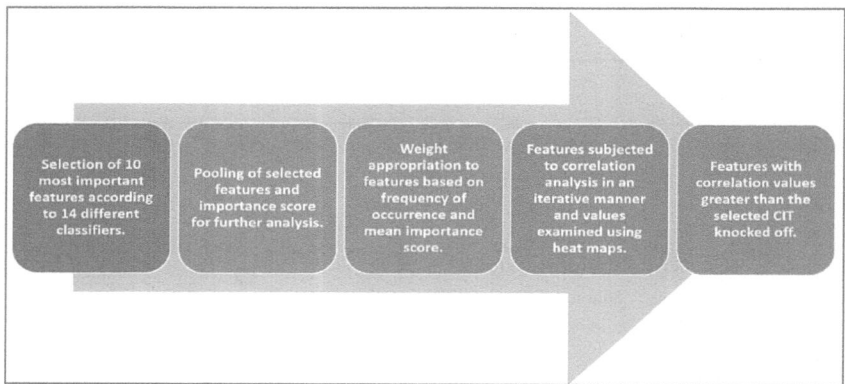

Fig. 2. Feature engineering pipeline

3   Results

To obtain a preliminary evaluation of the performance of the RFC, the top ten input features obtained from the MDI method were fed into the classifier. 75% of the dataset was used as the training set while the remaining 25% was used as the test set. The K-fold (K = 10) cross-validation technique was used to prevent overfitting and to provide a more reliable and robust evaluation of the model's performance by finding the mean value of the estimated results. Initial results obtained on the test set showed a root mean squared error (RMSE) of 0.25 and an accuracy of 0.94. To ensure that a maximally performing subset of hyperparameters for the RFC had been selected for the classification task, hyperparameter tuning was carried out on the dataset by using grid search cross validation (CV). Parameter values selected after this procedure were no different from the initial selection used to train the RFC. The dataset was trained again with the 'optimal' hyperparameter combination. Results obtained before and after hyperparameter optimisation remained basically the same. All patients from the control group were classified correctly. Only one out of the CHF patients was classified incorrectly.

In the case of Model 2, two different sets of input features arising from the feature engineering process described in Sect. 2.2 were fed into the model at separate instances, giving rise to Models 2a and 2b (i.e., at CIT $\leq$ 0.45 and CIT $\leq$ 0.40 respectively). The first set of inputs obtained with CIT $\leq$ 0.45 (see Table 1 for details) was used to

RMS: 0.25	precision	recall	f1-score	support
0	0.91	1.00	0.95	10
1	1.00	0.83	0.91	6
accuracy			0.94	16
macro avg	0.95	0.92	0.93	16
weighted avg	0.94	0.94	0.94	16
Accuracy: 0.9375				

	precision	recall	f1-score	support
0	1.00	0.89	0.94	9
1	0.88	1.00	0.93	7
accuracy			0.94	16
macro avg	0.94	0.94	0.94	16
weighted avg	0.95	0.94	0.94	16

(a)   (b)

Fig. 3. Performance metrics of Models 1 and 2a: (a) Model 2a with CIT ≤ 0.45, (b) Model 1 with hyperparameter tuning.

train Model 2a, resulting in a prediction accuracy of 0.9375 and a RMSE of 0.25, which was quite similar to results obtained for Model 1 although with some slight differences. Figure 3 shows the performance metrics of Models 1 and 2a. For Model 2b (where CIT ≤ 0.40), the model attained 100% accuracy. The confusion matrix shows that all labels were correctly identified unlike in Model 2a, which had a false positive in the predicted value. Notably, applying 10-fold CV did not change the test accuracy. The f1-score for Model 2b is 1.0, implying perfect precision and recall. Moreover, the receiver operating characteristics (ROC) curve output an area under the curve (AUC) score of 1.0.

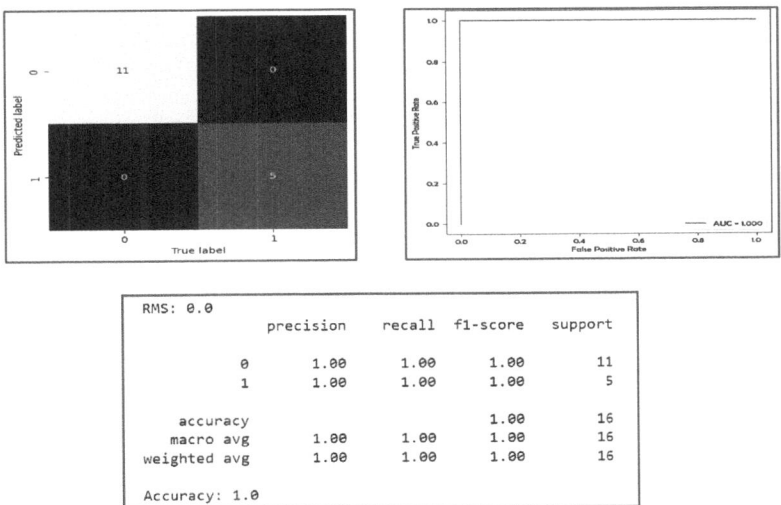

RMS: 0.0	precision	recall	f1-score	support
0	1.00	1.00	1.00	11
1	1.00	1.00	1.00	5
accuracy			1.00	16
macro avg	1.00	1.00	1.00	16
weighted avg	1.00	1.00	1.00	16
Accuracy: 1.0				

Fig. 4. Performance metrics of Model 2b

Additionally, stratified shuffle split cross-validation was employed to test the effect of class imbalance and data leaks on the model performance. It was interesting to note that the test accuracy for Model 2b remained at 100% for each stratified fold. From top left to bottom, Fig. 4 shows the evaluation results of Model 2b.

4 Discussion and Conclusion

From the results presented, Model 2 appears to provide consistent improvement on the model performance metrics (e.g., 100% accuracy) even when different cross-validation methods were used. This shows that this novel feature selection method (i.e., F-WIC) used to determine the input features to Model 2 has potential as an agnostic and generalised feature selection method across all classifiers to improve the performance of any model. The utility and effectiveness of this method will be further proved when used with much larger datasets with even many more input variables. The outcomes of this research can significantly impact clinical practice because the methods employed are easily understandable and explainable to clinical personnel. Also, the features having the most impact on the model prediction are the time domain features that are mostly used by cardiologists to diagnose patients with different forms of cardiovascular disease. Finally, the F-WIC approach can be applied generally to select optimal features for other disease diagnosis and categorization tasks.

Disclosure of Interests. The authors have no competing interests to declare that are relevant to the content of this article.

References

1. Silverman, M.E., Grove, D., Upshaw, C.B.: Why does the heart beat? The discovery of the electrical system of the heart. Circulation **113**(23), 2775–2781 (2006). https://doi.org/10.1161/CIRCULATIONAHA.106.616771
2. Shaffer, F., Ginsberg, J.P.: An overview of heart rate variability metrics and norms. Front. Public Heal. **5**(September), 1–17 (2017). https://doi.org/10.3389/fpubh.2017.00258
3. Malik, M., et al.: Heart rate variability: standards of measurement, physiological interpretation, and clinical use. Circulation **93**(5), 1043–1065 (1996). https://doi.org/10.1161/01.cir.93.5.1043
4. McCraty, R., Shaffer, F.: Heart rate variability: new perspectives on physiological mechanisms, assessment of self-regulatory capacity, and health risk. Glob. Adv. Heal. Med. **4**(1), 46–61 (2015). https://doi.org/10.7453/gahmj.2014.073
5. Pecchia, L., Melillo, P., Sansone, M., Bracale, M.: Heart rate variability in healthy people compared with patients with congestive heart failure (2009). https://doi.org/10.1109/ITAB.2009.5394352
6. Breiman, C.J., et al.: Stone, Classification and Regression Trees. Chapman & Hall/CRC, Taylor and Francis Group, LLC (2005)
7. Pecchia, L., Melillo, P., Sansone, M., Bracale, M.: Discrimination power of short-term heart rate variability measures for CHF assessment. IEEE Trans. Inf. Technol. Biomed. **15**(1), 40–46 (2011). https://doi.org/10.1109/TITB.2010.2091647
8. Pecchia, L., Melillo, P., Bracale, M.: Remote health monitoring of heart failure with data mining via CART method on HRV features. IEEE Trans. Biomed. Eng. **58**(3), 800–804 (2011). https://doi.org/10.1109/TBME.2010.2092776
9. Melillo, P., Fusco, R., Sansone, M., Bracale, M., Pecchia, L.: Discrimination power of long-term heart rate variability measures for chronic heart failure detection. Med. Biol. Eng. Comput. **49**(1), 67–74 (2011). https://doi.org/10.1007/s11517-010-0728-5
10. Melillo, P., De Luca, N., Bracale, M., Pecchia, L.: Classification tree for risk assessment in patients suffering from congestive heart failure via long-term heart rate variability. IEEE J. Biomed. Heal. Inform. **17**(3), 727–733 (2013). https://doi.org/10.1109/jbhi.2013.2244902

11. Chen, W., Zheng, L., Li, K., Wang, Q., Liu, G., Jiang, Q.: A novel and effective method for congestive heart failure detection and quantification using dynamic heart rate variability measurement. PLoS ONE **11**(11), 1–18 (2016). https://doi.org/10.1371/journal.pone.0165304

12. Alotaibi, F.S.: Implementation of machine learning model to predict heart failure disease. Int. J. Adv. Comput. Sci. Appl. **10**(6), 261–268 (2019). https://doi.org/10.14569/ijacsa.2019.010 0637

13. Hussain, L., et al.: Detecting congestive heart failure by extracting multimodal features and employing machine learning techniques. Biomed. Res. Int. **2020**(26), 23–34 (2020). https://doi.org/10.1155/2020/4281243

14. Samuel, O.W., Asogbon, G.M., Sangaiah, A.K., Fang, P., Li, G.: An integrated decision support system based on ANN and Fuzzy_AHP for heart failure risk prediction. Expert Syst. Appl. **68**, 163–172 (2017). https://doi.org/10.1016/j.eswa.2016.10.020

15. Plati, D.K., et al.: A machine learning approach for chronic heart failure diagnosis. Diagnostics **11**(10), 1863 (2021). https://doi.org/10.3390/diagnostics11101863

16. Choi, E., Schuetz, A., Stewart, W.F., Sun, J.: Using recurrent neural network models for early detection of heart failure onset. J. Am. Med. Inform. Assoc. **24**(2), 361–370 (2017). https://doi.org/10.1093/jamia/ocw112

17. Porumb, M., Iadanza, E., Massaro, S., Pecchia, L.: A convolutional neural network approach to detect congestive heart failure. Biomed. Signal Process. Control **55**, 101597 (2020). https://doi.org/10.1016/j.bspc.2019.101597

18. Li, D., Tao, Y., Zhao, J., Wu, H.: Classification of congestive heart failure from ECG segments with a multi-scale residual network. Symmetry **12**(12), 2019 (2020). https://doi.org/10.3390/SYM12122019

19. Rao, S., et al.: An explainable transformer-based deep learning model for the prediction of incident heart failure. IEEE J. Biomed. Health Inform. 26(7), 3362–3372 (2022). https://doi.org/10.1109/JBHI.2022.3148820

20. Goretti, F., Oronti, B., Milli, M., Iadanza, E.: Deep learning for predicting congestive heart failure. Electronics **11**(23), 3996 (2022). https://doi.org/10.3390/electronics11233996

21. Goldberger, A.L., et al.: PhysioBank, PhysioToolkit, and PhysioNet: components of a new research resource for complex physiologic signals. Circulation, **101**(23), e215–e220 (2000). https://doi.org/10.1161/01.cir.101.23.e215

22. Xia, H., et al.: Computer algorithms for evaluating the quality of ECGs in real time. Comput. Cardiol. **38**(June), 369–372 (2011)

23. Fisher, J.D.: New York heart association classification. Arch. Intern. Med. **129**(5), 836 (1972). https://doi.org/10.1001/ARCHINTE.1972.00320050160023

24. Gomes, P., Margaritoff, P., Silva, H.: pyHRV: development and evaluation of an open-source python toolbox for heart rate variability (HRV). In: Proceedings of International Conference on Electrical, Electronic and Computing Engineering (IcETRAN), pp. 822–828 (2018)

25. Gomes, P.M.C.: Development of an Open-Source Python Toolbox for Heart Rate Variability (HRV). University of Applied Sciences, Hamburg, Germany (2018)

MCL: Multi-level Consistency Learning for Medical Image Segmentation

Alou Diakite[1,2], Cheng Li[1], Lei Xie[3], Yuanjing Feng[3], Hairong Zheng[1], and Shanshan Wang[1(✉)]

[1] Paul C. Lauterbur Research Center for Biomedical Imaging, Shenzhen Institute of Advanced Technology, Chinese Academy of Sciences, Shenzhen 518055, China
`ss.wang@siat.ac.cn`
[2] University of Chinese Academy of Sciences, Beijing 100040, China
[3] Zhejiang University of Technology, Hangzhou 310023, China

Abstract. Semi-supervised learning (SSL) is crucial for advancing medical image segmentation by reducing the need for human annotations. However, current SSL-based consistency regularization techniques have limitations such as overlooking information in logits and potential overconfidence issues, which impede the segmentation performance. To address these challenges, we introduce a multi-level consistency learning (MCL) framework for enhancing the utilization of unlabeled data in medical image segmentation tasks. Our approach includes a dual-level consistency regularization that enforces consistency at both the instance and semantic levels, ensuring uniform class predictions for the same instance observed from different perspectives. Additionally, we introduce consistency between logits from different views to encourage both class prediction invariance within instances and prediction diversity across instances. We extensively evaluate the proposed framework on the Human Connectome Project (HCP) and the University of Pennsylvania glioblastoma (UPenn-GBM) datasets for brain tumor and visual pathway segmentation tasks, respectively, and compare its performance against six state-of-the-art methods. Our experimental findings demonstrate that the MCL approach surpasses existing methods, highlighting its efficacy in various medical image segmentation scenarios. The code will be made available upon acceptance.

Keywords: Medical Image Segmentation · Semi-Supervised Learning · Consistency Regularization

This research was partly supported by the National Natural Science Foundation of China (62222118, U22A2040), Guangdong Provincial Key Laboratory of Artificial Intelligence in Medical Image Analysis and Application (2022B1212010011), and Shenzhen Science and Technology Program (RCYX20210706092104034, JCYJ20220531100213029) and Youth Innovation Promotion Association CAS.

U. Anazodo et al. (Eds.): MIImA 2024/EMERGE 2024, CCIS 2240, pp. 32–41, 2025.
https://doi.org/10.1007/978-3-031-79103-1_4

1   Introduction

In recent years, deep learning has made significant progress in medical image analysis [1,2], particularly in tasks like medical image segmentation [3,4]. The success of these networks mostly relies on having access to large amounts of well-labeled data [5,6]. However, gathering such data in the medical field is time-consuming and labor-intensive. This has led to a growing interest in developing algorithms that can learn effectively with minimal annotations to achieve satisfactory performance.

Semi-supervised learning (SSL) algorithms have gained prominence in this context [7–9]. These algorithms aim to enhance segmentation models by combining limited labeled data and a larger pool of unlabeled data. Among various SSL techniques, consistency regularization [10] has become a fundamental approach. It enforces alignment between predictions on unlabeled samples and their perturbed versions, improving segmentation performance. Techniques like mean-teacher (MT) [11], uncertainty-aware mean-teacher (UA-MT) [12], and ambiguity-consensus mean-teacher (AC-MT) [13] use random perturbations to achieve this concept. However, these methods overlook information stored in logits and may cause overconfidence issues, leading to poor performance. Techniques such as Contrastive learning [14], similarity matching [15], cross-correlation matching [16], etc., have been explored to address these issues individually.

Inspired by the above studies, in this paper, we propose a novel SSL framework, MCL, which has been illustrated in Fig. 1. In MCL, we apply consistency regularization to match the similarity relationships of semantic and instance levels simultaneously for different views. Specifically, we first require the perturbed view to have a similar label prediction with a non-perturbed view; besides, we also encourage the perturbed view to have similar instance characteristics to the non-perturbed one for better feature alignment. Moreover, different from the previous studies that ignore logit-level information, we further constrained logits of different views to be consistent, promoting both invariance in class predictions for the same instance and diversity in class predictions across different instances. Experimental results on the Human Connectome Project (HCP) and the UPenn-GBM datasets demonstrate that our proposed MCL method surpasses existing state-of-the-art approaches, especially when the available labeled data is small.

2   Method

As shown in Fig. 1, the proposed MCL follows a standard mean-teacher framework where the student model learns from labeled and unlabeled data, guided by the teacher model using the exponential moving average (EMA) technique. We define the SSL medical image segmentation problem as following. Given a batch of B labeled samples $X = x_b : b \in (1, \ldots, B)$, we employed a standard U-Net -based encoder $F(\cdot)$ to extract the feature-level information from these samples, i.e. $h = E(x)$. Then, a U-Net-based decoder $D(\cdot)$ with a fully connected class prediction head $\phi(\cdot)$ is used to map h_b into label prediction, which can be written as $P_1 = \phi(h)$. The labeled samples X could be directly optimized by the joint binary cross-entropy L_{bce} and dice L_{dice} losses with the ground truth labels $Y = y_b : b \in (1, \ldots, B)$:

$$L_s = 0.5 * L_{bce}(y, P_1) + L_{dice}(y, P_1) \tag{1}$$

Similarly, let us define a batch of B unlabeled samples $U = u_b : b \in (1, \ldots, B)$. By following, we randomly apply noise perturbation $T(\cdot)$ to the raw unlabeled data to obtain two different views of the same sample and use similar steps as the labeled samples to obtain label prediction for a raw sample Q_1 (pseudo-label) and perturbed sample Q_2. Then, the semantic level consistency L_u^1 can be defined as the mean-squared error between these two predictions:

$$L_u^1 = L_{mse}(Q_1, Q_2) \tag{2}$$

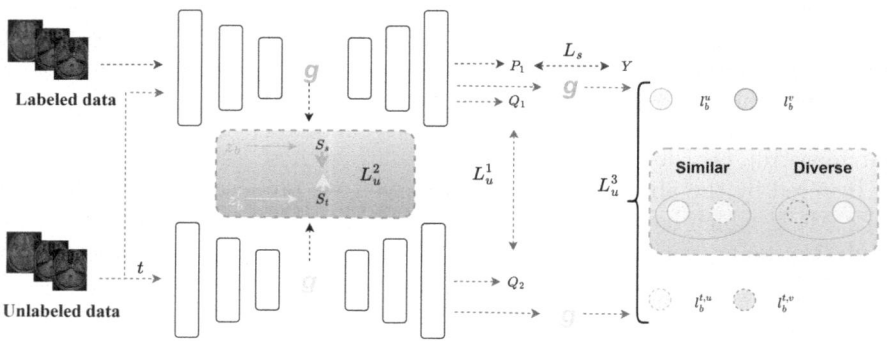

Fig. 1. Overview of our proposed MCL. MCL adopts a mean-teacher scheme to leverage both labeled and unlabeled data. u and v are class instances, and t is the perturbation operation. g is a non-linear projection head.

2.1   Instance-Level Consistency Learning

While semantic-level consistency holds promise in SSL learning, their pseudo-label is not reliable when the available labeled samples are very small. Therefore, In MCL, we also consider the instance-level consistency as we have discussed above. Specifically, we encourage the perturbed view to have an invariant feature distribution with the raw view. Let's say we have a non-linear projection function $g(\cdot)$ that transforms the representation h into a compact embedding $z_b = g(h_b)$ with lower dimensions. Following standard similarity matching methods [15], we use z_b and z_b^t to denote the embedding from the raw and perturbed view. Then, for a batch of non-perturbed unlabeled embeddings, we calculate the similarities distribution S_s between z and $i-th$ instance by utilizing a similarity function $sim(\cdot)$. A cosine angle similarity measure is adopted here to calculate the similarity distribution:

$$\{S_s\} = sim(z_b, z^i), \tag{3}$$

Similarly, we computed the similarity distribution S_t between z^t and $i - th$ instance as follows:

$$\{S_t\} = sim(z_b^t, z^i), \tag{4}$$

Note that we remove the diagonal values in S_s and S_t, representing the similarity value with itself as in this study [17]. This value is typically large and can have a dominant effect on the overall similarity distribution. Finally, the instance-level consistency optimization objective for our framework is:

$$L_u^2 = KL(\bar{S}, S_i), \tag{5}$$

where KL is the Kullback-Leibler divergence, $\bar{S}$ is the average similarity values of S_s and S_t, and $S_i \in \{S_s, S_t\}$. In our framework, we achieve feature-level knowledge sharing by enforcing better feature alignment between the perturbed and raw unlabeled data for feature-level semantic consistency as demonstrated in (5). This allows us to gather more invariant representation from unlabeled data, ultimately learning better generalizable representation.

2.2   Logit-Level Consistency Learning

Although the above consistency constraints can mitigate the overconfidence in consistency regularization-based methods, they overlook information stored in logits, leading to poor performance. Taking this into account, it is imperative to promote consistency among the class predictions from the same instance while also fostering diversity between the class predictions of different instances. Therefore, we further consider the consistency between the logits of the raw and perturbed images. Let's say we have a non-linear projection function $g(\cdot)$ that transforms the representation $\phi(h)$ into a compact embedding $l_b = g(\phi(h_b))$ with lower dimensions. We use l_b and l_b^t to denote the embedding from the raw and perturbed view. Then, for a batch of non-perturbed unlabeled embeddings, we calculate the cross-correlation matrix CM between l and l^t by utilizing a metric proposed in [17] as follows:

$$\{CM\} = \frac{\sum_b \big((l_b), (l_b)^t \big)}{N}, \tag{6}$$

where N represents the channel dimension. $\{CM\}$ is comprised between -1 and 1, where -1 indicates the highest degree of dissimilarity and 1 is the highest degree of similarity. Then, the logit-level consistency loss L_u^3 for our framework is defined as:

$$L_u^3 = \sum_b \big(1 - \{CM\}\big)^2 + \gamma \sum \sum_b \big(1 + \{CM\}\big)^2, \tag{7}$$

where γ is a positive constant that determines how much importance is placed on the first and second terms of the loss function. When the on-diagonal terms of the cross-correlation matrix are $+1$, it encourages similarity in the logit from the raw and perturbed view. Conversely, when the off-diagonal terms of the cross-correlation matrix are -1, it promotes dissimilarity in the sample predictions, as the dimensions (classes) of the output will be uncorrelated.

2.3   Overall Loss Function

The overall loss of MCL consists of three parts: the supervised segmentation loss L_s, the FSL loss L_{FSL}, and the ICCM loss L_{ICCM}. It is defined as:

$$L = L_s + L_u^1 + L_u^2 + L_u^3, \tag{8}$$

where L_s is the combination of binary cross-entropy and dice losses.

2.4   Network Architecture

The U-Net model core structure consists of an asymmetric encoder-decoder framework, enabling the learning of hierarchical features at multiple scales. The encoding part incorporates two successive 3×3 convolutional layers with a stride of 1, each supplemented by batch normalization (BN), rectified linear unit (ReLU), and a 2×2 max-pooling operation with a stride of 2 on 4 levels. With each down-sampling phase, the input image's spatial dimensions are halved while the feature channel count is doubled. The bottom level comprises two 3×3 convolutional layers, each with a BN but without a pooling layer. The decoding section restores the initial spatial dimensions by up-sampling the feature map, concatenating feature channels from the contracting path, and employing convolutional layers with BN and ReLU. In the case of the segmentation network, the output is a single-channel image with spatial dimensions of 128×160 voxels for the HCP data and 240×240 voxels for the Upenn-GBM data, where the channel holds voxel probabilities for the VP or brain tumor. We utilize this U-Net architecture as the backbone for all models to ensure a fair comparison with existing methods.

3   Experiments and Results

3.1   Dataset

We evaluate our MCL on the HCP [18] and UPenn-GBM [19] datasets. The HCP dataset provides 92 cases, among which 10 cases were used to assess the proposed method, and 82 cases (16 labeled and 66 unlabeled) were for training. The UPenn-GBM dataset provides 110 cases, among which 10 cases were used for testing, and 100 cases (5 labeled and 95 unlabeled) were used for training.

3.2 Implementation Details

Our MCL is built using PyTorch 2.0.1 and CUDA 11.8 on an NVIDIA GeForce RTX 3090 GPU. To prepare the training data, we normalize all volumes between 0 and 1, randomly crop HCP volumes to $128 \times 160 \times 128$ size, and convert 3D volumes to 2D slices as input. Each batch of 16 has 16 2D slices but not of the same 3D volume since they are randomly sampled. To avoid the imbalance issue of classes (foreground and background), we followed the conversion process in [20] and used dice loss in combination with binary cross-entropy loss. The model was trained with a batch size of 16, using the SGD optimizer with a learning rate of 0.0002, and eps of 1e-8 over 200 epochs. We balance loss terms by setting α and β to 0.5. During inference, we directly use the prediction results from the networks without any post-processing.

3.3 Quantitaive Evaluation and Comparison

Our MCL is assessed on four metrics: Dice Similarity Score (DSC), Recall (REC), Precision (PREC), and Average Symmetric Surface Distance (ASSD). We compare our approach with six state-of-the-art SSL methods, including URPC [21], DTC [22], UA-MT [12], MC-NET [8], CAML [7], AC-MT [13]. To ensure a fair comparison, we conduct all experiments on the same machine and report the mean and standard deviation of the test dataset from the final epoch.

Table 1. Comparison with state-of-the-art SSL methods on the HCP and UPenn-GBM datasets. Metrics are reported as *mean ± std*, and the best results are bolded.

Methods		Scans used		Metrics			
		Labeled	Unlabeled	DSC	REC	PREC	ASSD (mm)
HCP	URPC	16 (20%)	66 (80%)	0.76 ± 0.07	0.67 ± 0.1	**0.88 ± 0.03**	0.30 ± 0.1
	DTC	16 (20%)	66 (80%)	0.77 ± 0.06	0.70 ± 0.09	0.86 ± 0.02	0.28 ± 0.09
	UA-MT	16 (20%)	66 (80%)	0.78 ± 0.04	0.82 ± 0.04	0.80 ± 0.02	0.27 ± 0.06
	MC-NET	16 (20%)	66 (80%)	0.78 ± 0.05	0.75 ± 0.09	0.81 ± 0.03	0.29 ± 0.09
	CAML	16 (20%)	66 (80%)	0.79 ± 0.04	0.76 ± 0.08	0.83 ± 0.03	0.27 ± 0.09
	ACMT	16 (20%)	66 (80%)	0.81 ± 0.03	0.77 ± 0.05	0.86 ± 0.03	0.21 ± 0.05
	Ours	16 (20%)	66 (80%)	**0.84 ± 0.03**	**0.84 ± 0.05**	0.84 ± 0.02	**0.19 ± 0.05**
	FS	82 (100%)	0	0.86 ± 0.01	0.82 ± 0.03	0.90 ± 0.02	0.15 ± 0.08
UPenn-GBM	URPC	5 (5%)	95 (95%)	0.45 ± 0.13	0.34 ± 0.13	0.72 ± 0.14	4.21 ± 1.75
	DTC	5 (5%)	95 (95%)	0.52 ± 0.19	0.43 ± 0.17	**0.81 ± 0.07**	2.57 ± 1.04
	UA-MT	5 (5%)	95 (95%)	0.42 ± 0.18	0.32 ± 0.17	0.81 ± 0.15	4.85 ± 1.98
	MC-NET	5 (5%)	95 (95%)	0.52 ± 0.16	0.42 ± 0.18	0.79 ± 0.09	3.06 ± 0.93
	CAML	5 (5%)	95 (95%)	0.52 ± 0.20	0.44 ± 0.19	0.80 ± 0.07	2.88 ± 0.89
	ACMT	5 (5%)	95 (95%)	0.53 ± 0.18	0.45 ± 0.19	0.81 ± 0.06	2.78 ± 0.62
	Ours	5 (5%)	95 (95%)	**0.68 ± 0.08**	**0.65 ± 0.12**	0.72 ± 0.07	**2.21 ± 1.33**
	FS	100 (100%)	0	0.82 ± 0.06	0.82 ± 0.11	0.84 ± 0.04	0.82 ± 0.64

The results for HCP and UPenn-GBM are shown in Table 1. The fully supervised (FS) U-Net model was trained on the fully labeled data, and the results obtained from this training serve as the upper bounds. On the UPenn-GBM dataset, our method performed well with a DSC score of $\mathbf{0.68 \pm 0.08}$ and a REC score of $\mathbf{0.65 \pm 0.12}$. The PREC score was $\mathbf{0.72 \pm 0.07}$, lower than ACMT ($\mathbf{0.83 \pm 0.06}$). The ASSD was $\mathbf{2.21 \pm 1.33}$ mm, indicating reasonably accurate segmentation, with minor deviations from the ground truth surfaces. These results prove that our method can provide reliable segmentations even in challenging datasets like UPenn-GBM, highlighting its ability to capture essential brain tumor features even in challenging datasets like UPenn-GBM.

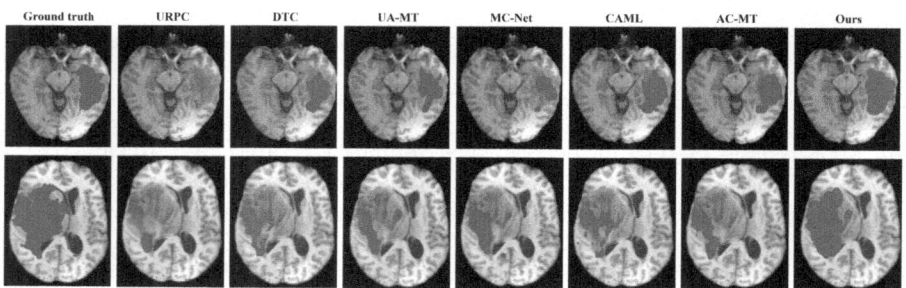

Fig. 2. Visualization of brain tumor segmentation on the UPenn-GBM dataset.

For the HCP dataset, our method excelled, achieving the highest DSC of $\mathbf{0.84 \pm 0.03}$ among SSL methods, indicating strong agreement with ground truth. Our method also showed a REC score of $\mathbf{0.84 \pm 0.05}$ and a PREC score of $\mathbf{0.84 \pm 0.02}$, slightly lower than ACMT ($\mathbf{0.86 \pm 0.03}$). The ASSD was $\mathbf{0.19 \pm 0.05}$ mm, demonstrating high segmentation accuracy. These results confirm our method's superiority in visual pathway segmentation on the HCP dataset.

Figures 2 and 3 visually compare the segmentation results from different methods on both datasets. One can see that MCL (Ours) demonstrated meaningful results across datasets, with its superiority being most evident in illustrated cases.

When comparing our method with the other evaluated methods, the consistent improvement achieved by our method on both datasets compared to most of the state-of-the-art methods indicates the robustness of our method in handling different datasets and tasks.

3.4   Ablation Study

In this section, we conducted ablation studies to evaluate the effectiveness of the (ILC) instance-level consistency (ILC) and logit-level consistency (LLC). We implement the MT model as our baseline ($L_s + L_u^1$). Three important conclusions can be made from these studies. The results are presented in Fig. 4. Firstly,

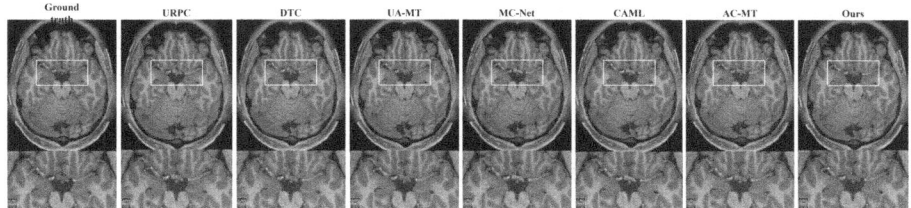

Fig. 3. Visualization of visual pathway (optic chiasm) segmentation on the HCP dataset.

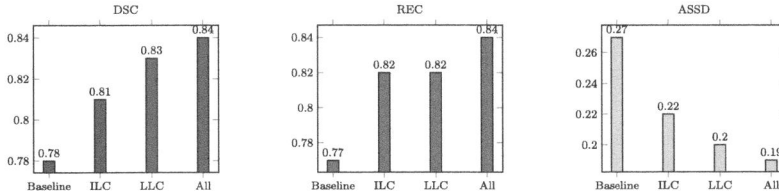

Fig. 4. Ablation study results on the HCP dataset.

integrating ILC into the baseline shows significant improvements, highlighting the significance of feature alignment between the perturbed and raw views for feature-level semantic consistency. Secondly, adding LLC to the baseline also results in notable performance enhancements compared to the baseline in all metrics, demonstrating the importance of encouraging both similarity and diversity across the same and different instances, respectively. Overall, our final model that introduces both ILC and LLC to the baseline achieves the best segmentation performance. This supports our approach to addressing the overconfidence issue and ignorance of logit information in consistency regularization-based methods.

4   Conclusion

In this paper, we introduced the MCL framework, a novel approach for semi-supervised medical image segmentation. Our key idea is to incorporate multi-level consistency into the learning process to improve performance. To this end, we first apply consistency regularization on both instance and semantic levels to encourage similarity of label prediction from the same instance, thus mitigating the overconfidence issue in consistency regularization-based methods. Furthermore, a logit-level consistency is applied between logits from raw and perturbed views to promote both invariance in the same instance's class predictions and diversity across different instances' class predictions. Experimental results on the HCP and UPenn-GBM datasets demonstrated the superiority of MCL over existing methods, showcasing its potential for improving accuracy in medical imaging applications with reduced annotation efforts.

References

1. Huang, W., et al.: A coarse-to-fine deformable transformation framework for unsupervised multi-contrast MR image registration with dual consistency constraint. IEEE Trans. Med. Imaging **40**(10), 2589–2599 (2021)
2. Li, C., Li, W., Liu, C., Zheng, H., Cai, J., Wang, S.: Artificial intelligence in multiparametric magnetic resonance imaging: A review. Med. Phys. **49**(10), e1024–e1054 (2022)
3. Litjens, G., et al.: A survey on deep learning in medical image analysis. Med. Image Anal. **42**, 60–88 (2017)
4. Zhou, Y., Huang, W., Dong, P., Xia, Y., Wang, S.: D-UNet: a dimension-fusion U shape network for chronic stroke lesion segmentation. IEEE/ACM Trans. Comput. Biol. Bioinf. **18**(3), 940–950 (2019)
5. Hu, X., Zeng, Y., Xu, X., Zhou, S., Liu, L.: Robust semi-supervised classification based on data augmented online elms with deep features. Knowl.-Based Syst. **229**, 107307 (2021)
6. Osman, Y.B.M., Li, C., Huang, W., Wang, S.: Collaborative learning for annotation-efficient volumetric MR image segmentation. J. Magn. Reson. Imaging (2023)
7. Gao, S., Zhang, Z., Ma, J., Li, Z., Zhang, S.: Correlation-aware mutual learning for semi-supervised medical image segmentation. In: International Conference on Medical Image Computing and Computer-Assisted Intervention, pp. 98–108. Springer (2023)
8. Wu, Y., Xu, M., Ge, Z., Cai, J., Zhang, L.: Semi-supervised left atrium segmentation with mutual consistency training. In: Medical Image Computing and Computer Assisted Intervention–MICCAI 2021: 24th International Conference, Strasbourg, France, 27 September –1 October 2021, Proceedings, Part II, pp. 297–306. Springer (2021)
9. Wang, S., et al.: Annotation-efficient deep learning for automatic medical image segmentation. Nat. Commun. **12**(1), 5915 (2021)
10. Liu, Y., Tian, Y., Chen, Y., Liu, F., Belagiannis, V., Carneiro, G.: Perturbed and strict mean teachers for semi-supervised semantic segmentation. In: Proceedings of the IEEE/CVF Conference on Computer Vision and Pattern Recognition, pp. 4258–4267 (2022)
11. Tarvainen, A., Valpola, H.: Mean teachers are better role models: weight-averaged consistency targets improve semi-supervised deep learning results. In: Advances in Neural Information Processing Systems, vol. 30 (2017)
12. Yu, L., Wang, S., Li, X., Fu, C.-W., Heng, P.-A.: Uncertainty-aware self-ensembling model for semi-supervised 3D left atrium segmentation. In: Shen, D., et al. (eds.) MICCAI 2019. LNCS, vol. 11765, pp. 605–613. Springer, Cham (2019). https://doi.org/10.1007/978-3-030-32245-8_67
13. Xu, Z., et al.: Ambiguity-selective consistency regularization for mean-teacher semi-supervised medical image segmentation. Med. Image Anal. **88**, 102880 (2023)
14. Huang, Z., Gai, D., Min, W., Wang, Q., Zhan, L.: Dual-stream-based dense local features contrastive learning for semi-supervised medical image segmentation. Biomed. Signal Process. Control **88**, 105636 (2024)
15. Zheng, M., You, S., Huang, L., Wang, F., Qian, C., Xu, C.: SimMatch: semi-supervised learning with similarity matching. In: Proceedings of the IEEE/CVF Conference on Computer Vision and Pattern Recognition, pp. 14471–14481 (2022)

16. Sun, B., Yang, Y., Zhang, L., Cheng, M.M., Hou, Q.: CorrMatch: Label propagation via correlation matching for semi-supervised semantic segmentation. In: Proceedings of the IEEE/CVF Conference on Computer Vision and Pattern Recognition, pp. 3097–3107 (2024)
17. Huang, W., Ye, M., Shi, Z., Du, B.: Generalizable heterogeneous federated cross-correlation and instance similarity learning. IEEE Trans. Pattern Anal. Mach. Intell. (2023)
18. Van Essen, D.C., et al.: The human connectome project: a data acquisition perspective. Neuroimage **62**(4), 2222–2231 (2012)
19. Bakas, S., et al.: The university of Pennsylvania glioblastoma (UPenn-GBM) cohort: advanced MRI, clinical, genomics, & radiomics. Scientific data **9**(1), 453 (2022)
20. Xie, L., et al.: Deep multimodal fusion network for the retinogeniculate visual pathway segmentation. In: 2023 42nd Chinese Control Conference (CCC), pp. 7946–7950. IEEE (2023)
21. Luo, X., et al.: Semi-supervised medical image segmentation via uncertainty rectified pyramid consistency. Med. Image Anal. **80**, 102517 (2022)
22. Luo, X., Chen, J., Song, T., Wang, G.: Semi-supervised medical image segmentation through dual-task consistency. In: Proceedings of the AAAI Conference on Artificial Intelligence, vol. 35, pp. 8801–8809 (2021)

Trustworthiness for Deep Learning Based Breast Cancer Detection Using Point-of-Care Ultrasound Imaging in Low-Resource Settings

Marisa Wodrich[1]([✉]), Jennie Karlsson[1], Kristina Lång[2,3], and Ida Arvidsson[1]

[1] Centre for Mathematical Sciences, Lund University, Lund, Sweden
marisa.wodrich@math.lth.se
[2] Department of Translational Medicine, Division of Diagnostic Radiology, Lund University, Lund, Sweden
[3] Unilabs Mammography Unit, Skåne University Hospital, Malmö, Sweden

Abstract. Poor survival for breast cancer in low- and middle-income countries is largely attributed to late-stage diagnosis and limited access to diagnostic tools. Therefore, we propose using point-of-care ultrasound (POCUS) as a low-cost diagnostic solution paired with a deep learning (DL) model for cancer detection. While using DL has shown great potential, it is crucial to ensure the trustworthiness of a model for diagnostic support in a setting with minimally trained healthcare workers, as a wrong prediction can lead to severe consequences. In the present study, we investigated different measures of trustworthiness from the field of uncertainty quantification, including deep ensembles, Bayesian neural networks and softmax score, for the application of breast cancer detection in POCUS imaging. The results show that all methods exhibit a correlation between uncertainty scores and correctness of prediction. The correlation was strongest when using an average ensemble and an entropy-based total predictive uncertainty. When excluding 30% of the test samples based on highest uncertainty scores, the area under the receiver operating characteristic curve (AUC) for cancer detection increases significantly from 95.6% to 98.9% (with 95% confidence intervals [93.3, 97.0] to [97.5, 99.9]), comparable to an expert radiologist's performance.

Keywords: Trustworthy AI · Uncertainty quantification · Breast cancer detection · Point-of-care ultrasound · low-resource settings

1 Introduction

Breast cancer is the most common type of cancer worldwide and the number one leading cause of cancer-related deaths in women [20]. While survival rates

Supplementary Information The online version contains supplementary material available at https://doi.org/10.1007/978-3-031-79103-1_5.

U. Anazodo et al. (Eds.): MImA 2024/EMERGE 2024, CCIS 2240, pp. 42–51, 2025.
https://doi.org/10.1007/978-3-031-79103-1_5

(SR) are high in many high-income countries, with five-year breast cancer SR above 90%, chances for a good outcome are poor in many low- and middle-income countries (LMIC), with SR of 66% in India, 40% in South Africa [16] and 12.1% in Kyadondo, Uganda [9]. The comparatively low survival rates in many LMIC are mainly caused by the lack of access to diagnostic tools, poor health infrastructure [4,17] and late-stage diagnosis [20]. Early detection and diagnosis are pivotal for reducing the ramifications resulting from breast cancer.

Many high-income countries have implemented large screening programs using mammography which has lead to a reduction in breast cancer morbidity and mortality [18,21]. Due to the high cost of mammography machines, as well as the need for trained personnel, the implementation of such a solution becomes unfeasible in many LMIC. A possible low-cost solution could be the use of point-of-care ultrasound (POCUS) as a diagnostic tool instead. POCUS devices are small, portable, cheap and easy to use, and therefore hold potential for being an accessible medical solution. To be applicable on a broader scale and with only minimally trained health care workers, a deep learning (DL) based classification algorithm could be used to interpret POCUS images in real-time [22]. It has previously been shown that ultrasound (US) images can be used for breast cancer detection in LMIC [19]. Furthermore, previous studies have shown that there is high potential for using specifically POCUS paired with DL for breast cancer classification [10] and applying this as a low-cost solution for classification of palpable breast lumps to replace mammography in LMIC [12].

DL has shown to have great potential for medical image assessment in the past [24]. However, DL algorithms tend to make overconfident decisions [11] which can be harmful and dangerous to trust [1]. In a field like medical diagnostics, where a wrong prediction can have severe consequences up to life-threatening outcomes, it becomes crucial to ensure the safety and trustworthiness of a classification algorithm. Interpreting a classifier's output together with quantified uncertainties in the prediction is considered to strengthen algorithmic transparency and trustworthiness [13], ultimately increasing its safety. Access to uncertainty scores can help determine if or how much a prediction can be trusted and can serve as a basis for deciding when computer-aided intervention can be used, and which cases require further examination or human intervention. The need for safety and trustworthiness of a breast cancer detection algorithm is extra high in LMIC settings where there is a limited number of highly trained radiologists that can confidently interpret breast ultrasound images alone.

In the scope of this work, different uncertainty quantification (UQ) methods were investigated for the purpose of detecting predictions that cannot be trusted and would need additional examination, see Fig. 1, ultimately leading to an increased trustworthiness and better performance of our classifier. Deep ensembles, Bayesian neural networks and softmax score were used to estimate uncertainties alongside predictions. Thorough evaluation was performed, by examining the classification performance when excluding different percentages of data based on estimated uncertainties.

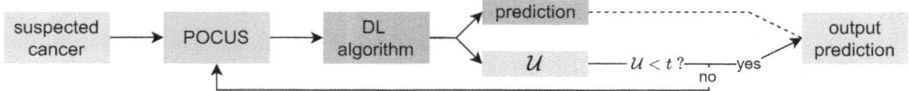

Fig. 1. Framework of our suggested approach. If the uncertainty score $\mathcal{U}$ for a prediction is below the trustworthiness threshold t, it is safe to believe the prediction.

2   Methods

2.1   Data

Both POCUS and conventional US images have been used in the scope of this work. All images were collected at Skåne University Hospital, Malmö, Sweden, and labeled by radiologists as either normal breast tissue, benign lesion or malignant lesion. The POCUS images were collected with a GE Vscan air CL probe [6]. The conventional US images were collected using the conventional US machines Logiq E9 and Logiq E10 [5]. The conventional US images were all used for training (total of 1160 images, of which 386 normal, 254 benign, 520 malignant), while POCUS images were divided into training (total of 814 images, of which 463 normal, 173 benign, 178 malignant) and testing (total of 531 images, of which 284 normal, 131 benign, 116 malignant). This study was performed in line with the principles of the Declaration of Helsinki. Approval was granted by the Swedish Ethical Review Authority of Region Skåne (2019-04607).

2.2   Base Classification Network

The base classification network used for this study is a convolutional neural network (CNN) with input size $180 \times 180 \times 1$. The CNN consists of five convolutional blocks followed by two dense blocks. A detailed description of the network architecture can be found in [10].

2.3   Uncertainty Quantification Methods

Deep Ensembles. Neural network ensembling [8], or deep ensembling, is a method that directly yields access to uncertainties alongside predictions. A deep ensemble consists of several independently trained neural networks (ensemble members) that are combined to make a more informed final decision. As a result, the deep ensemble is better at generalization [15] and more robust against overfitting [13]. Uncertainties can be quantified by looking at the separate predictions given by the ensemble members.

In the present study, an average ensemble using 20 ensemble members of the same architecture as the base classification network was used. To diversify the different ensemble members, the training parameters were randomized for each network: 0–15% of the training data was left out, the learning rate was between 0.0001 and 0.001, the optimizer was either Adam or RMSprop, the number of epochs was between 25 and 85, and the batch size was 8, 16, 32, 64 or 128. The uncertainty was measured in two different ways, using entropy-based total predictive uncertainty or variance-based uncertainty. The entropy-based total predictive uncertainty is defined as

$$\mathcal{U}_{entropy} = -\sum_{c=1}^{K} F_c(x|\theta)\log F_c(x|\theta), \tag{1}$$

with $F_c(x|\theta)$ being the prediction for class c of ensemble F, with this prediction being formed as the average over the different ensemble members

$$F_c(x|\theta) = \frac{1}{N}\sum_{m=1}^{N} f_c^{(m)}(x|\theta^{(m)}). \tag{2}$$

Here, $f^{(m)}$ is the m-th ensemble member and $\theta^{(m)}$ are the corresponding model parameters, thus the prediction F is based on the separate predictions of N models that are weighted equally.

The variance-based uncertainty is defined as the sum of the variances per class amongst all ensemble members,

$$\mathcal{U}_{var} = \sum_{c=1}^{K} \mathrm{Var}[F_c(x|\theta)] = \sum_{c=1}^{K} \frac{1}{N}\sum_{m=1}^{N} (f_c^{(m)}(x|\theta^{(m)}) - F_c(x|\theta))^2. \tag{3}$$

Bayesian Neural Networks. BNNs use probability distributions instead of fixed values as weights, containing information about the probabilities for a range of possible weight values [3]. Popular ways of building BNNs include variational inference, markov chain monte carlo and monte carlo dropout (MCD). Using sampling, uncertainties alongside the predictions can be estimated.

In the present study the MCD method was implemented. The base classification network was used as a backbone and the dropout layers were enabled during inference. A prediction was sampled 20 times and the average was used to determine the final prediction. As an uncertainty measure, entropy-based uncertainty was used. The calculations are based on Eq. 1 and 2. Here the 20 sampled prediction are treated the same as ensemble members for the ensemble method.

Softmax Score. The softmax score refers to the values outputted from the softmax activation function after the last layer of a neural network. While in theory a high softmax score should represent high internal model confidence, several studies have shown that this method is not the most reliable one [7,14]. Due to its simplicity, it however still remains of great interest.

2.4   Experiments

Breast Cancer Detection. The different UQ methods use the same classification network architecture for the task of breast cancer detection, but different training approaches defined by each UQ method. Since they serve as a base for the trustworthiness measures, a good performance of the cancer detection task needs to be ensured. All methods are compared and evaluated using the balanced accuracy score, binary balanced accuracy score (grouping benign and normal into one class) and the area under the receiver operating characteristic curve (AUC). The 95% confidence interval (CI) was calculated for the AUC using 1000 bootstrapped versions of the POCUS test set. Statistically significant differences in performance were determined using Mann-Whitney U-test, with p-values less than 0.05 considered significant.

Uncertainty Quantification Experiment. All methods were tested for the purpose of UQ and usefulness for increasing the trustworthiness. With each method, an uncertainty score was calculated for each prediction. The hypothesis is that there is a correlation between uncertainty scores and correctness of prediction, with prediction with low uncertainty score likely being correct and prediction with high uncertainty scores more likely being wrong. To evaluate the performance and quality of the UQ measures, the effect on cancer detection performance when excluding samples with high uncertainty scores was tested. Ultimately, the goal was to find a suitable threshold that excludes samples with high uncertainties, such that the predictions on the samples kept should be more trustworthy and safe to believe.

3   Results

The results of the breast cancer detection are shown in Table 1. The ensemble method performs significantly better than the other methods.

Table 1. Breast cancer detection performance of the different classification networks. All values are in percentage.

Method	ACC	binaryACC	AUC	AUC 95% CI
CNN	68.6	80.3	93.3	[91.1, 95.2]
Bayesian MCD	70.1	78.1	90.3	[87.4, 92.9]
Deep Ensemble	68.5	81.9	95.6	[93.8, 97.0]

The performance on the test data set when leaving out samples with high uncertainties is shown in Fig. 2. All methods show increased performance quality when samples with high uncertainties are left out. The desired effect is measured the strongest using the ensemble method with entropy-based uncertainty.

When 30% of the test data is excluded based on their uncertainties, the ensemble with entropy-based uncertainty performs the best. The results for that method for thresholds that would mark 0–30% of the test data as not trustworthy are shown in Table 2 including the 95% confidence intervals. Continuing with

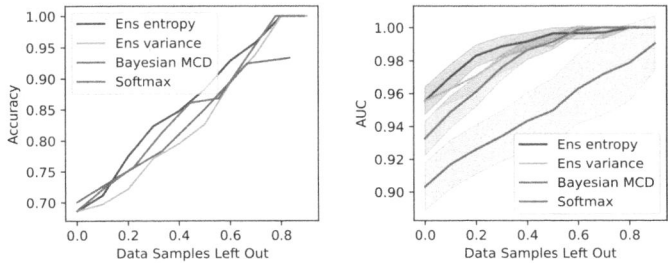

Fig. 2. Balanced accuracy (left) and AUC (right, mean ± one standard deviation) versus percentage of data samples left out. Samples with the highest uncertainty scores were gradually left out and the performance was calculated on the remaining data.

the same method, the distribution and histogram of uncertainty scores of correct and wrong predictions are shown in Fig. 3A. Based on that, the likelihood of a prediction being correct given its uncertainty score was calculated, also shown in Fig. 3A. Data samples with low, medium and high uncertainties are visualized in a two-dimensional space using principal component analysis (PCA) of the feature space representations from the first dense layer in the base CNN is shown in Fig. 3B. Examples of images from the subsets with highest and lowest uncertainties can be found in Fig. 4 and in the supplementary document.

Table 2. Performance on the remaining test data when different percentages of samples are left out, based on the uncertainties estimated by the ensemble method with entropy-based uncertainty. All values are in percentage.

Leave-out rate	ACC	binaryACC	AUC	AUC 95% CI
0	68.5	81.9	95.6	[93.3, 97.0]
10	71.3	85.4	97.0	[95.3, 98.4]
20	76.8	89.9	98.3	[96.8, 99.4]
30	81.5	93.9	98.9	[97.5, 99.9]

4   Discussion

The results show that overall, POCUS can be used as a low-cost imaging modality as a base for breast cancer detection in low-resource settings. All investigated classification networks achieved high performance at breast cancer detection, with AUCs between 90.3% and 95.6% (see Table 1). The achieved AUCs are comparable to other studies that are using conventional ultrasound [23], implying that POCUS is a promising tool for LMIC. While the ensemble achieves the best performance, it is also computationally the most costly one, with 20 times

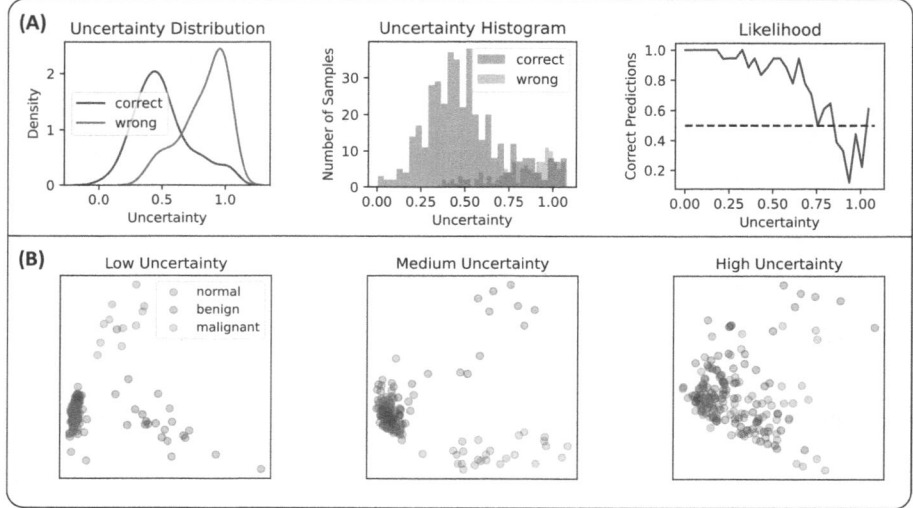

Fig. 3. (A) Distribution and histogram of uncertainties, based on the ensemble method with entropy-based uncertainty, for correctly and wrongly predicted test samples (left and middle), and likelihood of the prediction being correct (right). **(B)** PCA of different subsets of the test data sorted by their uncertainties. All subset are of the same size. The uncertainties were calculated with an average ensemble and entropy-based uncertainty. The PCA was performed on the feature space representations from the first dense layer in the base CNN of the respective POCUS images.

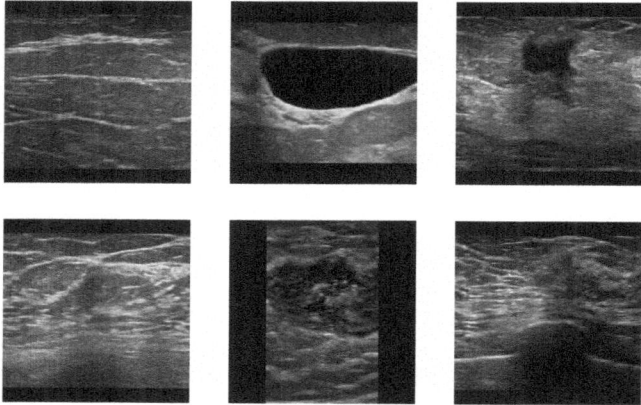

Fig. 4. Images with low uncertainties (upper row) compared to images with high uncertainties (lower row) for the classes normal, benign and malignant (left to right). The images with low uncertainties here were all classified correctly, while the images with high uncertainties were all misclassified.

higher inference and training time. However, inference is still fast (< 1s) and should not cause an issue even with limited computational resources.

In a medical application, eliminating the risk of wrong prediction which could lead to severe consequences is of great interest. For this purpose, several uncertainty quantification methods have been applied in this study, which are all showing great potential. All tested methods exhibit a correlation between uncertainty score and correctness of prediction. When leaving out samples with high uncertainty scores, the AUC and accuracy on the remaining samples increase (Fig. 2). The trustworthiness can thus be improved using any of those methods. For an ideal method, steep improvement in accuracy and AUC when excluding a small percentage of samples with high uncertainty scores is desired. This behavior is observed the strongest when using the ensemble with entropy-based uncertainty. Most images with low uncertainties can be clearly separated by class in the latent space, while images with high uncertainties are not clearly separable by class (Fig. 3B).

On average, wrong predictions tend to have higher uncertainties, and correct ones lower uncertainties. There is however an overlap and they are not fully distinguishable using the uncertainty scores, although the probability of a prediction being correct decreases for higher uncertainties (Fig. 3A). This suggests that a threshold for when not to trust a prediction is not the only way to improve the safety and trustworthiness of our breast cancer detection algorithm, but that a scale-based approach might also be suitable. In that case, a value describing the reliability could be given to the user alongside the prediction. It should be considered that this solution aims to be applied in LMIC with minimally-trained healthcare workers, where a simple threshold-based method might be more suitable. The threshold would decide if a prediction can be trusted, or if additional images need to be taken (see workflow in Fig. 1). Clinical studies in representative settings are needed to investigate which approach is the most appropriate.

Using a 30% threshold, our achieved sensitivity improves from 69.8% to 89.1% and the specificity from 94.0% to 99.7%. The performance is similar to the one from radiologists using conventional US images in a recent study [2], which had a better sensitivity of 98.5%, but a worse specificity of 90.8%.

5   Conclusion

Using POCUS as a low-cost modality that is feasible for LMIC has great potential, as the achieved results are similar to ones using conventional ultrasound. Using an AI-based approach yields good results, and when adding uncertainty thresholds, the performance increases to one similar to expert radiologists.

This study shows the importance and potential of improving trustworthiness of our classifier using UQ methods in a medical setting with minimally trained healthcare workers. To get a reliable and safe classifier, a threshold can be set at an uncertainty score where 30% of the test data would be rejected due to too large uncertainties, in which cases additional images would need to be captured. The best results were achieved using an average ensemble with entropy-based uncertainty. The AUC increased significantly from 95.6% with 95% CI [93.3, 97.0] to an AUC of 98.9% with 95% CI [97.5, 99.9]. These promising results serve as a proof-of-concept for using UQ methods to improve a classifier's trustworthiness. The clinical feasibility of using POCUS combined with our DL based algorithm needs to be further tested in a prospective setting, and collaborations with local healthcare expert are needed.

Acknowledgments. This work was supported by strategic research area eSSENCE and Analytic Imaging Diagnostics Arena (AIDA).

Disclosure of Interests. The authors have no competing interests to declare that are relevant to the content of this article.

References

1. Amodei, D., Olah, C., Steinhardt, J., Christiano, P., Schulman, J., Mané, D.: Concrete problems in AI safety. arXiv preprint arXiv:1606.06565 (2016)
2. Appelman, L., et al.: Us and digital breast tomosynthesis in women with focal breast complaints: results of the breast us trial (bust). Radiology **307**(4), e220361 (2023)
3. Blundell, C., Cornebise, J., Kavukcuoglu, K., Wierstra, D.: Weight uncertainty in neural network. In: International Conference on Machine Learning, pp. 1613–1622. PMLR (2015)
4. El Saghir, N.S., et al.: Breast cancer management in low resource countries (LRCS): consensus statement from the breast health global initiative. Breast **20**, S3–S11 (2011)
5. GE Healthcare: Logiq e10 ultrasound series. https://www.gehealthcare.com/products/ultrasound/logiq/logiq-e10. Accessed 15 Feb 2024
6. GE Healthcare: Vscan air. https://vscan.rocks/products-and-solutions/vscan-air-cl. Accessed 15 Feb 2024
7. Goodfellow, I.J., Shlens, J., Szegedy, C.: Explaining and harnessing adversarial examples. arXiv preprint arXiv:1412.6572 (2014)
8. Hansen, L.K., Salamon, P.: Neural network ensembles. IEEE Trans. Pattern Anal. Mach. Intell. **12**(10), 993–1001 (1990)
9. Joko-Fru, W.Y., et al.: Breast cancer survival in sub-saharan africa by age, stage at diagnosis and human development index: a population-based registry study. Int. J. Cancer **146**(5), 1208–1218 (2020)
10. Karlsson, J., et al.: Classification of point-of-care ultrasound in breast imaging using deep learning. In: Medical Imaging 2023: Computer-Aided Diagnosis, vol. 12465, pp. 192–200. SPIE (2023)
11. Lakshminarayanan, B., Pritzel, A., Blundell, C.: Simple and scalable predictive uncertainty estimation using deep ensembles. Adv. Neural Inf. Process. Syst. **30** (2017)
12. Love, S.M., et al.: Palpable breast lump triage by minimally trained operators in Mexico using computer-assisted diagnosis and low-cost ultrasound. J. Glob. Oncol. **4**, 1–9 (2018)
13. Nemani, V., et al.: Uncertainty quantification in machine learning for engineering design and health prognostics: a tutorial. arXiv preprint arXiv:2305.04933 (2023)

14. Nguyen, A., Yosinski, J., Clune, J.: Deep neural networks are easily fooled: high confidence predictions for unrecognizable images. In: Proceedings of the IEEE Conference on Computer Vision and Pattern Recognition, pp. 427–436 (2015)
15. Opitz, D., Maclin, R.: Popular ensemble methods: an empirical study. J. Artif. Intell. Res. **11**, 169–198 (1999)
16. Organization, W.H.: Global breast cancer initiative implementation framework: assessing, strengthening and scaling up of services for the early detection and management of breast cancer. World Health Organization (2023)
17. Pace, L.E., Shulman, L.N.: Breast cancer in sub-saharan africa: challenges and opportunities to reduce mortality. Oncologist **21**(6), 739–744 (2016)
18. Shapiro, S., Venet, W., Strax, P., Venet, L., Roeser, R.: Ten-to fourteen-year effect of screening on breast cancer mortality. J. Natl. Cancer Inst. **69**(2), 349–355 (1982)
19. Sood, R., et al.: Ultrasound for breast cancer detection globally: a systematic review and meta-analysis. J. Glob. Oncol. **5**, 1–17 (2019)
20. Sung, H., et al.: Global cancer statistics 2020: globocan estimates of incidence and mortality worldwide for 36 cancers in 185 countries. CA Cancer J. Clin. **71**(3), 209–249 (2021)
21. Tabar, L., Yen, M.F., Vitak, B., Chen, H.H.T., Smith, R.A., Duffy, S.W.: Mammography service screening and mortality in breast cancer patients: 20-year follow-up before and after introduction of screening. Lancet **361**(9367), 1405–1410 (2003)
22. Venkatayogi, N., et al.: From seeing to knowing with artificial intelligence: a scoping review of point-of-care ultrasound in low-resource settings. Appl. Sci. **13**(14), 8427 (2023)
23. Wu, G.G., Zhou, L.Q., Xu, J.W., Wang, J.Y., Wei, Q., Deng, Y.B., Cui, X.W., Dietrich, C.F.: Artificial intelligence in breast ultrasound. World J. Radiol. **11**(2), 19 (2019)
24. Zhou, S.K., et al.: A review of deep learning in medical imaging: imaging traits, technology trends, case studies with progress highlights, and future promises. Proc. IEEE **109**(5), 820–838 (2021)

Advancing the Reliability of Ultra-Low Field MRI Brain Volume Analysis Using CycleGAN

Peter Hsu[1,2]([✉]), Elisa Marchetto[1], Daniel Sodickson[1,2,3], Patricia Johnson[1,2,3], and Jelle Veraart[1,3]

[1] Bernard and Irene Schwartz Center for Biomedical Imaging, Department of Radiology, New York University Grossman School of Medicine, New York, NY, USA
`peter.hsu@nyulangone.org`
[2] Vilcek Institute of Graduate Biomedical Sciences, New York University Grossman School of Medicine, New York, NY, USA
[3] Center for Advanced Imaging Innovation and Research (CAI2R), Department of Radiology, New York University Grossman School of Medicine, New York, NY, USA

Abstract. The increasing prevalence of neurodegenerative diseases poses a significant threat to the well-being of the growing elderly population, with biological age being a major risk factor. This has increased the demand for cost-effective and informative neuroimaging modalities and analysis tools. Specifically, measuring brain volume is of critical importance as abnormal atrophy patterns are strong indicators of disease onset. Ultra-low field (ULF) MRI provides an innovative pathway to more accessible neuroimaging by mitigating various logistical, financial, and safety considerations associated with clinical MRI. However, the image quality of ULF-MRI impacts the reliability of brain volume analysis. Advancements in deep learning (DL) have proven capable of enhancing the image quality and analysis of medical images. Yet, these tools have not been fully realized for ULF-MRI, largely due to data scarcity as the technology is still relatively new. As a result, existing DL techniques for ULF image enhancement are trained with synthetically generated images, leading to potential "domain shift" issues when applied to real images. Here, we introduce a CycleGAN framework that learns with real ULF and high-field (HF) MRIs to improve the image enhancement process compared to existing methods. We demonstrate that this approach increases the accuracy of brain volume measurements based on improved correlations with paired clinical data and higher test-retest reliability across repeat measurements. Ultimately, our proposal has the potential to enhance clinical and research workflows through the increased accessibility and reliability of ULF-MRI.

Keywords: Ultra-Low Field MRI · CycleGAN · Vision Transformers · Brain Volume Analysis · Transfer Learning

Supplementary Information The online version contains supplementary material available at https://doi.org/10.1007/978-3-031-79103-1_6.

1 Introduction

Improving healthcare systems for the elderly across the globe is essential for the well-being of the growing aging population. By 2050, 16.5% of the world population will be 65 and older [1]. This is particularly important because biological age is one of the largest risk factors for developing psychiatric and neurodegenerative diseases, with Alzheimer's Disease (AD) and related dementias representing a common endpoint of various age-related pathologies and a growing cause of mortality [2,3]. There is a critical need for more accessible neuroimaging technology to break the barriers to equity in diagnoses and treatment. Access to software tools, such as those for volumetric analysis, is also necessary as abnormal atrophy patterns are a strong indicator of potential disease onset [4,5]. Today, clinical trials, or neuroscientific research in general, have significant demographic biases relative to the global population due to the limited accessibility of neuroimaging technology [6]. Therefore, major knowledge gaps exist in our understanding of treatment efficacy for neurodegenerative diseases such as AD in the global population. More accessible imaging technology, such as ultra-low field (ULF) MRI (<100 mT), holds the promise to bridge healthcare disparities and accelerate the study of biological brain aging across demographics and genders, both in health and disease [7]. However, technical challenges related to ULF image quality must be overcome to maximize its utility [8]. In this paper, we capitalize on emerging AI technologies to address these limitations of ULF-MRI, promoting its further development.

ULF-MRI has recently emerged as a promising, more accessible imaging modality as it mitigates the drawbacks of clinical MRI with significant cost reductions, flexibility for diverse patient conditions, and fewer operational constraints [7,8]. These factors enable more frequent deployment in local and remote clinics, thereby mitigating challenges related to accessibility, recruitment, and retention. This has also led to significant interest from both commercial and academic research sectors, signified by growing developments of novel ULF systems [9–11]. By expanding access to neuroimaging care, ULF-MRI holds the potential to improve health equity, especially for underserved and vulnerable populations. Yet, there are two critical barriers that impede the efficacy and utility of ULF-MRI. First, the signal-to-noise ratio (SNR) is significantly reduced compared to clinical MRI, leading to decreased spatial resolution and lengthened scan time to achieve adequate image quality. Second, field-dependent variations in MR contrast challenge its clinical interpretability and its compatibility with routinely used neuroimaging analysis software [12].

Recent advances in AI, or deep learning (DL), have promoted the development of techniques for improving SNR, synthesizing contrasts, enhancing spatial resolution, and accelerating the acquisitions of medical images [13–15]. Most of these techniques have been developed and evaluated for clinical MRI data, but they hold the potential to be translated to ULF-MRI. Unfortunately, while clinical MRI data is abundant, both in clinical data archives as well as public data repositories, such data collections are missing for ULF-MRI as this technology remains under active development [16–18]. Given the need for voluminous training data, researchers are actively exploring the utility of synthetic data that emulates ULF image quality and contrast [19–21]. However, training an image enhancement model with real-world ULF-MRI data rather than simulated data will improve robustness to biological variability, leading to augmented images that reflect true tissue properties and promote the use of ULF-MRI in clinical and research settings [7, 22].

We propose the use of a CycleGAN framework with real ULF and 3T training data to provide more accurate DL image enhancement [23]. This method learns to match distributions between ULF and 3T cohorts without needing paired subject data. In addition, we utilize 3D vision transformers in our framework to maximize spatial information processing for more efficient DL enhancement [24]. These have a distinct advantage over the classic U-Net architecture as spatial relationships are directly encoded and processed throughout the network layers rather than attempting to preserve them through skip connections [25].

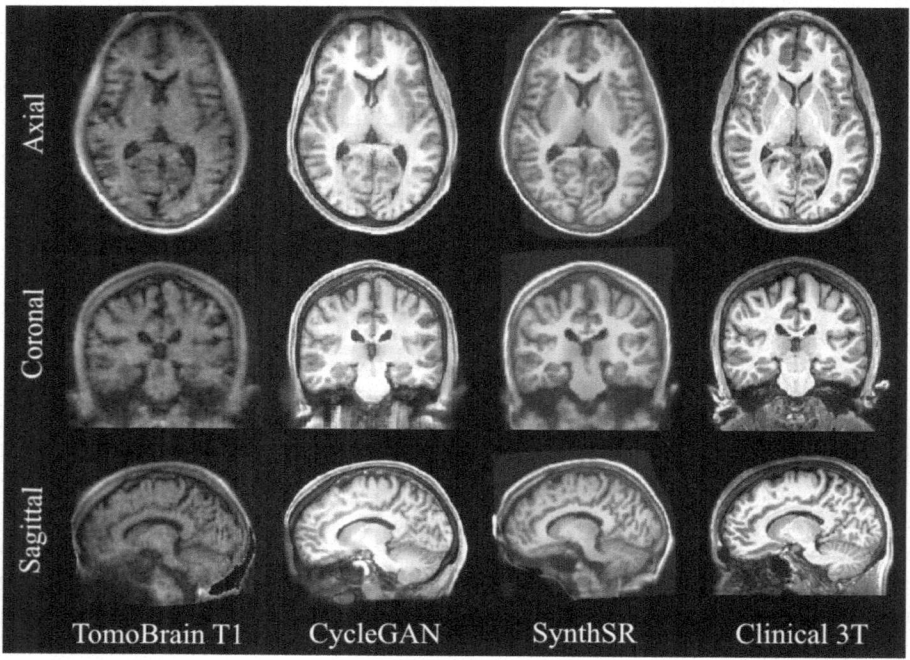

Fig. 1. Comparison of image quality between a composite T1w ULF image (here coined *TomoBrain*), our proposed CycleGAN output, the SynthSR model output, and a 3T clinical MP-RAGE for reference.

In this work, we aim to advance the volumetric analysis of brain structures using DL-enhanced ULF-MRI (Fig. 1). We evaluate the precision and accuracy of brain volumetrics relative to 3T clinical MRI scans that represent the gold standard, and in comparison to SynthSR, the current state-of-the-art tool that is trained on synthetically-generated images [19]. We demonstrate that the test-retest reliability and the agreement with clinical MRI data improves when using our proposed technology.

2 Methods

2.1 MRI Data

Study-Specific MRI Data. ULF images from 50 healthy control subjects (25 female) aged 23–72 years were acquired with a Hyperfine *Swoop*® scanner (64mT; v8.2.0 - v8.6.1) [9] (Table 1). Data were acquired using a 3D fast spin echo (FSE) imaging sequence with T1-weighted (T1w) and T2-weighted (T2w) contrast. For each contrast, data were acquired in axial, coronal, and sagittal orthogonal imaging directions. Total scan time was approximately 35 min.

From these 50 subjects, 29 subjects had previously received a T1w MP-RAGE scan on a clinical 3T MRI scanner, no more than 12 months prior to their ULF-MRI scan. These prior scans were acquired with a clinical protocol but with variations in the acquisition parameters. Additionally, 13 of the 50 subjects received a repeat scan on the ULF-MRI scanner, with an identical protocol for evaluating test-retest reproducibility.

Public Data Collections. T1w and T2w data from the M4Raw (N=468, 0.3T) and Human Connectome Project (HCP) (N=1113, 3T) repositories were collected for model training [26,27]. Additional HCP cases (N=45) were collected to assess the test-retest reliability of measurements on 3T MRI data.

2.2 MRI Pre-processing

Orthogonal T1w and T2w ULF acquisitions were combined into isotropic 1.5mm resolution template images through the advanced normalization tools (ANTs) multivariate template construction. We refer to the resulting T1w and T2w images as *TomoBrain* images [28,29]. The *TomoBrain* images have improved SNR and resolution as they leverage the overlapping and non-overlapping parts of the outer k-space of individual orthogonal imaging directions. Rigid registration was performed to align the T1w and T2w *TomoBrains* and HD-BET was used to create brainmasks [30,31]. Field inhomogeneities were further reduced using ANTs N4 bias correction. The same workflow for skull-stripping and bias correction was applied to the clinical 3T data.

For DL training and testing, all images were ultimately resampled to 1 mm isotropic resolution and padded to fit dimensions of $256 \times 256 \times 256$.

Table 1. A. Acquisition parameters of ULF T1w and T2w images including Echo Time (TE), Repetition Time (TR), Echo Train Length (ETL), and Scan Time. **B.** Acquisition information for our retrospectively collected MRI data: ULF T1w, ULF T2w, and 3T MP-RAGE.

A. Contrast	TE	TR	TI	ETL	Scan Time (per direction)
ULF T1w	6ms	1500ms	300ms	24	5:36min
ULF T2w	209ms	2000ms	N/A	80	5:46min
B. Contrast	Resolution (mm^2)		Slice Thickness (mm)		Field-of-View (FOV)
ULF T1w	1.6x1.6		5.0		179 x 220 x 180
ULF T2w	1.5x1.5		5.0		180 x 219 x 180
3T MP-RAGE	0.8x0.8 - 2.0x2.0		0.8 - 2.0		(166.4 x 240 x 256) - (256 x 256 x 192)

2.3   Model Framework and Network Architecture

Our proposed DL framework utilizes a modified Residual Vision Transformer (ResViT) model architecture for the generator models of a CycleGAN (Supplementary Fig. 1) [24]. This enabled the processing of 3D image volumes. Our chosen architecture combines the strengths of ResNet convolutional layers to process local image features and transformer attention mechanisms to integrate global context [32,33]. Contextual processing is furthered by additional residual connections that preserve higher-order information throughout the network. Model input consisted of ULF T1w and T2w *TomoBrain* images and model output consisted of corresponding 3T T1w and T2w images. The discriminator models were 3-layer convolutional networks.

2.4   Model Training

Training Data. Our training dataset consists of 423 low-field (LF) images from M4Raw, 35 of our collected ULF images, and 1000 HF images from HCP.

Our training process had two stages. Stage 1 used T1w and T2w images from the M4Raw and HCP datasets to pre-train our model for 400 epochs. Stage 2 utilized our T1w and T2w ULF and HCP images to fine-tune our model for 1200 epochs. This process leverages the benefits of transfer learning to enhance model performance in data-sparse environments [34,35].

We trained using the AdamW optimizer ($lr = 2 \times 10^{-4}$) with a learning rate decay (2×10^{-6}) for the final 100 and 200 epochs of stage 1 and stage 2 model training, respectively. Three loss functions were used: cycle consistency and identity loss for the generators, and GAN loss for the discriminators. Cycle consistency and identity loss use L1 loss, and GAN loss uses binary cross entropy loss. Patch-based ($128 \times 128 \times 128$) learning with a batch size of 1 was implemented to reduce computational constraints. Our model was trained with a single A100 GPU using a SLURM HPC.

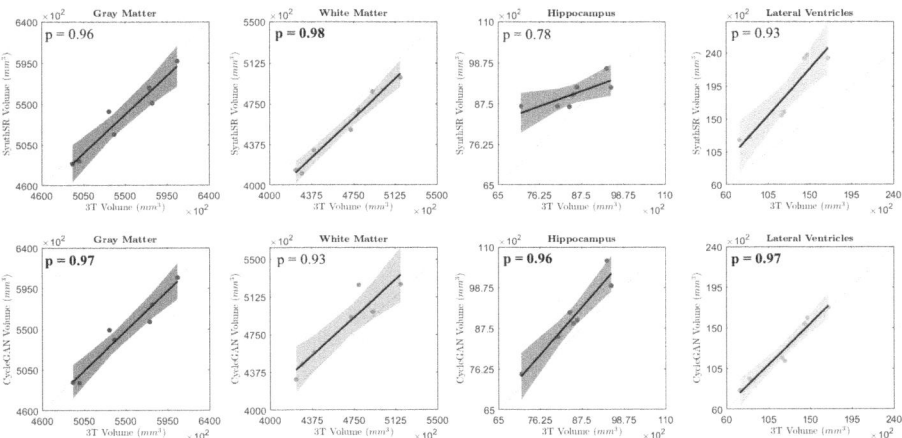

Fig. 2. Pearson correlation with 95% confidence interval bounds between estimated volumes of DL-enhanced ULF images (y-axis) and paired clinical 3T images (x-axis). Comparisons are made between the output of SynthSR (top row) and our proposed CycleGAN (bottom row). The highest performing results are denoted with bold.

2.5 Volumetric Analysis

The primary metrics of interest are the volumes of gray matter (GM), white matter (WM), hippocampus (HC), lateral ventricles (LV), and total intracranial volume (ICV). We utilized SynthSeg+, a contrast-agnostic AI-based segmentation tool, to provide brain volume estimates from T1w DL-enhanced ULF images and T1w clinical 3T images (ground truth) [36]. This tool has the potential to improve the reliability and consistency of segmentations on MRI data of varying image quality and contrasts.

2.6 Statistical Analysis

Test Data. The testing dataset consists of 13 subjects with repeat ULF-MRI scans. Of these subjects, 8 had paired 3T MP-RAGE data. After a visual assessment, one paired case was excluded as an outlier from evaluation as several brain structures were not captured in the initial ULF scan.

Clinical Agreement. We evaluated the agreement between subject-matched DL-enhanced ULF-MRI and clinical 3T MRI in terms of brain volumes using Pearson correlation and Lin's Concordance Correlation Coefficient (CCC) [37]. These quantify how well observations conform relative to a clinical gold standard.

Test-Retest Reliability. We quantified the test-retest reproducibility of two measurements obtained from repeat ULF scans that were enhanced by DL models. This assessment utilized Lin's Concordance Correlation Coefficient (CCC) and the coefficient of variation (CoV) derived from the absolute mean difference of the estimated volumes [38].

3   Results

3.1   Clinical Agreement

In Fig. 2, we show a scatter plot of clinically-relevant brain morphological features that were derived from subject-matched DL-augmented ULF- and clinical MRIs. The 95% confidence interval for the linear regression is shown alongside the Pearson correlation coefficients. Lin's CCC is also computed between paired cases (Table 2).

Table 2. Lin's CCC between paired cases of DL-enhanced ULF images and 3T MRIs. Comparisons are made between estimated brain volumes from the output of SynthSR (row 1) and our CycleGAN (row 2). The highest performance is denoted with bold.

	ICV	GM	WM	HC	LV
SynthSR	0.9199	0.9195	**0.9081**	0.4586	0.4355
CycleGAN	**0.9281**	**0.9687**	0.8303	**0.7812**	**0.9667**

3.2   Test-Retest Reliability

The reproducibility in estimating brain morphological features is shown in Table 3, for both SynthSR and our proposed CycleGAN approach. We quantified test-retest reliability using Lin's CCC and variability using the CoV, expressed in percentage. For comparison, we also present these metrics from a set of 3T MRI data using the HCP test-retest dataset.

Table 3. Test-retest comparisons of HCP 3T reference data (row 1), SynthSR (row 2), and our CycleGAN (row 3) across ICV, GM, WM, HC, and LV volumes. Metrics include Lin's CCC and the CoV in percentage for each brain volume. The highest performing DL model results are denoted with bold.

	ICV		GM		WM		HC		LV	
	CCC	CoV	CCC	CoV	CCC	CoV	CCC	CoV	CCC	CoV
HCP 3T	0.9993	0.26	0.9988	0.41	0.9993	0.34	0.9934	0.74	0.9971	2.76
SynthSR	0.9475	1.95	0.9725	1.65	0.9497	2.03	0.7212	3.79	0.9266	8.26
CycleGAN	**0.9884**	**0.69**	**0.9944**	**0.69**	**0.9965**	**0.44**	**0.9139**	**2.68**	**0.9982**	**2.21**

4 Discussion

The training and validation of AI tools for ULF image enhancement benefit from the acquisition of a subject-matched data collection across field strengths. However, such an endeavor is time-consuming and expensive, prolonging the development of emerging technologies. Previous strategies proposed to mitigate such challenges by using synthetic data in model training. We propose a Cycle-GAN model that is pre-trained with publicly available data to reduce the need for voluminous training data compared to training such a model from scratch.

The acquisition of ULF-MRI is typically performed using anisotropic voxel sizes (e.g. $1.5 \times 1.5 \times 5.0$ mm^3) to recover SNR. We here acquire such images in three orthogonal imaging directions (axial, sagittal, and coronal) and merge them in a composite *TomoBrain* image with an isotropic 1.5mm voxel size. These resulting images are used for DL augmentation. While the image acquisition is significantly longer (approximately 30 mins for T1w and T2w), we leverage complementary k-space information to improve the spatial resolution of the data, thereby lowering the resolution gap between the input and generated output of our model. This is conceptually different from SynthSR, or similar strategies, which use data from a single imaging direction. In follow-up studies, we will accelerate our protocol by training our network on subsets of the input data, varying both the contrasts and directions, to maximize performance with minimal scan time [39,40].

Our CycleGAN showcases significantly improved test-retest reliability in the quantification of brain volumes compared to state-of-the-art approaches - with CoV values below 1% in most regions. We hypothesize that the gain in precision can be attributed to the data-driven upsampling of the ULF-MRI data by merging orthogonal measurements. While test-retest reliability is "moderate" to "excellent" following the suggested interpretation of Lin's CCC, we observe an expected loss in precision compared to 3T MRI data, primarily in smaller anatomical regions such as the hippocampus. In research studies, any loss in precision can be compensated by an increased sample size. This is likely to be more attainable given the increased accessibility and scalability of ULF technology. Overall, given our current test-retest values, we hypothesize that the sensitivity of research studies will improve when leveraging ULF-MRI to promote more diverse and larger cohort studies.

The estimated brain volumes that are derived from our subject-matched DL-augmented ULF and clinical MRI are highly correlated (p = 0.93–0.97), but significant discrepancies are observed in the hippocampus. The limited sample size, spatial resolution, and field-dependent differences in MR contrast might be culprits that will be investigated in additional experiments. Moreover, we will further evaluate the impact of segmentation tools used for volumetric analysis, e.g. SynthSeg+ compared to FreeSurfer and/or FastSurfer [41,42].

We conclude that training DL models to augment ULF-MRI in terms of SNR, spatial resolution, and contrast using *real-world* data is feasible, specifically when leveraging public data collections for pre-training. Overall, we observed increased precision and accuracy in the estimation of brain morphological features compared to current state-of-the-art techniques that are trained on synthetic data.

Acknowledgments. This work has grant support through NIH P41 EB017183. This work was performed under the rubric of the Center for Advanced Imaging Innovation and Research (CAI2R, www.cai2r.net), an NIBIB National Center for Biomedical Imaging and Bioengineering.

Disclosure of Interests. This research was conducted as part of a research agreement with Hyperfine Inc. This does not carry any financial incentives and there were no influences to the study design, data collection/analysis, or preparation of this manuscript.

References

1. United Nations Department of Economic and Social Affairs, Population Division: World Population Prospects 2022: Summary of Results. UN DESA/POP/2022/TR/NO. 3 (2022)
2. Brown, R.C., et al.: Neurodegenerative diseases: an overview of environmental risk factors. Environ. Health Perspect. **113**(9), 1250 (2005)
3. Fox, N.C., Schott, J.M.: Imaging cerebral atrophy: normal aging to Alzheimer's disease. Lancet **363**(9406), 392–394 (2004)
4. Bell-McGinty, S., et al.: Differential cortical atrophy in subgroups of mild cognitive impairment. Arch. Neurol. **62**(9), 1393–1397 (2005)
5. Frenzel, S., et al.: A biomarker for Alzheimer's disease based on patterns of regional brain atrophy. Front. Psych. **10**, 953 (2020)
6. van Dyck, C.H., et al.: Lecanemab in early Alzheimer's disease. N. Engl. J. Med. **388**(1), 9–21 (2023)
7. Kimberly, W. T., et al.: Brain imaging with portable low-field MRI. Nat. Rev. Bioeng. **1**(9), Article 9 (2023)
8. Parasuram, N.R., et al.: Future of neurology & technology: neuroimaging made accessible using low-field, portable MRI. Neurology **100**(22), 1067–1071 (2023)
9. Chetcuti, K., et al.: Implementation of a low-field portable MRI scanner in a resource-constrained environment: our experience in Malawi. AJNR Am. J. Neuroradiol. **43**(5), 670–674 (2022)
10. Liu, Y., et al.: A low-cost and shielding-free ultra-low-field brain MRI scanner. Nat. Commun. **12**(1), 7238 (2021)
11. Guallart-Naval, T., et al.: Portable magnetic resonance imaging of patients indoors, outdoors and at home. Sci. Rep. **12**(1), Article 1 (2022)
12. Arnold, T.C., et al.: Low-field MRI: clinical promise and challenges. J. Magn. Reson. Imaging **57**(1), 25–44 (2023)
13. Preetha, C.J., et al.: Deep-learning-based synthesis of post-contrast T1-weighted MRI for tumor response assessment in neuro-oncology: a multicentre, retrospective cohort study. Lancet Digit. Health **3**(12), e784–e794 (2021)
14. Pham, C.-H., et al.: Brain MRI super-resolution using deep 3D convolutional networks. In: 2017 IEEE 14th International Symposium on Biomedical Imaging (ISBI 2017), pp. 197–200 (2017)
15. Johnson, P.M., et al.: Deep learning reconstruction enables prospectively accelerated clinical knee MRI. Radiology **307**(2), e220425 (2023)

16. Petersen, R.C., et al.: Alzheimer's disease neuroimaging initiative (ADNI). Neurology **74**(3), 201–209 (2010)
17. Knoll, F., et al.: fastMRI: a publicly available raw k-space and DICOM dataset of knee images for accelerated MR image reconstruction using machine learning. Radiol. Artif. Intell. **2**(1), e190007 (2020)
18. Littlejohns, T.J., et al.: The UK Biobank imaging enhancement of 100,000 participants: rationale, data collection, management and future directions. Nat. Commun. **11**(1), 2624 (2020)
19. Iglesias, J. E., et al.: SynthSR: A public AI tool to turn heterogeneous clinical brain scans into high-resolution T1-weighted images for 3D morphometry. Sci. Adv. **9**(5) (2023)
20. Lin, H., et al.: Low-field magnetic resonance image enhancement via stochastic image quality transfer. Med. Image Anal. **87**, 102807 (2023)
21. Lau, V., et al.: Pushing the limits of low-cost ultra-low-field MRI by dual-acquisition deep learning 3D superresolution. Magn. Reson. Med. **90**(2), 400–416 (2023)
22. Chen, R.J., et al.: Synthetic data in machine learning for medicine and healthcare. Nat. Biomed. Eng. **5**(6), Article 6 (2021)
23. Zhu, J.-Y., et al.: Unpaired image-to-image translation using cycle-consistent adversarial networks. arXiv preprint arXiv:1703.10593 (2020)
24. Dalmaz, O., et al.: ResViT: residual vision transformers for multi-modal medical image synthesis. IEEE Trans. Med. Imaging **41**(10), 2598–2614 (2022)
25. Ronneberger, O., et al.: U-net: convolutional networks for biomedical image segmentation. arXiv preprint arXiv:1505.04597 (2015)
26. Lyu, M., et al.: M4Raw: a multi-contrast, multi-repetition, multi-channel MRI k-space dataset for low-field MRI research. Sci. Data **10**(1), Article 1 (2023)
27. Van Essen, D.C., et al.: The WU-Minn human connectome project: an overview. Neuroimage **80**, 62–79 (2013)
28. Avants, B.B., et al.: A reproducible evaluation of ANTs similarity metric performance in brain image registration. Neuroimage **54**(3), 2033–2044 (2011)
29. Deoni, S.C.L., et al.: Simultaneous high-resolution T2-weighted imaging and quantitative T2 mapping at low magnetic field strengths using a multiple TE and multi-orientation acquisition approach. Magn. Reson. Med. **88**(3), 1273–1281 (2022)
30. Tournier, J.-D., et al.: MRtrix3: a fast, flexible and open software framework for medical image processing and visualization. NeuroImage **202**, 116137 (2019)
31. Isensee, F., et al.: Automated brain extraction of multi-sequence MRI using artificial neural networks. Hum. Brain Mapp. **40**(17), 4952–4964 (2019)
32. He, K., et al.: Deep residual learning for image recognition. arXiv preprint arXiv:1512.03385 (2015)
33. Dosovitskiy, A., et al.: An image is worth 16×16 words: transformers for image recognition at scale. arXiv preprint arXiv:2010.11929 (2021)
34. Valverde, J.M., et al.: Transfer learning in magnetic resonance brain imaging: a systematic review. J. Imaging **7**(4), 66 (2021)
35. Minaee, S., et al.: Deep-COVID: predicting COVID-19 from chest X-ray images using deep transfer learning. Med. Image Anal. **65**, 101794 (2020)
36. Billot, B., et al.: Robust machine learning segmentation for large-scale analysis of heterogeneous clinical brain MRI datasets. Proc. Natl. Acad. Sci. **120**(9), e2216399120 (2023)
37. McBride, G. B.: A proposal for strength-of-agreement criteria for Lin's concordance correlation coefficient. NIWA client report: HAM2005-062 45, 307-310 (2005)

38. Veraart, J., et al.: The variability of MR axon radii estimates in the human white matter. Hum. Brain Mapp. **42**(7), 2201–2213 (2021)
39. Marquez, E.S., et al.: Deep cascade learning. IEEE Trans. Neural Netw. Learn. Syst. **29**(11), 5475–5485 (2018)
40. Meyes, R., et al.: Ablation studies in artificial neural networks. arXiv preprint arXiv:1901.08644 (2019)
41. Fischl, B.: FreeSurfer. Neuroimage **62**(2), 774–781 (2012)
42. Henschel, L., et al.: FastSurfer—a fast and accurate deep learning based neuroimaging pipeline. Neuroimage **219**, 117012 (2020)

Deep Learning Based Non-invasive Meningitis Screening Using High-Resolution Ultrasound in Neonates and Infants from Mozambique, Spain and Morocco

Beatrice M. Jobst[1]([✉]), Francesc Carandell[1], Sara Ajanovic[2], Hassan Sial[3],
Javier Jiménez[1], Rita Quesada[1], Fabião Santos[1], Manuela Lopez-Azorín[4],
Eva Valverde[5], Marta Ybarra[5], M. Carmen Bravo[5], David Muñoz[6], Thais Agut[6],
Barbara Salas[6], Nuria Carreras[7], Ana Alarcón[7], Martín Iriondo[7], Carles Luaces[6],
Muhammad Sidat[8], Mastalina Zandamela[9], Paula Rodrigues[10], Luzidina Martins[11],
Uneisse Cassia[8], Justina Bramugy[9], Anelsio Cossa[9], Campos Mucasse[9],
W. Chris Buck[12], Sara Arias[2], Chaymae El Abbass[13,14], Houssain Tligui[13,14],
Amina Barkat[13,14], Najat Amalik[13,14], Imane Zizi[13,14], Alberto Ibáñez[15],
Montserrat Parrilla[15], Luis Elvira[15], Cristina Calvo[16,17,18,19], Adelina Pellicer[5,19],
Fernando Cabañas[4,20], Quique Bassat[2,9,21,22,23], and Paula Petrone[3]

[1] Kriba, Barcelona Science Park, 08028 Barcelona, Spain
Beatrice.jobst@kriba.ai
[2] Barcelona Institute for Global Health (ISGlobal) - Hospital Clínic, Universitat de Barcelona, 08036 Barcelona, Spain
[3] Biomedical Data Science Team, Barcelona Institute for Global Health (ISGlobal), 08003 Barcelona, Spain
[4] Department of Pediatrics and Neonatology, Quirónsalud Madrid University Hospital, 28223 Madrid, Spain
[5] Neonatology Department, La Paz University Hospital - IdiPaz (Hospital La Paz Institute for Health Research), 28046 Madrid, Spain
[6] Emergency Department, Sant Joan de Déu Hospital, Institut de Recerca Sant Joan de Déu, Universitat de Barcelona, 08950 Barcelona, Spain
[7] Neonatology Department, Sant Joan de Déu Hospital, Institut de Recerca Sant Joan de Déu, Universitat de Barcelona, 08950 Barcelona, Spain
[8] Faculdade de Medicina, Universidade Eduardo Mondlane, Maputo 257, Mozambique
[9] Centro de Investigação em Saúde de Manhiça (CISM), Manhiça 1929, Mozambique
[10] Maputo Central Hospital, Maputo, Maputo 1100, Mozambique
[11] Hospital Geral de Mavalane, Maputo 1100, Mozambique
[12] David Geffen School of Medicine, University of California Los Angeles, Los Angeles 90095, USA
[13] Centre National de Référence en Néonatologie Et Nutrition - Hôpital d'enfants-Centre Hospitalier Universitaire Ibn Sina, 6527 Rabat, Morocco
[14] Équipe de Recherche en Santé Et Nutrition du Couple Mère-Enfant, Faculté de Médecine et Pharmacie, Université Mohammed V Rabat, 6203 Rabat, Morocco
[15] Instituto de Tecnologías Físicas y de la Información (CSIC), 28006 Madrid, Spain

U. Anazodo et al. (Eds.): MImA 2024/EMERGE 2024, CCIS 2240, pp. 63–72, 2025.
https://doi.org/10.1007/978-3-031-79103-1_7

[16] Pediatrics and Infectious Diseases Department, La Paz University Hospital, Fundación IdiPaz. 28046, Madrid, Spain

[17] Biomedical Research Network Centre for Infectious Diseases (CIBERINFEC), Carlos III Health Institute, 28029 Madrid, Spain

[18] Translational Research Network in Pediatric Infectious Diseases (RITIP), 28046 Madrid, Spain

[19] Universidad Autonoma de Madrid, 28049 Madrid, Spain

[20] Biomedical Research Foundation, La Paz University Hospital-IDIPAZ, 28046 Madrid, Spain

[21] ICREA, 08010 Barcelona, Spain

[22] Pediatrics Department, Hospital Sant Joan de Déu, Universitat de Barcelona, 08950 Barcelona, Spain

[23] CIBER Epidemiology and Public Health, Carlos III Health Institute, 28029 Madrid, Spain

Abstract. Meningitis is a life-threatening disease, resulting in severe neurological damage or death if not treated in time. The standard diagnostic method is an invasive lumbar puncture (LP), not exempt of complications and not always feasible, especially in resource-poor settings. To overcome these challenges, we developed Neosonics®, a novel non-invasive high-resolution ultrasound (HRUS) technology that detects backscatter signals from white blood cells (WBCs) in cerebrospinal fluid (CSF) below the infant fontanel. Using deep learning (DL) for image analysis, this technology classifies patients based on WBC levels, providing a rapid, non-invasive screening method for infant meningitis and with this, limiting indications for LPs for positive cases only, if not contraindicated. A convolutional neural network was trained and validated using CSF HRUS patient data gathered from cohorts from three different countries, Mozambique, Spain and Morocco. Then, a soft voting ensemble-learning technique was applied to give a classification on a patient-level. The DL model showed on a single-image level a sensitivity (SE) and specificity (SP) of 71.1% and 87.0% for the Mozambican cohort, SE = 73.3% and SP = 75.5% for the Spanish cohort, and SE = 72.0% and SP = 88.7% for the Moroccan cohort. On a patient-level, the overall SE and SP was 94.4% and 94.8%, respectively. We demonstrate in this study the significance of combining HRUS and DL for the non-invasive detection of infant meningitis, offering an efficient, cost-effective solution suitable also for limited resource settings and remote areas.

Keywords: Neonatal and Infant Meningitis · High-Resolution Ultrasound · Deep Learning

1   Introduction

Meningitis is a life-threatening disease, particularly dangerous for newborns and infants. Early and accurate diagnosis is crucial to prevent severe neurological damage and high mortality rates [1]. In 2019, there were an estimated 2.51 million new cases and 336,000 deaths worldwide, with nearly half of these affecting children under five [2].

Q. Bassat and P. Petrone—co-supervising author

The standard diagnostic method, lumbar punctures (LPs), involves collecting cerebrospinal fluid (CSF) for laboratory analysis, to detect elevated white blood cell (WBC) counts or pathogens. LPs can cause complications like infections, bleeding, nerve damage, or respiratory arrest, especially in young infants, requiring specialized equipment, skilled personnel, and laboratory infrastructure [3]. This poses significant challenges in resource-poor settings, leading to delays in diagnosis and treatment leading to empirical management of the disease and consequently unnecessary antibiotic use [4]. In high-income countries on the other hand, LPs are performed proactively on high-risk infants, but due to the low incidence of the disease, less than five percent yield positive results [5–7], underscoring the need for less invasive diagnostic procedures.

To address these challenges, we developed Neosonics®, a novel non-invasive high-resolution ultrasound (HRUS) technology that detects WBC backscatter signals in CSF below the infant fontanel. Using deep learning (DL) for image analysis, this technology classifies patients based on WBC levels, offering a rapid, non-invasive screening method for infant meningitis and thus limiting indications for LPs for positive cases only, if not contraindicated.

Our team previously developed a deep learning methodology to predict meningitis from HRUS images acquired by the Neosonics® device [8–10]. In this study, we train and validate a convolutional neural network for the screening of meningitis using patient data gathered from three countries, Mozambique (MZB), Spain (ESP) and Morocco (MOR). This approach further demonstrates the adaptability of the algorithm across different ethnicities and backgrounds. We apply Explainable AI (XAI) techniques to enhance the transparency of the AI-driven decision-making process, allowing bias detection, models' response interpretation and promotion of ethical AI usage. For this, we use Gradient-weighted Class Activation Mapping (Grad-CAM) and statistical image analysis to improve the screening process transparency and outcomes.

This study demonstrates the potential of combining HRUS and DL to help modulate the clinician's meningitis suspicion level and thus limit the use of invasive LPs in neonatal and infant care only to positive patients, providing an efficient, non-invasive and cost-effective solution suitable also for limited resource settings and remote areas.

2   Materials and Methods

2.1   Data Acquisition and Ground Truth Generation

From June 2020 to June 2023, we prospectively recruited neonates and infants with an open anterior fontanelle and suspected meningitis in 5 different hospitals in 3 countries: Hospital Central de Maputo in MZB; La Paz University Hospital, Quirónsalud University Hospital and Sant Joan de Déu University Hospital in ESP; and Hôpital d'Enfants-Centre Hospitalier Universitaire Ibn Sina in MOR. The study in MZB was approved by the Mozambican National Bioethics Committee (608/CNBS/23); in ESP by the Spanish ethical committee (HULP 5183) and the Spanish Agency of Medicines and Medical Devices (AEMPS 711/18/EC); and in MOR by the Moroccan Ethics Committee (C70–20) and by the Directorate of Medicines and Pharmacy from Ministry of Health and Social Protection of the Kingdom of Morocco (N°01 RB/DMP/18/AIC).

Patient inclusion criteria were having an open anterior fontanel, being younger than 24 months, requiring a lumbar puncture (LP) due to suspected meningitis, having an LP performed within 24 h of image acquisition, and obtaining consent from legal guardians. Exclusion criteria were central nervous system malformations, history of traumatic events, or suspected/confirmed intracranial bleeding. The data were integrated as three different cohorts, one corresponding to each country (MZB, ESP and MOR).

For ground truth generation, WBC count was performed using the Neubauer chamber in MZB, the Fuchs-Rosenthal chamber in ESP and the Malassez chamber in MOR, with a standard correction formula applied in hematic punctures (reducing 1 WBC for each 1000 red blood cells (RBCs)). Samples with $\geq$ 100,000 RBCs were excluded [11]. A WBC count of 30 or more WBC/mm3 in CSF was set as the positivity threshold for the purpose of this study [12].

The Neosonics® device, a compact ultrasound system with a central frequency of 20 MHz, can detect backscatter signals of WBCs in suspension and analyze serous body fluids' composition with higher sensitivity than conventional ultrasound systems. For the HRUS data collection, we applied the Neosonics® ultrasonic probe positioned over the open anterior fontanelle region of the neonates' or infants' head to non-invasively record HRUS images of the underlying CSF.

2.2 Data Preprocessing

All images were processed using standard image processing methods using the SciPy Python library [13]. First, a Butterworth bandpass filter was applied, then the signal envelopes were calculated using the Hilbert transform and a min-max normalization was applied to a uniform range. Additionally, for model training and testing, a one-directional Sobel filter was applied to the images, to enhance the visibility of the cell patterns in case of their presence.

2.3 Deep Learning Model Training and Patient Level Prediction

For each participant an HRUS scan was performed, where multiple HRUS images were acquired. Good quality images were manually selected (good US sensor coupling, no blood vessels present, correctly located recording area) [8] and HR images were labelled as Control (<30 cells/mm3) or Case ($\geq$30 cells/mm3) [12]. We trained a deep learning (DL) model with these images to classify between Controls and Cases based on a ResNet50 architecture [14] pretrained on ImageNet. The final convolutional block was fine-tuned, and fully connected layers were replaced with a global average pooling layer and a dense layer for binary classification. We used binary cross-entropy loss and the Adam optimizer with a learning rate varying between [5E-6, 5E-7].

Each image frame was assigned a probability of being Case or Control by the model. Then, a soft-voting ensemble-learning technique was applied, by averaging the predicted probability scores over all images per patient for each class to obtain one probability score for each class per patient. The higher class score was defined as the final patient level classification result.

To ensure robust validation, we used a cross-cohort training and testing strategy. Specifically, we trained the model on images from two cohorts and tested it on the third cohort, where data was imbalanced and thus less suited for training. Within the two cohorts used for training, a leave-one-patient-out methodology was adopted, ensuring that each patient was tested individually. Additionally, we tested the algorithm on an independent cohort to evaluate its validity across different ethnicities and backgrounds.

2.4 Explainable AI (XAI)

To better understand the deep learning model's decision-making processes, we used Gradient-weighted Class Activation Mapping (GradCAM) [15] using the pytorch-gradcam library. We applied the GradCAM function, utilizing the final convolutional layer of the DL model, generating heatmaps for each image in the range of [0, 255]. After qualitative visual inspection, we calculated the mean GradCAM score per image and computed the median over these scores to obtain a patient-level GradCAM score for quantitative evaluation. This was compared with the model's predicted output using the Pearson correlation coefficient.

We also performed statistical comparisons of image characteristics between Control and Case images. We computed the relative pixel intensity frequency over all images for each group: Pixels values ranged from [0, 255]. We summed the frequency of each pixel value over all Control and Case images, respectively, separately for each cohort. The resulting values were divided by 256 (pixel value range length) to obtain the relative pixel intensity frequency. We tested for difference between Case and Control groups using a Wilcoxon Rank sum test, considering p-values < 0.05 as significant.

3   Results

We recruited 68 patients from a public teaching hospital in MZB, 30 patients from 3 Spanish University Hospitals and 127 patients from a Moroccan University Hospital. After exclusions, 2237 good quality HR images were used to train the algorithm from 32 patients (16 patients from MZB (median age: 47.5 days) and 16 from ESP (median age: 23 days)). Two patients had a repeated LP with an associated HRUS scan, thus resulting in 34 scans (10 Cases (3 MZB, 7 ESP), 24 Controls (14 MZB, 10 ESP)).

The algorithm was trained in a leave-one-out fashion within these two patient cohorts, to be able to report results for each patient. Finally, a model was trained including the two cohorts (MZB + ESP) and tested on the third Moroccan cohort, consisting of 1545 images from 42 scans (8 Cases, 34 Controls; median age: 3.5 days) [8].

3.1 Image- and Patient-Level Results of the DL Model

At an image level, the DL model showed a sensitivity (SE) and specificity (SP) of 71.1% and 87.0%, respectively, for the Mozambican cohort; SE was 73.3% and SP 75.5% for the Spanish cohort; and for the Moroccan cohort SE and SP were 72.0% and 88.7%. The confusion matrices for single image level results separated per cohorts are shown in Fig. 1A.

The number of images analyzed per patient in this study ranged from 6 to 233 (25.39 $\pm$ 42.72 on average). The patient-level classification integrated information from all the patient's images by calculating the median probability for both binary model outputs. Likewise, the higher median probability determined the final classification. In Fig. 1B the image level probabilities with the population level median are shown for the 3 cohorts. We achieved a patient-level SE and SP of both 100% in the Mozambican cohort, an SE of 100% and an SP of 80% (2 false positives) in the Spanish cohort, and an SE of 87.5% (1 false negative, ground truth = 40WBC/mm^3, hematic LP) and an SP of 97.1% (1 false positive) in the Moroccan cohort. The overall sensitivity and specificity on a patient-level was 94.4% and 94.8%, respectively.

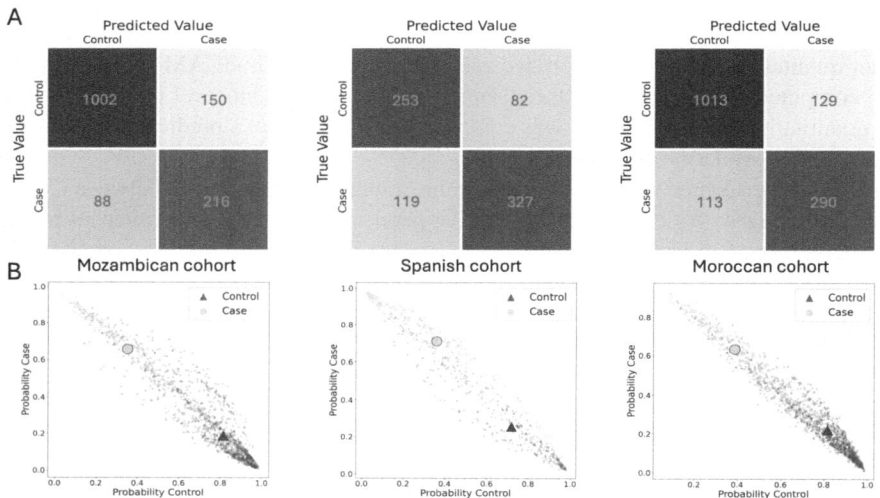

Fig. 1. Image-level model results for each cohort. **A.** Image-level confusion matrices for Control/Case classification. **B.** Probability distributions of image frames. Yellow dots show the probability of an image as being classified as Case and blue triangles as Control. The cohort level Control/Case class median probability is represented with a larger blue triangle/yellow dot. (color figure online)

3.2 Explainable AI (XAI) Results

For better interpretability of the DL model, we applied Gradient-weighted Class Activation mapping (GradCAM) [9, 15]. GradCAM visualizations highlighted areas important for the DL model's prediction. In Fig. 2 example Control and Case images are shown from each cohort (MZB, ESP, MOR from top to bottom) with their respective GradCAM output heatmaps for each output class. These visualizations show the importance of large parts of the images for the decisive class, clearly relating to the absence or presence of WBC patterns (Fig. 2).

For quantitative characterization, we analyzed the mean GradCAM scores of the images and computed the median average GradCAM score for each patient [9] for the Case class to compare scores between the two classes. This analysis revealed a significant

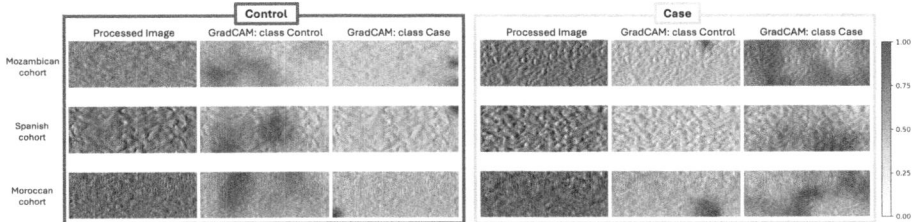

Fig. 2. GradCAM heatmap visualizations. For each cohort (MZB, ESP, MOR from top to bottom) a Control and a Case example are presented. The processed image is the model input. The highlighted features through GradCAM heatmaps are shown for each output class, blue/white indicating regions of highest/least influence on the model's prediction. (color figure online)

correlation between the mean GradCAM scores and the median classification probability (correlation coefficient > 0.98 in all 3 cohorts) and shows a high separation between Controls and Cases (Fig. 3).

Statistical image comparison showed significant differences in pixel intensity frequency between Control and Case images in all 3 cohorts (from left to right: MZB, ESP, MOR), with Case images represented by higher pixel values (Fig. 4).

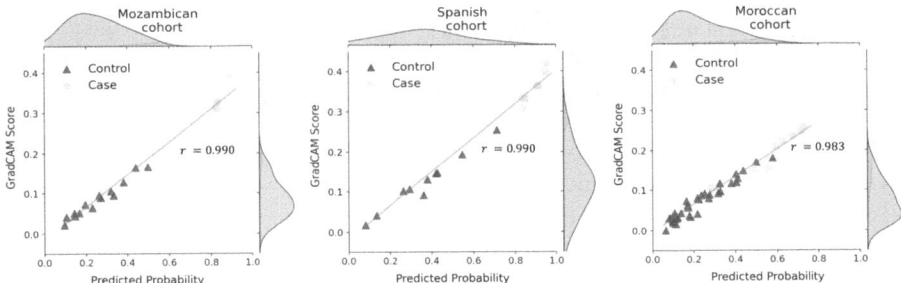

Fig. 3. Scatter plot showing the relationship between the median predicted probability for the model's class Case and the median average GradCAM score for each patient in each of the studied cohorts (from left to right): MZB, ESP and MOR. Each plot is accompanied by its regression line and correlation coefficient.

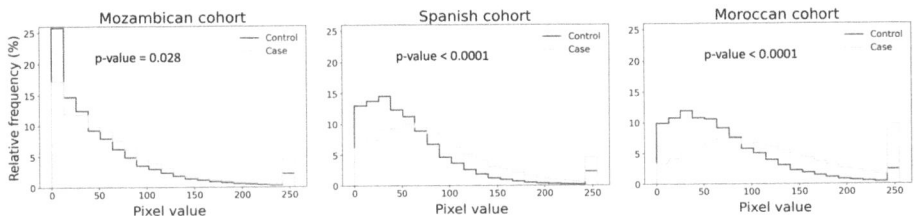

Fig. 4. Histogram comparing the relative frequency for each of the pixel values (in the 0–255 range) between Control and Case patients in each cohort (from left to right): MZB, ESP and MOR. p-values were computed applying a Wilcoxon Rank sum test.

4 Discussion and Conclusions

In this study, we developed, validated and explained a deep learning algorithm to predict meningitis in newborns and infants using data from 3 different cohorts across various countries and ethnicities. The model, based on a pretrained ResNet50 architecture, was trained based on a leave-one-patient-out approach within two cohorts and was additionally tested on an independent cohort to minimize bias and maximize performance across different populations.

Our method achieved on an image-level an SE and SP between 70%–90% (Fig. 1), which, applying a soft-voting technique, resulted in an excellent patient-level performance: In MZB, both SE and SP were 100%; in ESP, SE was 100% and SP 80%; and in MOR, SE was 87.5% and SP 97.1%.

Out of 74 patients (76 LPs with 76 associated scans), four patients were misclassified, including a false negative with a traumatic LP and a borderline WBC count [8] and additionally a thick fontanelle tissue allowing the US signal to pass through with more difficulty. The false positive cases were patients with small liquid space which makes WBC detection more difficult. Given the primary application of this technology for meningitis screening, the main performance metric to consider is sensitivity, which is very high across cohorts, indicating an effective screening of positive patients.

Explainable AI techniques helped interpret the model's predictions. GradCAM visualizations highlighted pixel clusters significant for Case classification (Fig. 2), correlating well with classification probabilities (Fig. 3).

Statistical analyses showed higher relative pixel intensity in Case images, especially in Spain and Morocco (Fig. 4), demonstrating a tendency towards greater saturation. Univariate analysis of the different pixel intensity distributions between groups is not enough though for accurate classification. The deep learning algorithm goes beyond intensity patterns and incorporates complex image gradient information important for the efficient detection of WBCs in CSF.

This study has limitations. First, obtaining gold-standard measurements involves manual counting with a microscope, which may introduce errors, especially at low concentrations. Second, the threshold of 30 WBC/mm^3 for diagnosing meningitis applies only to newborns up to 28 days old. For older infants (1 to 12 months), a threshold of > 5 WBC/mm^3 is used [16, 17]. Future studies will integrate patient age into the classification algorithm and develop a regression model to estimate WBC values.

This study demonstrates the effectiveness of the Neosonics® device for meningitis screening, offering a non-invasive, cost-effective solution, particularly valuable in low-resource settings. Training and testing across diverse cohorts enhances the model's validity. The use of XAI offers a more comprehensive interpretation of the model's results, which is paramount when supporting clinical decisions.

Acknowledgments. Kriba acknowledges support from the European Union's Horizon Europe research and innovation programme, project code 190155553 - NEOSONICS. All authors with ISGlobal affiliation acknowledge support from the grant CEX2018-000806-S funded by MCIN/AEI/ https://doi.org/10.13039/501100011033, and support from the Generalitat de Catalunya through the CERCA Program. Mozambique: This work was supported by the Bill &

Melinda Gates Foundation INV-048197. Morocco: This project was supported in part by Spanish Agency of Cooperation (AECID) under the reference number 2019/ACDE/001196. Project PI16/01822 from PI ENRIQUE BASSAT ORELLANA and project PI16/00738 funded by Instituto de Salud Carlos III, co-funded by the European Union (FEDER) "Una manera de hacer Europa".

Disclosure of Interests. BJ, FCar, JJ, RQ, and FS are part of the technical team developing the presented technology. HS and PP are external consultants for Kriba on algorithm development. SA, CC, AP, FC, QB and PP are clinical collaborators for the development of Kriba's technology.

References

1. Schiess, N., Groce, N.E., Dua, T.: The impact and burden of neurological sequelae following bacterial meningitis: a narrative review. Microorganisms. **9** (2021). https://doi.org/10.3390/MICROORGANISMS9050900

2. GBD 2019 meningitis antimicrobial resistance collaborators.: global, regional, and national burden of meningitis and its aetiologies, 1990–2019: a systematic analysis for the Global Burden of Disease Study 2019. Lancet Neurol. **22**, 685–711 (2023). https://doi.org/10.1016/S1474-4422(23)00195-3

3. Glatstein, M.M., Zucker-Toledano, M., Arik, A., Scolnik, D., Oren, A., Reif, S.: Incidence of traumatic lumbar puncture: experience of a large, tertiary care pediatric hospital. Clin. Pediatr. (Phila) **50**, 1005–1009 (2011). https://doi.org/10.1177/0009922811410309

4. Dalai, R., Dutta, S., Pal, A., Sundaram, V., Jayashree, M.: Is lumbar puncture avoidable in low-risk neonates with suspected sepsis? Am. J. Perinatol. **39**, 099–105 (2022). https://doi.org/10.1055/S-0040-1714397

5. Vickers, A., Donnelly, J.P., Moore, J.X., Barnum, S.R., Schein, T.N., Wang, H.E.: Epidemiology of lumbar punctures in hospitalized patients in the United States. PLoS One. **13** (2018). https://doi.org/10.1371/JOURNAL.PONE.0208622

6. Flidel-Rimon, O., Leibovitz, E., Eventov Friedman, S., Juster-Reicher, A., Shinwell, E.S.: Is lumbar puncture (LP) required in every workup for suspected late-onset sepsis in neonates? Acta Paediatr. **100**, 303–304 (2011). https://doi.org/10.1111/J.1651-2227.2010.02012.X

7. Bedetti, L., et al.: Pitfalls in the diagnosis of meningitis in neonates and young infants: the role of lumbar puncture. J. Matern. Fetal. Neonatal. Med. **32**, 4029–4035 (2019). https://doi.org/10.1080/14767058.2018.1481031

8. Ajanovic, S., et al.: UNITED study group: non-invasive meningitis screening in neonates and infants from Spain, morocco, and Mozambique: a proof-of-concept study. https://doi.org/10.2139/SSRN.4883993

9. Sial, H., et al.: Novel AI-Driven Infant Meningitis Screening from High Resolution Ultrasound Imaging (2024). [Manuscript submitted for publication]

10. Jiménez, J., et al.: System and method for non-invasive white blood cell counting in serous body fluids (2023)

11. Kannarkat, G.T., Darrow, J., Moghekar, A.: Reassessing accuracy of blood cell correction factor for traumatic lumbar puncture. J. Neurol. Sci. **432**, 120097 (2022). https://doi.org/10.1016/J.JNS.2021.120097

12. Martín-Ancel, A., García-Alix, A., Salas, S., Del Castillo, F., Cabañas, F., Quero, J.: Cerebrospinal fluid leucocyte counts in healthy neonates. Arch Dis Child Fetal Neonatal Ed **91**, F357 (2006). https://doi.org/10.1136/ADC.2005.082826

13. Virtanen, P., et al.: SciPy 1.0: fundamental algorithms for scientific computing in Python. Nat. Methods **17**(3), 261–272 (2020). https://doi.org/10.1038/s41592-019-0686-2

14. He, K., Zhang, X., Ren, S., Sun, J.: Deep residual learning for image recognition. In: Proceedings of the IEEE Computer Society Conference on Computer Vision and Pattern Recognition, pp. 770–778 (2016). https://doi.org/10.1109/CVPR.2016.90
15. Selvaraju, R.R., Cogswell, M., Das, A., Vedantam, R., Parikh, D., Batra, D.: Grad-CAM: visual explanations from deep networks via gradient-based localization. Int. J. Comput. Vis. **128**, 336–359 (2016). https://doi.org/10.1007/s11263-019-01228-7
16. Oostenbrink, R., Moons, K.G.M., Twijnstra, M.J., Grobbee, D.E., Moll, H.A.: Children with meningeal signs: predicting who needs empiric antibiotic treatment. Arch. Pediatr. Adolesc. Med. **156**, 1189–1194 (2002). https://doi.org/10.1001/ARCHPEDI.156.12.1189
17. Griffiths, M.J., McGill, F., Solomon, T.: Management of acute meningitis. Clin. Med. **18**, 164 (2018). https://doi.org/10.7861/CLINMEDICINE.18-2-164

Automated Segmentation of Ischemic Stroke Lesions in Non-contrast Computed Tomography Images for Enhanced Treatment and Prognosis

Toufiq Musah[1]([✉]), Prince Ebenezer Adjei[1,2], and Kojo Obed Otoo[3]

[1] Department of Computer Engineering, Kwame Nkrumah University of Science and Technology, Kumasi, Ghana
`toufiqmusah32@gmail.com`
[2] Kumasi Centre for Collaborative Research in Tropical Medicine, Kumasi, Ghana
[3] Radiology Department, Komfo Anokye Teaching Hospital, Kumasi, Ghana

Abstract. Stroke is the second leading cause of death worldwide, and is increasingly prevalent in low- and middle-income countries (LMICs). Timely interventions can significantly influence stroke survivability and the quality of life after treatment. However, the standard and most widely available imaging method for confirming strokes and their sub-types, the NCCT, is more challenging and time-consuming to employ in cases of ischemic stroke. For this reason, we developed an automated method for ischemic stroke lesion segmentation in NCCTs using the nnU-Net framework, aimed at enhancing early treatment and improving the prognosis of ischemic stroke patients. We achieved Dice scores of 0.596 and Intersection over Union (IoU) scores of 0.501 on the sampled dataset. After adjusting for outliers, these scores improved to 0.752 for the Dice score and 0.643 for the IoU. Proper delineation of the region of infarction can help clinicians better assess the potential impact of the infarction, and guide treatment procedures.

Keywords: Ischemic Stroke · Acute infarcts · Non-Contrast CT · nnU-Net

1 Introduction

There is an estimated 15 million cases of stroke annually, with one-third of these cases leading to death, and another one-third leading to some form of disability [3]. The rate of prevalence is more apparent in Low- and Middle-Income Countries (LMICs), where existing healthcare infrastructures remain unprepared to deal with the surge in cases [10,20]. Timely intervention is a critical factor in the survivability and post-treatment quality of life of patients [18]. Expert radiologist interpretations of the standard Non-Contrast Computed Tomography (NCCT) images are far more accurate for hemorrhagic stroke (95%), when compared to

U. Anazodo et al. (Eds.): MImA 2024/EMERGE 2024, CCIS 2240, pp. 73–80, 2025.
https://doi.org/10.1007/978-3-031-79103-1_8

ischemic stroke where a diagnostic ruling is given for major onsets only two-thirds of the time [11,18]. CT imaging is the most accessible modality for brain diagnostic imaging in LMICs [2].

The process of diagnosing stroke, and subsequently segmenting the lesion region using radiological images can guide treatment and predict rehabilitation outcomes [4,14]. Manual segmentation can be highly time-consuming and error-prone, with significant inter-observer variability depending on the observer's level of expertise [25]. The process of segmenting lesions in NCCT images is a relatively difficult task (due to low tissue contrast) as compared to MRI [17], especially when done manually. It is important that robust automated lesion segmentation processes for varying modalities (in this case, NCCT) be made available, especially in regions with limited imaging modality accessibility.

Several works explore the segmentation of stroke lesions, employing a variety of approaches. Earlier methods based on classical machine learning approaches such as Maier et al. [15], utilized the Support Vector Machine (SVM) algorithm to segment ischemic stroke lesions in multi-spectral magnetic resonance volumes consisting of T1, T2, FLAIR, DW, and ADC sequences. They obtained an average dice coefficient score of 0.74 on the testing data, using leave-one-out cross-validation. Classical methods often require the development and use of feature selection methods to reduce the redundancy caused by highly correlated features. Commonly used feature selection method include the Minimum Redundancy Maximum Relevance (mRMR) measure [19], and the [13]. The SLNet stroke lesion segmentation network was proposed by [23], which takes as input, synthesized DWI images obtained from Computed Tomography Angiography scans. They achieved a best dice coefficient of 0.622 ± 0.154 . Liang et al. [12] proposed the Symmetry-Enhanced Attention Network (SEAN) to automatically segment acute ischemic infarcts in NCCTs. The input image is first transformed into the standard space in an unsupervised manner by an 'Alignment Network'. In the transformed image with its bilateral symmetry, long range dependencies are better captured by SEAN, which has proven to outperform other existing symmetry based methods, with a dice score of 0.5784. The nnU-Net was developed by Fabian et al. [7], and is described as an out-of-the-box tool for biomedical image segmentation. It is a well-validated segmentation method, and has achieved state-of-the-art (SOTA) results in various biomedical image segmentation benchmark datasets such as the Medical Segmentation Decathlon (MSD) [1], Brain Tumor Segmentation (BraTS) Challenge [16], and the Ischemic Stroke Lesion Segmentation Challenge (ISLES) [6].

In this paper, we achieve promising results in ischemic lesion segmentation in NCCTs, by using out-of-the-box, well-validated methods. Initial image processing involved normalizing the input NCCT scans to ensure uniformity in image quality and resolution. We employed a Residual Encoder U-Net architecture from the nnU-Net framework, known for its robust performance in medical image segmentation tasks. The network was trained using a curated dataset consisting of NCCT images which were annotated using DWI MRI scans of the subjects as reference. The results obtained reiterate the potential of using validated methods to automate the difficult task of ischemic lesion segmentation in NCCT scans, further extending their utility in the stroke diagnosis and treatment process. Comparisons with existing techniques highlight and underscore areas for future improvements, including robust data collection and annotation processes, as well as the integration of deep learning models with other diagnostic tools for a comprehensive assessment of stroke severity and prognosis.

2   Materials and Methods

2.1   Dataset

The dataset used for this study is the Acute Ischemic stroke Dataset (AISD) [12], comprising of Non-Contrast-enhanced Computed Tomography (NCCT), and diffusion-weighted MRI (DWI) scans from 398 subjects. The NCCT scans are obtained less than 24 h from the onset of ischemia symptoms, and have a slice thickness of 5mm. The lesion regions in the NCCT are manually segmented by a doctor using the DWI scans (taken within 24 h after the NCCT imaging procedure) as reference, and then double-reviewed by a senior doctor. The scans are multi-labelled for various forms of ischemias present as shown in Fig. 1. This study only considers acute infarctions, as they are more pertinent in onset strokes, and require immediate response in diagnosis and treatment.

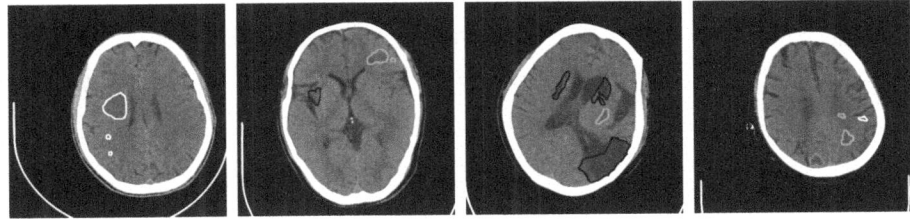

Fig. 1. Segmentation masks indicating different types of infarcts in the scans. Each color represents a specific type of infarct: Red - Clear acute infarcts, Blue - Remote infarcts, Green - Blurred acute infarcts, Magenta - Infarct. (Color figure online)

2.2   Data Preprocessing

The data preprocessing pipeline was automatically selected and applied by nnU-Net, using its new ResEnc presets [9]. The first step involved CT Normalisation, which clips the image intensities between the 0.5th and 99.5th percentiles to remove outlier pixels, followed by zero-mean and unit-variance normalization based on the intensity statistics of the dataset. It standardises the images which may have been produced using varying protocols and or machinery, while preserving relevant CT characteristics. A small epsilon value of 1×10^{-8} was used to prevent division by zero when standardising.

The individual 5mm slices were handled as 2D images and resampled to an isotropic spacing of 1mm x 1mm, with patch sizes of 512×512 pixels. For image data, cubic interpolation (order 3) was used for in-plane resampling, while linear interpolation (order 1) was applied to segmentation masks to preserve label integrity.

2.3   Model Architecture

In using the out-of-the-box nnU-Net framework for both preprocessing and training, the selected architecture was the Residual Encoder U-Net, which was first proposed by [8], in a 3D configuration. It comprised of 8 stages in the encoder (with the following feature maps per stage: [32, 64, 128, 256, 512, 512, 512, 512]), and 7 stages in the decoder. In the encoder, the first three stages had 1, 3, and 4 blocks respectively, whiles the last five stages each had 6 blocks. The decoder only had 1 block per stage. This design allows for increasing feature complexity as the network deepens.

Each residual convolution block in the encoder comprised of 5 layers as seen in Fig. 2. An input 2D convolution layer, followed by an instance normalisation, a LeakyReLU activation function, another 2D convolution layer, and finally an instance normalisation of the input features before handing it off to the next block.

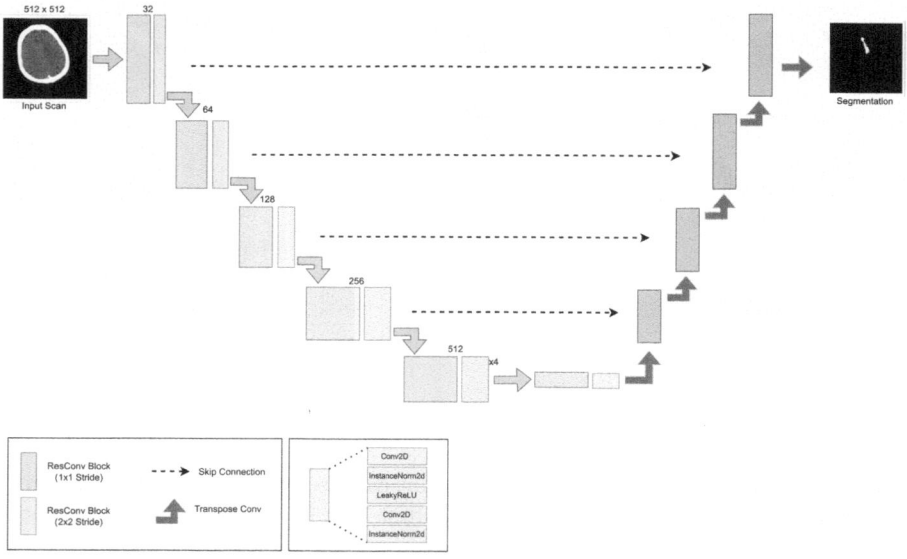

Fig. 2. Residual Encoder U-Net Architecture

2.4 Training Recipe

The network was trained for 50 epochs using a linearly decaying learning rate starting from 0.01. We used a batch size of 13 and calculated dice losses per batch. Five-fold cross-validation was performed, with testing conducted on two fold sets. The entire training process took approximately 4 h using an NVIDIA T4 GPU.

3 Results and Discussion

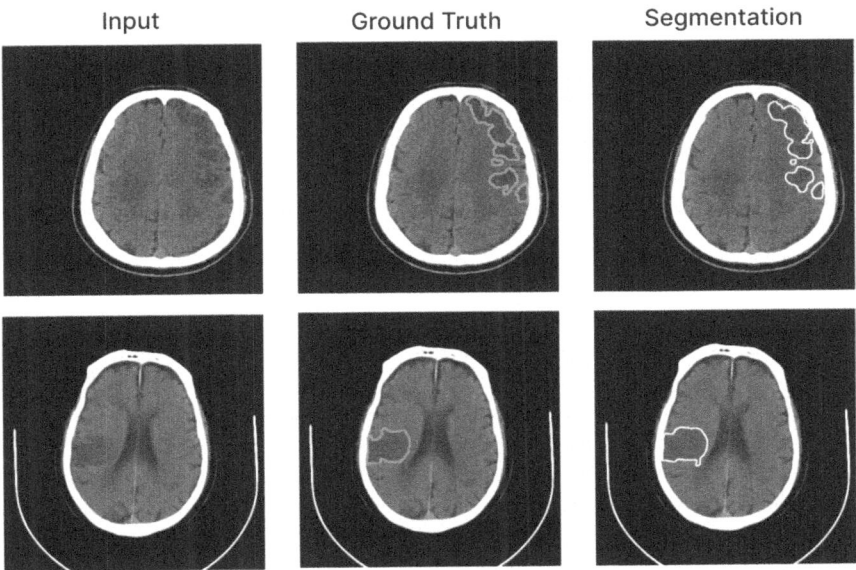

Fig. 3. Comparing the ground truth segmentation results in blue, to the network's segmentation results in red. (Color figure online)

Acute infarct lesions in the non-contrast computed tomography (NCCT) scans were segmented utilizing the trained Residual Encoder U-Net. The segmentation performance was subsequently evaluated by employing the Dice similarity coefficient and Jaccard's Index, also known as the Intersection Over Union (IoU) metric.

A 5-fold cross-validation approach was used in the evaluation, to assess the consistency of the network. This approach allowed us to validate the trained models against diverse subsets of the data, mitigating potential biases associated with any single fold. Out of the 5 folds, the network was validated on two folds with the highest results being reported

The mean Dice score achieved was 0.596, with a Jaccard's index of 0.510 as seen in Table 1. Compared to the Symmetry-Enhanced Attention Network's [12] reported dice score of 0.578 on the same dataset, these results can be viewed as promising in enabling the segmentation of acute infarctions from NCCT scans. Scores ranged from as low as 0.00, to as high as 0.923, which was attributable to the existence of several outliers in the dataset. The outliers included volume regions located outside the typical areas associated with stroke, extending to regions such as the top and base of the skull. Without the outliers, the resulting dice score is as high as 0.752, with a Jaccard's index of 0.643.

Table 1. Comparison of Segmentation Performance Metrics

Metric	Value	Description
Mean Dice Score	0.596	Overall mean score
Jaccard's Index	0.510	Overall mean index
Dice Score Range	0.00–0.983	Extremes due to outliers
Adjusted Mean Dice Score	0.752	Excluding outliers
Adjusted Jaccard's Index	0.643	Excluding outliers
Comparative Dice Score	0.578	SEAN [12]

The relatively low Dice scores observed in the segmentation task can largely be attributed to the inherent challenges in tissue differentiation within NCCT scans [17], compounded by the presence of several outliers in the dataset. Despite these challenges, NCCT remains an essential diagnostic tool for distinguishing between hemorrhagic and ischemic strokes, with the latter presenting more complexities in accurate detection. The results are promising and underscore the need for further data collection to enhance the effectiveness of segmentation algorithms in infarct segmentation using NCCT scans. The additional data could potentially refine the training process, leading to improvements in model performance and, consequently, more accurate clinical assessments for better patient outcomes.

In the diagnosis and treatment process of stroke, the NCCT scan's utility is often limited to excluding the presence of a hemorrhagic stroke or intracerebral hemorrhages [21]. Computed Tomography Perfusion (CTP) and or Angiography (CTA) is the next imaging procedure carried out to estimate the volume of the ischemia [22] and obtain map of the affected vessels which can influence the treatment procedure [24]. The stated process, which makes up a part of the 'code stroke' guide [5] may be further optimised in the early phase of NCCT imaging to enable quick decision making. The utility of NCCT can be extended in this regarded if stroke lesions are properly segmented, allowing for its use in estimating the volume and location of an infarct and further find occluded arteries, aiding in the treatment planning process.

4   Conclusion

In this paper, we employ nnU-Net's residual encoder U-Net architecture in the automatic segmentation of acute stroke infarctions in NCCT scans. Our investigations revealed that while NCCT poses significant challenges in ischemic lesion identification due to low tissue differentiation, the deployed demonstrated potential in segmenting the lesions with a reasonable degree of accuracy. Notably, the Dice scores, although relatively low, showed promising avenues for future improvements.

Accurate delineation of the infarction region is pivotal for clinical decision-making. It enables clinicians to evaluate the extent and severity of the stroke, providing essential insights into the affected regions and potential impact on brain function. Detailed infarct delineation supports targeted treatment strategies, such as the administration of thrombolytics for ischemic strokes or surgical interventions in more severe cases. While challenges persist, the improvements in automated segmentation of stroke lesions in NCCT scans hold considerable promise for improving stroke diagnosis and treatment, ultimately enhancing patient care and outcomes.

References

1. Antonelli, M., et al.: The medical segmentation decathlon. Nat. Commun. **13**(1), 4128 (2022)
2. Bour, B.K., et al.: National inventory of authorized diagnostic imaging equipment in Ghana: data as of september 2020. Pan African Med. J. **41**(1) (2022)
3. Feigin, V.L., et al.: Global, regional, and national burden of neurological disorders during 1990–2015: a systematic analysis for the global burden of disease study 2015. Lancet Neurol. **16**(11), 877–897 (2017)
4. Feng, W., et al.: Corticospinal tract lesion load: an imaging biomarker for stroke motor outcomes. Ann. Neurol. **78**(6), 860–870 (2015)
5. Gomez, C.R., Malkoff, M.D., Sauer, C.M., Tulyapronchote, R., Burch, C.M., Banet, G.A.: Code stroke: an attempt to shorten in hospital therapeutic delays. Stroke **25**(10), 1920–1923 (1994)
6. Hernandez Petzsche, M.R., et al.: Isles 2022: a multi-center magnetic resonance imaging stroke lesion segmentation dataset. Sci. Data **9**(1), 762 (2022)
7. Isensee, F., Jaeger, P.F., Kohl, S.A., Petersen, J., Maier-Hein, K.H.: nnu-net: a self-configuring method for deep learning-based biomedical image segmentation. Nat. Methods **18**(2), 203–211 (2021)
8. Isensee, F., Maier-Hein, K.H.: An attempt at beating the 3d u-net. arXiv preprint arXiv:1908.02182 (2019)
9. Isensee, F., et al.: nnu-net revisited: a call for rigorous validation in 3d medical image segmentation. arXiv preprint arXiv:2404.09556 (2024)
10. Kengne, A.P., Anderson, C.S.: The neglected burden of stroke in sub-saharan africa. Int. J. Stroke **1**(4), 180–190 (2006)
11. Kidwell, C.S., et al.: Comparison of mri and ct for detection of acute intracerebral hemorrhage. JAMA **292**(15), 1823–1830 (2004)

12. Liang, K., et al.: Symmetry-enhanced attention network for acute ischemic infarct segmentation with non-contrast CT images. In: de Bruijne, M., et al. (eds.) MICCAI 2021. LNCS, vol. 12907, pp. 432–441. Springer, Cham (2021). https://doi.org/10.1007/978-3-030-87234-2_41

13. Kursa, M.B., Rudnicki, W.R.: Feature selection with the boruta package. J. Stat. Softw. **36**, 1–13 (2010)

14. Lo, B.P., Donnelly, M.R., Barisano, G., Liew, S.L.: A standardized protocol for manually segmenting stroke lesions on high-resolution t1-weighted mr images. Front. Neuroimaging **1**, 1098604 (2023)

15. Maier, O., Wilms, M., von der Gablentz, J., Krämer, U., Handels, H.: Ischemic stroke lesion segmentation in multi-spectral mr images with support vector machine classifiers. In: Medical Imaging 2014: Computer-Aided Diagnosis, vol. 9035, pp. 21–32. SPIE (2014)

16. Menze, B.H., et al.: The multimodal brain tumor image segmentation benchmark (brats). IEEE Trans. Med. Imaging **34**(10), 1993–2024 (2014)

17. Muir, K.W., Buchan, A., von Kummer, R., Rother, J., Baron, J.C.: Imaging of acute stroke. Lancet Neurol. **5**(9), 755–768 (2006)

18. Musuka, T.D., Wilton, S.B., Traboulsi, M., Hill, M.D.: Diagnosis and management of acute ischemic stroke: speed is critical. CMAJ **187**(12), 887–893 (2015)

19. Peng, H., Long, F., Ding, C.: Feature selection based on mutual information criteria of max-dependency, max-relevance, and min-redundancy. IEEE Trans. Pattern Anal. Mach. Intell. **27**(8), 1226–1238 (2005)

20. Prust, M.L., Forman, R., Ovbiagele, B.: Addressing disparities in the global epidemiology of stroke. Nat. Rev. Neurol., 1–15 (2024)

21. Shafaat, O., Sotoudeh, H.: Stroke imaging. In: StatPearls [Internet]. StatPearls Publishing (2023)

22. Sotoudeh, H., Bag, A.K., Brooks, M.D.: "code-stroke" ct perfusion; challenges and pitfalls. Acad. Radiol. **26**(11), 1565–1579 (2019)

23. Wang, G., Song, T., Dong, Q., Cui, M., Huang, N., Zhang, S.: Automatic ischemic stroke lesion segmentation from computed tomography perfusion images by image synthesis and attention-based deep neural networks. Med. Image Anal. **65**, 101787 (2020)

24. Wildermuth, S., Knauth, M., Brandt, T., Winter, R., Sartor, K., Hacke, W.: Role of ct angiography in patient selection for thrombolytic therapy in acute hemispheric stroke. Stroke **29**(5), 935–938 (1998)

25. Yepes-Calderon, F., Gordon McComb, J.: Manual segmentation errors in medical imaging. proposing a reliable gold standard. In: Florez, H., Leon, M., Diaz-Nafria, J.M., Belli, S. (eds.) ICAI 2019. CCIS, vol. 1051, pp. 230–241. Springer, Cham (2019). https://doi.org/10.1007/978-3-030-32475-9_17

Spatial Attention-Enhanced Diffusion Model for Multiple Sclerosis MRI Synthesis

Khaoula Alaoui Belghiti[1]($\boxtimes$) ⓘ, Islem Rekik[2] ⓘ, Sahar Selim[3] ⓘ,
Mikram Mounia[1] ⓘ, and Maryem Rhanoui[1,4] ⓘ

[1] School of Information Sciences, Rabat, Morocco
`khaoula.alaoui-belghiti@esi.ac.ma`
[2] BASIRA Lab, Department of Computing, Imperial College London,
London, UK
[3] School of Information Technology and Computer Science, Nile University, Giza,
Egypt
[4] Laboratory Health Systemic Process (P2S), UR4129, University Claude Bernard
Lyon 1, University of Lyon, Lyon, France

Abstract. In the domain of neurological disorders such as Multiple Sclerosis (MS), MRI scans serve as pivotal tool for diagnosis and treatment evaluation. However, the availability of public datasets on Multiple Sclerosis is limited and insufficient. Although deep learning models have proven effective in aiding the diagnosis and analysis of disease progression, their accuracy is dependent on access to large and diverse datasets. To address this issue, we propose a novel approach by augmenting a UNet-based Denoising Diffusion Probabilistic Model (DDPM) with a spatial attention mechanism, benefiting from the interpretability of the generated attention maps to enhance the model's convergence as well as the generated MRIs. Our modified architecture demonstrates superior performance, surpassing state-of-the-art GANs and diffusion models on existing public datasets as well as African Local dataset, with the lowest loss values 0.13 and 0.04 for L1 and L2 respectively, and a Fréchet inception distance FID of 0.20 between generated and ground truth scans. Showcasing its effectiveness in capturing the intricacies of MS lesions and the unique characteristics of MRI scans in the African context. Thereby broadening its applicability and relevance in diverse clinical settings. While the generated synthetic dataset offer a practical solution to the limited availability of real-world data and addresses the privacy concerns inherent in medical data sharing.

Keywords: Multiple Sclerosis · MRI Generation · Diffusion Model · Spatial Attention

1 Introduction

Multiple sclerosis is a neurodegenerative disease with spatial and temporal expansion through the brain affecting the daily activities of the patients [1].

U. Anazodo et al. (Eds.): MImA 2024/EMERGE 2024, CCIS 2240, pp. 81–90, 2025.
https://doi.org/10.1007/978-3-031-79103-1_9

Unfortunately this is an autoimmune disease with no specific cause, making the lesions highly unpredictable, Therefore, the neurologists rely on symptom based treatment, where the treatment evaluation heavily depend on Magnetic Resonance Imaging (MRI) scans [2]. However despite the critical role of MRI scans in MS diagnosis, the availability of public datasets related to MS is notably limited. This scarcity hinders the development and refinement of deep learning models, which heavily depend on large and diverse datasets for optimal performance. Some studies focused on using heterogeneous datasets for tasks such as classification and segmentation of MS [3], it is insufficient to produce highly scalable and generalizable models. To address this challenge, deep generative models have emerged as powerful tools for uncovering intricate patterns within medical imaging data. Generative Adversarial Networks (GANs) [4] became popular in generating high-quality synthetic medical images [5]. Studies such as [6] have demonstrated the efficacy of GANs in generating realistic MS MRI scans, also in [7] using a Unet structure MS image generator as well as in [8] with lesion filling techniques, However, GAN-based models often suffer from training instability and mode collapse, limiting their applicability in clinical settings, also these approaches are based on using healthy brain MRI datasets and transferring the lesion structures to generate synthetic MS MRIs, even though these techniques have proven to be useful in such scenarios, it has unstable training dynamics and limited diversity in small lesion generation, due to the adversarial training strategy, which is necessary to represent the different disease stages and also applying lesions to healthy scans oversees the brain pathological structure and pattern that falls within the MS MRI characteristics. Therefore Using a complete MS dataset to capture the underlying patterns.

Diffusion models on the other hand have proven their efficacy in generating high quality images in different scales [9], and have gained more and more attention as they perform as well on medical images, especially denoising diffusion models [10] for image synthesis [11,12], anomaly detection [13] and medical MS MRI segmentation [14]. Yet so far no one have used the diffusion models for MS MRI generation to generate high quality scans. Furthermore, attention mechanisms [15] have been integrated into deep learning architectures to improve model interpretability and performance. While spatial attention mechanisms, in particular, have been shown to enhance the discriminate capabilities of deep learning models by focusing on relevant regions within an image [16].

Benefiting from the effectiveness and high quality of denoising diffusion probabilistic models, and using a complete MS patients datasets, we propose a novel approach by augmenting a UNet-based [17] DDPM with a spatial attention mechanism as middle block. In order to generate spatial attention maps that helps enhancing significant areas while diminishing less relevant ones in the MRI scan. Therefore we focus more on the brain structure and the small lesions of MS, with high convergence and low computation resources leveraging the interpretability offered by attention maps. Moreover, we extend the validation of the proposed solution to include an African Local dataset. This demonstrates its efficacy in capturing the intricacies of MS lesions and the unique characteristics of MRI scans within the African context, given the special phenotype of MS [18], with more small and random spread of lesions and its aggressive transition to advanced stages, as well as low quality MRI scans. By broadening the applicability and relevance of our approach across diverse clinical settings, we aim to address the pressing need for robust synthetic datasets, free of privacy concerns in global healthcare.

2   Methodology

In this section we outline the techniques used to develop our approach, starting by the acquisition and prepossessing of public datasets and local dataset we collected, followed by the background model and the proposed mechanism.

2.1   Datasets

For the model's training and evaluation we have used two public datasets along with an African dataset, with only fluid-attenuated inversion recovery (FLAIR) [19] modality MRI scans, given that this modality is widely used for multiple sclerosis diagnosis and progression analysis as it represents clearly the lesions structure mostly appearing in the white matter. This modality have also been proven to be the most effective among others to segment multiple sclerosis lesions using state of the art deep learning models [20].

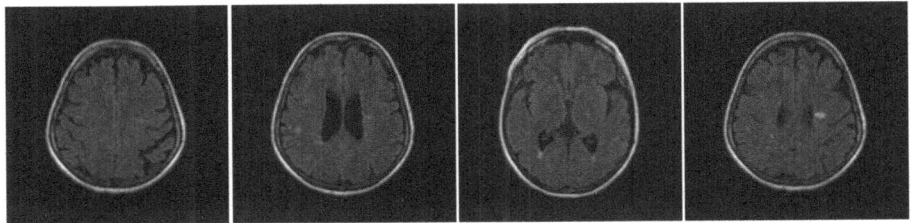

Fig. 1. FLAIR MRI Scan samples of different patients from the African Dataset.

Public Datasets. We have collected 3D MRI scans from two public datasets:

MSSEG-2. MS new lesions segmentation challenge by MICCAI 2021 [21], providing two time points of MRI scans from 100 MS patients, with an interval between the baseline scan and the follow up along with new lesion masks, addressed for the detection and segmentation of the new appearing MS lesions. We extracted the 2D lesions from both time points and used them for our approach training.

Brain MRI MS Dataset. The dataset provided publicly in [22] is an extensive dataset with imaging and clinical data from 60 MS patients, providing 3 MRI modalities with segmentation mask, for MS lesion segmentation. From which we also extracted the 2D slices using only the FLAIR scans, accurately representing the MS lesions. Then with the combined samples we resized the images to 128*128, and performed other transforms such as normalization, crop, and augmented with random flips. After collection and pre processing we preserved only the slices showing MS lesions to reduce the data imbalance and not confuse the generative model with unrepresentative samples.

African Dataset. With special characteristics of lesion distribution and shape, MS is showing aggressively in Africa [23], increasing the need for its pattern analysis. So to validate the new architecture on a more random and small lesion scans as shown in Fig. 1, we have acquired an African dataset from a Local University Hospital. We collected MRI scans from approximately 160 patients with 1 to 8 MRI scans per each, the samples are in dicom format. Which required extensive pre processing for patient privacy concerns, after selecting the brain region representing scans, and excluding the noise images, we picked the FLAIR scans, center cropped the images and then resized to 128 * 128 with Min-max normalisation to follow the same structure used on the public dataset.

2.2 Denoising Diffusion Probabilistic Model (DDPM)

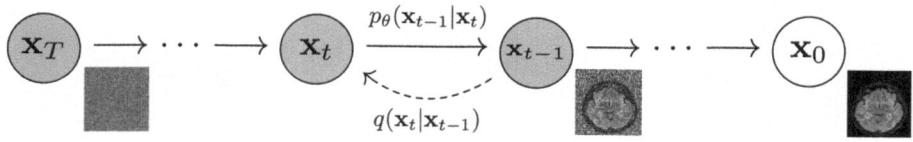

Fig. 2. The noise diffusion process in Denoising Diffusion Probabilistic Models.

Denoising diffusion probabilistic models are a class of non-autoregressive generative models, that have proved their efficacy and accurate generated samples exceeding GANs, and competing with transformers as they benefit from the sinusoidal embedding adding a positional information to the images, inspired from [24]; defining the positional embedding as in (1):

$$\text{PE}(pos, 2i) = \sin\left(\frac{pos}{10000^{2i/d_{\text{model}}}}\right) \& \text{PE}(pos, 2i+1) = \cos\left(\frac{pos}{10000^{2i/d_{\text{model}}}}\right) \ (1)$$

where pos represents the position and i is the dimension. d_{model} considered the size word embedding.

Mainly, the DDPMs are trained to predict slightly less noisy images from a noisy input. During inference, they can be utilized to iteratively transform random noise to generate an image, where $p_\theta(x_{t-1}|x_t)$ predicts a slightly less noisy image, as shown in Fig. 2. Through a forward and reversed diffusion (Markov chain).

2.3 Spatial Attention Denoising Diffusion Probabilistic Model (SA-DDPM)

Following the DDPM structure we introduce an enhanced version where we replace the middle block from only residual and self-attention layers to using spatial attention on the MRI scans, the model is based on a U-Net [17] structure with identical input and output dimensions. Where the downsampling and upsampling blocks are sequences of residual and self attention blocks, respectively downsample and then upsample the input image, and adds skip connections between layers having the same resolution helping with gradient flow as shown in Fig. 3.

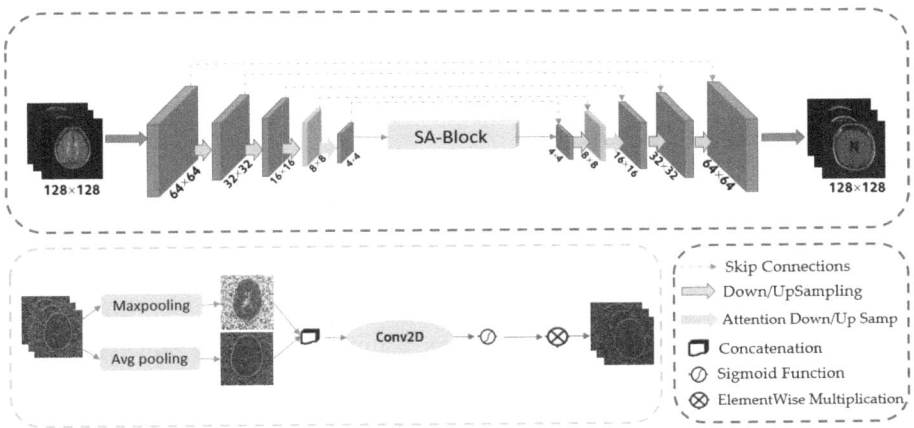

Fig. 3. The U-Net based SA-DDPM Architecture (Top), highlighting the integrated Spatial Attention Block (Bottom Left).

The attention block represented in Fig. 3 processes the output from the final downsampling block, denoted as $\mathbf{X} \in \mathbb{R}^{H \times W \times C}$, where H, W, and C represent the height, width, and channels of the feature map, respectively. The block applies a composite function to generate a spatial attention map and enhance significant areas while diminishing less relevant ones.

The composite function is defined as:

$$\mathbf{Y} = \mathbf{X} \odot \sigma \left(\text{Conv2D} \left(\text{MaxP}(\mathbf{X}) + \text{AvgP}(\mathbf{X}), \mathbf{W}, \mathbf{b} \right) \right)$$

where: $\mathbf{Y} \in \mathbb{R}^{H \times W \times C}$ is the output of the attention block, $\odot$ denotes the element-wise multiplication operation, σ denotes the Sigmoid function, Conv2D is a 2D convolutional layer with weights $\mathbf{W}$ and bias $\mathbf{b}$, MaxP and AvgP are max pooling and average pooling operations. The output feature map with applied spatial attention is then fed into the upsampling block of the model.

2.4   Evaluation Metrics

To evaluate our approach's performance and compare it to other generative models, we opted for quantitative and qualitative metrics. For quantitative evaluation we have used the common full reference image quality metrics such as pixel-Level metrics with L1 Loss (Mean Absolute Error) and L2 Loss (Mean Squared Error), Structural Similarity Index (SSIM), including feature-level evaluation with the Learned Perceptual Image Patch Similarity (LPIPS), and distribution level with Fréchet Inception Distance (FID).

3   Results and Discussion

The proposed architecture, leveraging a UNet-based Denoising Diffusion Probabilistic Model (DDPM) enhanced with a spatial attention block, exhibits compelling performance in generating MRI scans for Multiple Sclerosis (MS) patients through extensive experimentation and comparative analysis as illustrated in the following subsections;

3.1   Experimental Setup

As we divided the dataset into training and test sets, we utilized T4 GPUs from the cloud service provider Colab for both the new and comparative models, opting for a high RAM environment with approximately 15 Gb. The model was trained over 500 epochs with 1000 timesteps, employing the AdamW optimizer and a batch size of 16.

This setup was chosen to optimize the training process, balancing computational resources with the need for efficient learning and model convergence. For the model configuration we have used the predefined diffusion models from the Diffusers library by Hugging Face, and created the new model as a subclass incorporating the previously defined spatial attention block, using Pytorch framework. To assess the model performance over the combined public datasets, we have applied State Of The Art GANs and basic unconditional DDPM model, along with a Transformer based diffusion model [25] using SwinViT Model as a backbone.

3.2   Quantitative Evaluation

We present the evaluation results of our proposed model alongside those of comparative models, including VAE-GAN [26], DC-GAN [27], Transformer-based Diffusion Model (TDM) [25], and UNet-based Diffusion Model (Unet-Diffuser) [10], using 500 samples generated by each model, as listed in Table 1.

Where the enhanced DDPM consistently outperforms all comparative models across evaluation metrics. Specifically, it achieves the lowest L1 (0.13) and L2 (0.04) losses, indicating superior pixel-level reconstruction accuracy compared to the other architectures. Additionally, the model exhibits higher SSIM score

Table 1. Evaluation results of generated sets by each comparative model

	L1	L2	SSIM	LPIPS	FID
VAE-GAN	0.19	0.08	0.18	2.36	6.08
DC-GAN	0.28	0.19	0.17	2.60	5.26
TDM	0.16	0.07	0.20	2.49	3.38
Unet-Diffuser	0.21	0.09	0.19	2.63	2.12
Ours	**0.13**	**0.04**	**0.27**	**2.67**	**0.20**

(0.27) compared to the GAN based models, indicating better preservation of structural similarities between generated and ground truth samples. Moreover, the new model demonstrates competitive performance in terms of LPIPS. Most notably, SA-DDPM achieves the lowest FID score among all comparative models, indicating the closest statistical similarity between the distributions of feature embeddings extracted from the generated and real MRI scans.

The proposed architecture not only achieves lower FID scores on the public datasets, but also when implemented with African samples, with a FID distance of 0.12, indicative of superior image fidelity and realism, but also demonstrates better generalization capabilities across diverse imaging quality conditions and patient populations. These results underscore the superiority of our proposed architecture in MS MRI generation, demonstrating its potential to advance clinical practice and dealing with data scarcity issue especially in the context of neurodegenerative diseases such as multiple sclerosis.

3.3 Qualitative Evaluation

Visual Assessment: Visual inspection of the generated images supports these findings, showcasing sharper details, reduced artifacts, and improved structural coherence in the generated MRI scans as in Fig. 4, in both public (First row) and African (Second row) datasets.

This suggests that the proposed approach produces MRI scans that are not only visually realistic but also statistically indistinguishable from real MS patient scans. Either from high quality scanner settings as in the public datasets, or from lower standards and unique disease characteristics as in the African context.

Professional Assessment. To further validate the clinical realism and diagnostic utility of the synthetic MRI scans generated by our proposed model, we conducted a questionnaire-based assessment involving neurologists and radiologists. The participants were presented with a set of synthetic images and were asked to indicate which scans appeared to be fake and which were likely to belong to real Multiple Sclerosis (MS) patients.

Our assessment yielded an overall accuracy of 84%, indicating that 84% of the synthetic MRI scans generated by the new model were deemed to be representative of real MS MRI scans by healthcare professionals. This underscores

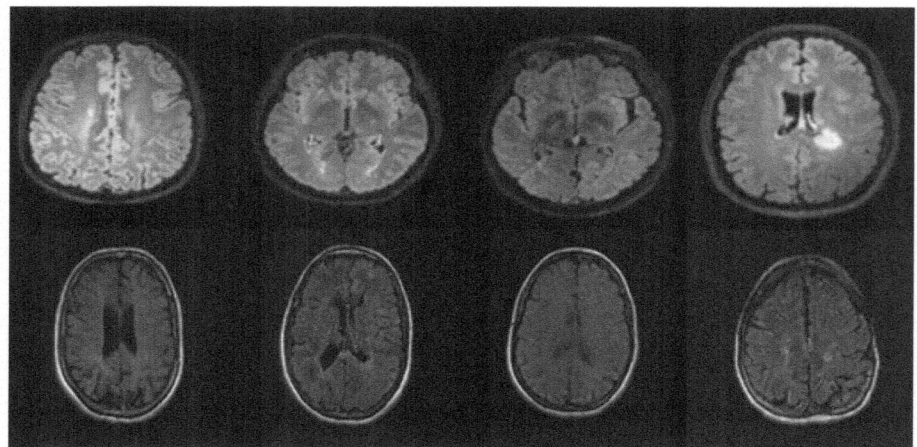

Fig. 4. Generated scans of synthetic MS MRI by the enhanced DDPM model: based on public datasets (First row), and African Dataset (Second row)

the effectiveness of the model in producing clinically plausible MRI images that closely resemble those obtained from actual MS patients, in both global and African Context.

4   Conclusion

While existing deep learning models have shown promise in MS imaging processing and segmentation, there is still a demand for robust and interpretable generative models that can effectively capture the complex features of MS lesions across diverse populations and varied imaging conditions.

In this paper, we propose a novel approach by augmenting a UNet-based Denoising Diffusion Probabilistic Model (DDPM) with a spatial attention middle block, leveraging the interpretability of attention maps to enhance model performance and generate high-quality synthetic MRI scans for MS diagnosis. The spatial attention mechanism played a pivotal role in enhancing the convergence of the model, reaching lower loss values in less training time compared to other diffusion models, which helps in lower compute settings. And also improves the quality of the generated images by selectively attending to relevant image regions. Through extensive experimentation, the enhanced model have proven to be superior compared to existing state of the art architectures, by generating high quality MRI scans representing MS lesions of different sizes and patterns based on two public dataset. While also capturing different disease patterns and imaging contexts from the African Dataset.

The generated synthetic MRI scans can augment the limited availability of real-world data, facilitating the development and validation of advanced diagnostic algorithms and treatment planning tools for MS. Future research directions may include further optimization of the model architecture, while exploring the generation of two time points scans for disease progression analysis tasks.

References

1. Dobson, R., Gavin, G.: Multiple sclerosis - a review. Eur. J. Neurol. **26**(1), 27–40 (2019)
2. Mendelsohn, Z., et al.: Commercial volumetric MRI reporting tools in multiple sclerosis: a systematic review of the evidence. Neuroradiology **65**(1), 5–24 (2023)
3. Wu, Y., Wu, Z., Shi, H., Picker, B., Chong, W., Cai, J.: CoactSeg: learning from heterogeneous data for new multiple sclerosis lesion segmentation. In: Greenspan, H., et al. (eds.) MICCAI 2023. LNCS, vol. 14227, pp. 1–13. Springer, Cham (2023). https://doi.org/10.1007/978-3-031-43993-3_1
4. Goodfellow, I., et al.: Generative adversarial networks. Commun. ACM **63**(11), 139–144 (2020)
5. Creswell, A., et al.: Generative adversarial networks: an overview. IEEE Signal Process. Mag. **35**(1), 53–65 (2018)
6. Basaran, B.D., et al.: Subject-specific lesion generation and pseudo-healthy synthesis for multiple sclerosis brain images. Springer, Cham (2022)
7. Salem, M., et al.: Multiple sclerosis lesion synthesis in MRI using an encoder-decoder U-NET. IEEE Access **7**, 25171–25184 (2019)
8. Prados, F., et al.: A multi-time-point modality-agnostic patch-based method for lesion filling in multiple sclerosis. Neuroimage **139**, 376–384 (2016)
9. Croitoru, F. et al.: Diffusion models in vision: a survey. IEEE Trans. Pattern Anal. Mach. Intell. (2023)
10. Ho, J., Jain, A., Abbeel, P.: Denoising diffusion probabilistic models. Adv. Neural. Inf. Process. Syst. **33**, 6840–6851 (2020)
11. Dorjsembe, Z., et al.: Conditional diffusion models for semantic 3D medical image synthesis. arXiv preprint 2305.18453 (2023)
12. Pandey, K. et al.: VAEs meet diffusion models: efficient and high-fidelity generation. In: NeurIPS 2021 Workshop on Deep Generative Models and Downstream Applications (2021)
13. Yu, J. H., O., H., Y., J.: Adversarial denoising diffusion model for unsupervised anomaly detection. In: Deep Generative Models for Health Workshop NeurIPS (2023)
14. Chowdary, G. J., Yin, Z.: Diffusion transformer U-net for medical image segmentation. In: Greenspan, H., et al. (eds.) MICCAI 2023. LNCS, vol. 14223, pp. 1–13. Springer, Cham (2023). https://doi.org/10.1007/978-3-031-43901-8_59
15. Guo, M., et al.: Attention mechanisms in computer vision: a survey. Comput. Vis. Media **8**(3), 331–368 (2022)
16. Zhu, X. et al.: An empirical study of spatial attention mechanisms in deep networks. In: Proceedings of the IEEE CVF International Conference on Computer Vision (2019)
17. Ronneberger, O., Fischer, P., Brox, T.: U-net: convolutional networks for biomedical image segmentation. In: Navab, N., Hornegger, J., Wells, W.M., Frangi, A.F. (eds.) MICCAI 2015. LNCS, vol. 9351, pp. 234–241. Springer, Cham (2015). https://doi.org/10.1007/978-3-319-24574-4_28
18. Yamout, B.I., et al.: Epidemiology and phenotypes of multiple sclerosis in the Middle East North Africa (MENA) region. Multiple Sclerosis J.-Exp. Transl. Clin. **6**(1), 2055217319841881 (2020)

19. Bakshi, R., Ariyaratana, S., Benedict, R.H.B., Jacobs, L.: Fluid-attenuated inversion recovery magnetic resonance imaging detects cortical and juxtacortical multiple sclerosis lesions. Arch. Neurol. **58**(5), 742–748 (2001)

20. Liu, H. et al.: ModDrop++: a dynamic filter network with intra-subject co-training for multiple sclerosis lesion segmentation with missing modalities. In: Wang, L., et al. (eds.) MICCAI 2022. LNCS, vol. 13435, pp. 1–13. Springer, Cham (2022). https://doi.org/10.1007/978-3-031-16443-9_43

21. Commowick, O., et al.: MSSEG-2 challenge proceedings: multiple sclerosis new lesions segmentation challenge using a data management and processing infrastructure. In: Medical Image Computing and Computer Assisted Intervention MICCAI (2021)

22. Muslim, A. M. et al.: Brain MRI dataset of multiple sclerosis with consensus manual lesion segmentation and patient meta information. Data Brief **42**, 108139 (2022)

23. Yamout, B. I., Assaad, W., Tamim, H., Mrabet, S., Goueider, R.: Epidemiology and phenotypes of multiple sclerosis in the Middle East North Africa (MENA) region. Multiple Sclerosis J.-Exp. Transl. Clin. **1**(6) (2020)

24. Vaswani, A., et al.: Attention is all you need. In: Advances in Neural Information Processing Systems, vol. 30 (2017)

25. Pan, S., et al.: 2D medical image synthesis using transformer-based denoising diffusion probabilistic model. Phys. Med. Biol. **68**(10), 105004 (2023)

26. Niu, Z., Ke, Y., Xiaofei, W.: LSTM-based VAE-GAN for time-series anomaly detection. Sensors **20**(13), 3738 (2020)

27. Radford, A., Luke, M., Soumith, C.: Unsupervised representation learning with deep convolutional generative adversarial networks. arXiv preprint arXiv:1511.06434 (2015)

An Automated Pipeline
for the Identification of Liver Tissue
in Ultrasound Video

Eloise Ockenden[1,2], Simon Mpooya[3], J. Alison Noble[2],
and Goylette F. Chami[1(✉)]

[1] Nuffield Department of Population Health, University of Oxford, Oxford, UK
goylette.chami@ndph.ox.ac.uk
[2] Institute of Biomedical Engineering, University of Oxford, Oxford, UK
[3] Division of Vector-Borne and Neglected Tropical Diseases Control, Uganda
Ministry of Health, 15 Bombo Road, Kampala, Uganda

Abstract. Liver diseases are a leading cause of death worldwide, with an estimated 2 million deaths each year. Causes of liver disease are difficult to ascertain, especially in sub-Saharan Africa where there is a high prevalence of infectious diseases such as hepatitis B and schistosomiasis, along with alcohol use. Point-of-care ultrasound often is used in low-resource settings for diagnosis of liver disease due to its portability and low cost. For classification models that can automatically stage liver disease from ultrasound video, the region of interest is liver tissue. A fully-automated pipeline for liver tissue identification in ultrasound video is presented. Ultrasound video data was collected using a low-cost, portable ultrasound machine in rural areas of Uganda. The pipeline first detects the diaphragm in each ultrasound video frame, then segments the diaphragm to ultimately use this segmentation to infer the position of liver tissue in each frame. This pipeline outperforms directly segmenting liver tissue with an intersection over union of 0.84 compared to 0.62. This pipeline also shows improved results with respect to the ease of clinical interpretation and anticipated clinical utility.

Keywords: Ultrasound · Video · Liver · Diaphragm · Segmentation · Infectious Disease · Africa · Schistosomiasis

1 Introduction

Liver diseases are a leading cause of death in sub-Saharan Africa, with commonly recorded causes of hepatitis B and alcohol use [7,19]. A less recorded, but widely prevalent cause of chronic, severe liver disease in areas of sub-Saharan Africa without access to safe water and adequate sanitation is schistosomiasis. Chronic schistosomiasis infection can cause liver fibrosis, due to immune responses to eggs from intestinal species of this parasitic blood fluke that are trapped in vessels in the portal system [1,15]. This type of liver fibrosis is called schistosomal periportal fibrosis (PPF). In its early stages, schistosomal PPF is visible at the early segmental branches of the portal system then advances towards the main portal vein, and finally extends across the whole liver [3, 18].

Ultrasound imaging is the most-used, and only recommended method by the World Health Organisation (WHO) for the diagnosis of schistosomal PPF, used due to the high echogenicity of the fibrosis [12, 17]. To reach rural populations affected by schistosomiasis and provide point-of-care services where health systems lack diagnostic capacity, low-cost, portable ultrasound machines are used [5]. It has been shown that the quality of images attained by low-cost, portable ultrasound machines may be lower than the quality obtained from cart-based ultrasound machines [9, 21]. To enable liver tissue identification in this context, deep learning approaches might need to exploit features or anatomy that are known to appear with clearer boundaries.

Automated liver tissue identification from ultrasound video has not been widely explored, especially considering challenges posed in low-resource settings. Segmentation for liver ultrasound has focused on cancer, in particular segmenting focal lesions as a pre-processing step before classification [16]. There has been work on the segmentation of liver tissue in ultrasound images, using data collected in hospital settings. A proposed method has been to locate the liver capsule and use this as a boundary for liver tissue, before cirrhosis staging [14]. However, this approach was used only for ultrasound images that did not image deep into the body, which do not capture enough liver tissue for staging complex diseases that present in both diffuse and focal forms, such as schistosomal PPF. Recently, another method has been proposed, achieving very good quantitative results by using multi-head self attention to segment liver tissue and related structures in ultrasound images [22]. Despite this, there remains no literature on the identification of liver tissue in ultrasound video. Ultrasound video may be a more appropriate data form in low-resource settings. The acquisition of specific views of the liver to produce curated images requires expert knowledge that may not be available. Alternatively, predefined sweep procedures for video acquisition requires less expertise and training. Thus, there remains a need to investigate liver tissue identification using ultrasound video collected in low-resource settings.

In this paper, a pipeline is presented to automatically identify when and where liver tissue appears in ultrasound video, using ultrasound examinations of individuals in a rural, sub-Saharan African context.

2   Methods

2.1   Context and Data

This study was conducted within SchistoTrack, which is a community-based cohort in rural Uganda [2]. Data from an annual follow-up of 3219 participants in 2023 were used. Ultrasound videos were collected from each participant following a predefined protocol conducted by local sonographers who were highly experienced in assessing liver disease. The procedure for acquiring the video used for the development of this pipeline was to start with a good view of the gall bladder with the probe transverse, and then to move the probe to the cephalic position. Each video clip was 5 s long, at 20 frames per second (fps). Phillips C5-2 curvilinear transducers were used with the Lumify Application connected to tablets with Android 9 Pie.

Adults who had fasted before the scan and had no ultrasound findings including liver fibrosis, cirrhosis, fatty liver, hepatitis B and other abnormalities, were used for the development of the pipeline. Of the 728 adults who fit these criteria, a random sample of 110 adults were selected, each with one ultrasound video. The participants' age range was 18–84 years, with a mean (s. d.) of 42.0 (15.6) and 77% (85) were female. Every other frame of each video was labelled with whether or not the diaphragm was visible. Additionally, masks of the diaphragm were created for every other frame of ultrasound video for a randomly selected 25 videos. For diaphragm detection, 90 videos were used for training, 10 for validation, and 10 for testing. For automated diaphragm segmentation, 18 videos were used for training, two for validation, and five for testing.

2.2 Diaphragm Detection

As the first step of the pipeline, diaphragm detection was conducted. In B-mode liver ultrasound imaging, the diaphragm is viewed as a echogenic line between the lung and the liver [10]. The diaphragm appears echogenic and clear even in cases where liver tissue as a whole may have an ill-defined boundary. The model used for diaphragm detection was a fine-tuned PulseNet encoder with support vector machines (SVM) as the classification head. The PulseNet encoder used the SonoNet model [4], fine-tuned on data from the PULSE study to classify 13 fetal anatomy views [8,11]. In this study, the encoder was further fine-tuned for diaphragm detection. Predictions were considered in the context of temporal dependence between frames of video. If one isolated frame was predicted to have no diaphragm while neighbouring frames were predicted otherwise, the isolated frame was reclassified as containing the diaphragm. Similarly, if three or fewer consecutive frames were predicted to have a diaphragm, while their surrounding frames were predicted to have no diaphragm, these frames were reclassified to have no diaphragm. This temporal smoothing was based on the appearance of the diaphragm in the training set where the diaphragm never appeared in less than three frames consecutively nor disappeared for only one isolated frame.

Hyperparameters and Fine-Tuning. Bayesian hyperparameter tuning was used for fine-tuning the PulseNet encoder for diaphragm detection. The temperature parameter, used for estimating uncertainty in the dataset, was tuned to a value of 1.25. The learning rate was tuned to a value of 0.022. The model was fine-tuned for 100 epochs with early stopping implemented if the validation accuracy stayed within a window of 0.02 for 10 consecutive epochs. The batch size was 32.

2.3   Diaphragm Segmentation for Liver Tissue Identification

Following diaphragm detection, segmentation masks for the diaphragm were propagated across video frames using adaptive memory to exploit the temporal dependence between frames [23]. The original model was trained on fetal ultrasound video for segmentation of the maternal bladder and placenta, and was fine-tuned here using 20 annotated videos of the diaphragm. During inference, to initialise the model for a video, a segmentation mask of the target object (in this case, the diaphragm) for the first frame of ultrasound video was required. To fully automate the use of this model, a bank of ultrasound images was created from the training set. The bank contained liver views with diverse representations of the diaphragm that had been labelled with masks of the diaphragm. An ultrasound image was selected from this bank that best matched the first frame of ultrasound video, which was identified using the structural similarity index measure (SSIM) [20]. The labelled ultrasound image from the bank of ultrasound images and masks that had the highest SSIM value with the first frame of ultrasound video was used to initialise the model.

In ultrasound images, liver tissue appears in the area above the diaphragm in the field of view. Therefore, a boundary for liver tissue was identified by finding the end-points of the mask of the diaphragm, and linking these up to the closest points on the left and right edges of the field of view. A special case was where the diaphragm stretched less than halfway across the field of view. This special case could introduce an error where the right-hand closest point in Euclidean space was towards the top of the field of view. Therefore, using this point for partitioning created a crude under-estimate of liver tissue by cutting out the lower right side of the field of view. Accounting for the fan shape of the ultrasound field of view, the closest point in Euclidean space was corrected in cases with partial diaphragm observation by moving the point used to create the partition towards the bottom of the field of view, by a factor of $\frac{|p-c|}{2}$, where p is the original closest point, and c is the midpoint of the right-hand edge of the field of view.

2.4   Full Pipeline

The fully-automated liver tissue identification pipeline is presented in Fig. 1.

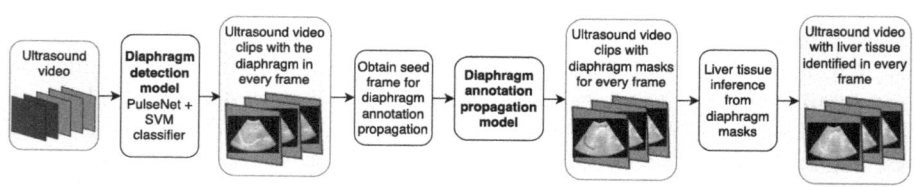

Fig. 1. Automated pipeline to identify liver tissue from ultrasound video.

Performance Metrics and Benchmarks. Performance was evaluated using sensitivity, specificity, and F1 score for diaphragm detection. For the final liver tissue identification, intersection over union (IoU) was used as a quantitative evaluation metric, in addition to the investigation of failure cases and clinical utility. For benchmarking diaphragm detection, ResNet-50 with cross entropy (CE) loss and a ResNet-50 using focal loss to account for the class imbalance [13] were used. The ResNet models were pre-trained on the ImageNet dataset [6]. As a benchmark for the full pipeline, the annotation propagation model detailed in [23] was applied without the full pipeline to directly segment liver tissue.

3 Experiments and Results

3.1 Diaphragm Detection

For diaphragm detection, 90 videos were used for training, 10 for validation, and 10 for testing. The total number of frames across videos in which the diaphragm appears and does not appear in each of these sets is detailed in Table 1. Not all videos had frames without a diaphragm. For the training set videos, the frames without diaphragm appeared in 45/90 videos. For the validation and test set videos, the frames without the diaphragm appeared in 6/10 and 4/10 videos, respectively.

Table 1. Number of frames in each class for diaphragm detection.

	Frames with diaphragm	Frames without diaphragm	Total frames
Train	3822 (87%)	565 (13%)	4387
Validation	393 (80%)	100 (20%)	493
Test	451 (95%)	24 (5%)	475

Table 2. Results for diaphragm detection. Sensitivity and specificity are reported with respect to detecting frames with no diaphragm.

	Specificity		Sensitivity		F1 score	
	Validation	Test	Validation	Test	Validation	Test
ResNet-50 + CE Loss (benchmark)	0.964	0.965	0.500	0.417	0.859	0.938
ResNet-50 + Focal Loss	0.967	0.97	0.370	0.000	0.825	0.911
PulseNet + SVM	**0.987**	**0.982**	**0.720**	**0.625**	**0.930**	**0.964**

Table 2 shows the sensitivities, specificities, and F1-scores for the PulseNet + SVM model as compared to the benchmark ResNet-50 models. The PulseNet encoder + SVM classifier were both more specific and sensitive in both sets, despite seeing an expected dip in performance on the test set. These results include temporal corrections as previously described. In particular, PulseNet + SVM was much more sensitive than the ResNet models, thereby conserving data. Figure 2 shows a set of predictions, before and after post-processing, as compared to the ground truth, taken from the test set. Common errors in prediction for the PulseNet model were in the transition between the diaphragm disappearing or appearing from view, or when the diaphragm was particularly blurry or unclear.

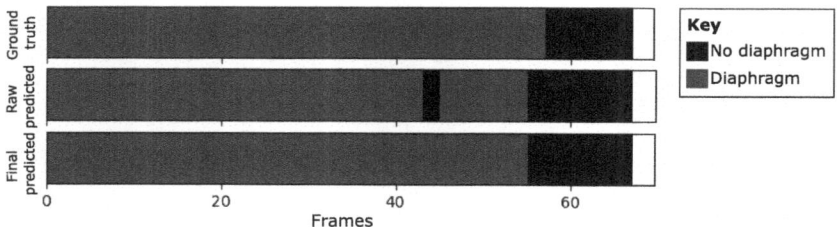

Fig. 2. Diaphragm detection for an example video. Frame-by-frame diaphragm detection predictions, both before and after post-processing, compared to the ground truth for an example video with a total number of 66 frames.

3.2 Identification of Liver Tissue

Table 3. Liver tissue identification results. Comparison in intersection over union (IoU) of directly segmenting liver tissue and inferring liver tissue from diaphragm position. The mean IoU is weighted by the number of frames in each video.

	IoU	
	Direct liver tissue segmentation	Liver tissue identification pipeline
Video 1	0.742	0.840
Video 2	0.459	0.828
Video 3	0.639	0.781
Video 4	0.549	0.912
Video 5	0.657	0.841
Mean (s.d.)	**0.622 (0.160)**	**0.837 (0.067)**

There was an improvement (Table 3) in the intersection over union (IoU) when identifying liver tissue using the full pipeline using the diaphragm as an anchor, as compared to directly segmenting liver tissue. The results presented in this table use the SSIM selected ultrasound image to initialise the segmentation model. For comparison, the model was rerun using the first frame of ultrasound video for initialisation; the SSIM method was nearly equivalent to using the first frame with a difference in the mean IoU of <1%. Figure 3 shows examples of failure cases for each model. Figures 3(a) and (b) show that the direct liver segmentation model incorrectly used vessel walls as the liver tissue boundary. Figure 3(c) shows an example where both lobes of the liver are in view, and the poor prediction of this tissue by both models.

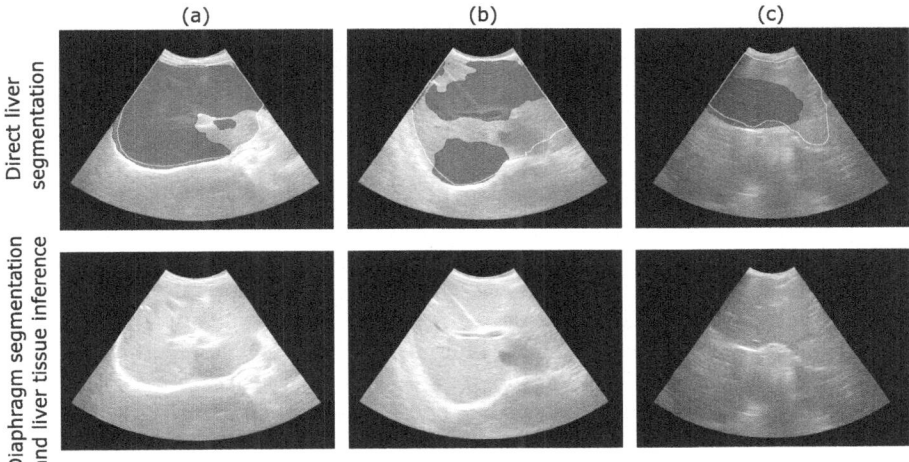

Fig. 3. Failure cases. Examples of where the diaphragm segmentation and inference improves liver partitioning.

4   Conclusion

A fully automated pipeline for identification of liver tissue in ultrasound video has been presented. This pipeline enables automatic annotation of the diaphragm and liver tissue in large volumes of ultrasound video data, as a processing step before liver disease diagnosis and staging. The liver identification pipeline presented here is more fit-for-purpose than direct liver segmentation. Errors encountered using direct liver segmentation include the detection of other echogenic structures, in particular vessel walls, as the boundary for liver tissue. Not only is this anatomically incorrect, but the results from direct segmentation are not useful as a pre-processing step for schistosomal PPF diagnosis or similarly complex liver diseases requiring whole views of the liver for accurate staging. In particular, schistosomal PPF develops along vessel walls, so using these as a boundary will discard important information for staging. Importantly, the proposed pipeline is suitable for processing data from low-cost, portable ultrasound in rural, resource-limited areas and generalisable to sub-Saharan African populations.

Limitations. There were several underlying assumptions that underpin this work. Firstly, there could be cases, if free-hand probe sweeps are instead used, where the diaphragm does not appear in an ultrasound video frame but there is liver tissue in that frame. However, this can be avoided by following the same acquisition protocol used in this study. Further to this, the method to infer the position of liver tissue from the diaphragm might provide a crude estimate of liver tissue area in a given frame of ultrasound video, especially when the diaphragm does not stretch fully across the field of view. This issue would be an important consideration if the pipeline were to be used for liver tissue area measurement or volume estimation but does not undermine the purpose of this pipeline; which is to identify the region of interest for liver disease classification models.

Future Work. The generalisability of this pipeline to diseased populations, to children, and to videos with known quality issues should be investigated in future studies. Subsequently, the key next steps for this pipeline include an evaluation as a pre-processing step before the use of a classification model for liver disease staging, including for schistosomal PPF. Importantly, a full pipeline for liver tissue identification and morbidity prediction for diseases like schistosomal PPF could serve as a capacity building tool for trainee sonographers, a risk prediction tool for clinical decision support, and a triage system for screening patients where staff are limited in areas of sub-Saharan Africa.

Acknowledgments. We would like to acknowledge Betty Nabatte as the Ugandan project manager; Victor Anguajibi, Timothy Mugume and Benjamin Ntegeka as sonographers for the SchistoTrack study; and the study teams and participants. We would like to acknowledge He Zhao for providing code for the annotation propagation model and advice, and Christopher Ho for assisting with the preparation of ultrasound videos.

ESO receives a DPhil studentship (2593890) associated with the EPSRC CDT in Health Data Science (EP/S02428X/1). GFC receives funding from the Welcome Trust Institutional Strategic Support Fund (204826/Z/16/Z) and John Fell Fund as part of the SchistoTrack Project, Robertson Foundation Fellowship, and UKRI EPSRC Award (EP/X021793/1). AN acknowledges EPSRC grants EP/X040186/1, ERC grant PULSE (ERC-2015-AdG-694581) and the NIHR Oxford Biomedical Research Centre Imaging Theme. For the purpose of Open Access, the author has applied a CC BY public copyright licence to any Author Accepted Manuscript version arising from this submission.

Ethics Approvals. Data collection and use were reviewed and approved by Oxford Tropical Research Ethics Committee (OxTREC 509-21), Vector Control Division Research Ethics Committee of the Uganda Ministry of Health (VCDREC146), and Uganda National Council of Science and Technology (UNCST HS 1664ES).

Disclosure of Interests. The authors have no competing interests to declare that are relevant to the content of this article.

References

1. Schistosomiasis. https://www.who.int/news-room/fact-sheets/detail/schistosomiasis. Accessed 16 May 2024
2. SchistoTrack. https://www.bdi.ox.ac.uk/research/schistotrack. Accessed 07 June 2024
3. Andrade, Z.A.: Schistosomiasis and liver fibrosis. Parasite Immunol. **31**(11), 656–663 (2009). https://doi.org/10.1111/j.1365-3024.2009.01157.x
4. Baumgartner, C.F., et al.: SonoNet: real-time detection and localisation of fetal standard scan planes in freehand ultrasound. IEEE Trans. Med. Imaging **36**(11), 2204–2215 (2017). https://doi.org/10.1109/TMI.2017.2712367
5. Becker, D.M., Tafoya, C.A., Becker, S.L., Kruger, G.H., Tafoya, M.J., Becker, T.K.: The use of portable ultrasound devices in low- and middle-income countries: a systematic review of the literature. Trop. Med. Int. Health **21**(3), 294–311 (2016). https://doi.org/10.1111/tmi.12657
6. Deng, J., Dong, W., Socher, R., Li, L.J., Li, K., Fei-Fei, L.: ImageNet: a large-scale hierarchical image database. In: 2009 IEEE Conference on Computer Vision and Pattern Recognition, pp. 248–255 (2009). https://doi.org/10.1109/CVPR.2009.5206848
7. Devarbhavi, H., Asrani, S.K., Arab, J.P., Nartey, Y.A., Pose, E., Kamath, P.S.: Global burden of liver disease: 2023 update. J. Hepatol. **79**(2), 516–537 (2023). https://doi.org/10.1016/j.jhep.2023.03.017
8. Drukker, L., et al.: Transforming obstetric ultrasound into data science using eye tracking, voice recording, transducer motion and ultrasound video. Sci. Rep. **11**(1), 14109 (2021). https://doi.org/10.1038/s41598-021-92829-1
9. Eggleston, A.J., Farrington, E., McDonald, S., Aziz, S.: Portable ultrasound technologies for estimating gestational age in pregnant women: a scoping review and analysis of commercially available models. BMJ Open **12**(11), e065181 (2022). https://doi.org/10.1136/bmjopen-2022-065181
10. Fayssoil, A., et al.: Diaphragm: pathophysiology and ultrasound imaging in neuromuscular disorders. J. Neuromuscul. Dis. **5**(1), 1–10 (2018). https://doi.org/10.3233/JND-170276
11. Fu, Z., Jiao, J., Yasrab, R., Drukker, L., Papageorghiou, A.T., Noble, J.A.: Anatomy-aware contrastive representation learning for fetal ultrasound (2022). https://doi.org/10.48550/arXiv.2208.10642
12. Hashim, A., Berzigotti, A.: Noninvasive assessment of schistosoma-related periportal fibrosis. J. Ultrasound Med. **40**(11), 2273–2287 (2021)
13. Lin, T.Y., Goyal, P., Girshick, R., He, K., Dollár, P.: Focal loss for dense object detection (2018). https://doi.org/10.48550/arXiv.1708.02002
14. Liu, Y., Liu, X., Wang, S., Song, J., Zhang, J.: A novel method for accurate extraction of liver capsule and auxiliary diagnosis of liver cirrhosis based on highfrequency ultrasound images. Comput. Biol. Med. **125**, 104002 (2020). https://doi.org/10.1016/j.compbiomed.2020.104002
15. McManus, D.P., Dunne, D.W., Sacko, M., Utzinger, J., Vennervald, B.J., Zhou, X.N.: Schistosomiasis (primer). Nat. Rev.: Dis. Primers **4**(1), 13 (2018)
16. Meiburger, K.M., Acharya, U.R., Molinari, F.: Automated localization and segmentation techniques for B-mode ultrasound images: a review. Comput. Biol. Med. **92**, 210–235 (2018). https://doi.org/10.1016/j.compbiomed.2017.11.018
17. Ockenden, E.S., Frischer, S.R., Cheng, H., Noble, J.A., Chami, G.F.: The role of point-of-care ultrasound in the assessment of schistosomiasis-induced liver fibrosis: a systematic scoping review. PLOS Neglected Trop. Dis. **18**(3), e0012033(2024)

18. Richter, J., Hatz, C., Campagne, G., Bergquist, N.R., Jenkins, J.M.: Ultrasound in schistosomiasis: a practical guide to the standard use of ultrasonography for assessment of schistosomiasis-related morbidity: Second international workshop, 22–26 October 1996, Niamey, Niger. Technical report TDR/STR/SCH/00.1, World Health Organization (2000)

19. Spearman, C.W., Sonderup, M.W.: Health disparities in liver disease in sub-Saharan Africa. Liver Int. **35**(9), 2063–2071 (2015). https://doi.org/10.1111/liv.12884

20. Wang, Z., Bovik, A., Sheikh, H., Simoncelli, E.: Image quality assessment: from error visibility to structural similarity. IEEE Trans. Image Process. **13**(4), 600–612 (2004). https://doi.org/10.1109/TIP.2003.819861

21. Wilkinson, J.N., Saxhaug, L.M.: Handheld ultrasound in training - the future is getting smaller! J. Intensive Care Soc. **22**(3), 220–229 (2021). https://doi.org/10.1177/1751143720914216

22. Zhang, L., et al.: SEG-LUS: a novel ultrasound segmentation method for liver and its accessory structures based on multi-head self-attention. Comput. Med. Imaging Graph. **113**, 102338 (2024). https://doi.org/10.1016/j.compmedimag.2024.102338

23. Zhao, H., Men, Q., Gleed, A., Papageorghiou, A.T., Noble, J.A.: Ultrasound video segmentation with adaptive temporal memory. In: Kainz, B., Noble, A., Schnabel, J., Khanal, B., Müller, J.P., Day, T. (eds.) ASMUS 2023. LNCS, pp. 3–12. Springer, Cham (2023). https://doi.org/10.1007/978-3-031-44521-7_1

Democratizing AI in Africa: Federated Learning for Low-Resource Edge Devices

Jorge Fabila[1(✉)], Víctor M. Campello[1], Carlos Martín-Isla[1],
Johnes Obungoloch[2], Kinyera Leo[2], Amodoi Ronald[2], and Karim Lekadir[1,3]

[1] Dept. de Matemàtiques i Informàtica, Universitat de Barcelona, Barcelona, Spain
jorge_fabila@ub.edu
[2] Mbarara University of Science and Technology, Mbarara, Uganda
[3] Institució Catalana de Recerca i Estudis Avançats (ICREA), Barcelona, Spain

Abstract. Africa faces significant challenges in healthcare delivery due to limited infrastructure and access to advanced medical technologies. This study explores the use of federated learning to overcome these barriers, focusing on perinatal health. We trained a fetal plane classifier using perinatal data from five African countries: Algeria, Ghana, Egypt, Malawi, and Uganda, along with data from Spanish hospitals. To incorporate the lack of computational resources in the analysis, we considered a heterogeneous set of devices, including a Raspberry Pi and several laptops, for model training. We demonstrate comparative performance between a centralized and a federated model, despite the compute limitations, and a significant improvement in model generalizability when compared to models trained only locally. These results show the potential for a future implementation at a large scale of a federated learning platform to bridge the accessibility gap and improve model generalizability with very little requirements.

Keywords: Federated Learning · Low resources · Edge devices · Fetal ultrasound

1 Introduction

Today's advancements in Artificial Intelligence (AI) are proposed at a blistering pace and rely greatly on the availability of large-scale and varied datasets. As a result, the training of AI models is becoming restricted to users with access to such datasets and a high-performance computing cluster with a large storage that can train a model in a reasonable amount of time.

These requirements might threaten the democratization of AI tools, and specifically, its use on low-resource settings, worsening the accessibility gap [13]. One particular region where reducing this gap is paramount is Africa, a continent that faces deep challenges such as widespread poverty, limited infrastructure, and insufficient access to quality medical services. In fact, AI presents a unique opportunity to improve healthcare access in Africa. For instance, by providing support

U. Anazodo et al. (Eds.): MImA 2024/EMERGE 2024, CCIS 2240, pp. 101–109, 2025.
https://doi.org/10.1007/978-3-031-79103-1_11

to healthcare workers with acquiring medical images or extracting biomarkers or risk scores in the absence of the appropriate imaging modality or clinical expert. Several studies have assessed the viability of such an approach, for example, for oral cancer screening in the absence of clinical experts in India [14] or for simplifying image acquisition of fetal ultrasound and estimating gestational age from non-standard planes [15].

Federated learning (FL) has been proposed to help circumvent the requirement of having access to a large database [16]. With this technique, the desired model design and weights are shared across a set of nodes in each participating institution and trained with a corresponding local dataset. When finished, the model weights are sent back to a central node that combines the different weights received to maximize final performance. This process is usually repeated several times until the improvements stall. By following this approach one avoids transferring the different datasets to one central node, guaranteeing a stronger data privacy. Several works have tested this learning technique in simulated scenarios [2–4] and recently, one study conducted a real-world validation with Raspberry Pi devices in three institutions in the United Kingdom [8], but only using classical machine learning. Another study trained a federated model across 20 clients using COVID-19 cases [19], although no details were provided about the infrastructure and all nodes are assumed to have a GPU. In all cases, the different nodes had either the same hardware devices or had a GPU. To our knowledge, no existing work has considered a real heterogeneous setting where devices with different capabilities are involved in the training process.

In this study, we aim to develop the first scientific framework for inclusive imaging AI in resource-limited settings bringing the current state-of-the-art methods (mostly developed for high-income settings) towards new imaging AI algorithms that are fundamentally inclusive, i.e. affordable for resource-limited clinical centres, scalable to under-represented population groups, and accessible to minimally trained clinical workers. With this we aim to bridge the healthcare accessibility gap, providing essential support for working with medical imaging to those who need it most.

In this work, we focus on the enhancement of obstetric ultrasound screening in rural Africa, with the aim of empowering midwives with AI-based tools through the implementation of an FL platform that aims to overcome the hardware constrains in some healthcare facilities. With our FL platform we enable collaboration across borders by training a fetal ultrasound plane classifier with data from Algeria, Ghana, Egypt, Malawi, and Uganda. We demonstrate the added value of collective training for creating robust and generalizable AI models even when working with edge devices, such as a Raspberry Pi device.

2 Methodology

2.1 Datasets

For this study, we used two publicly available fetal ultrasound datasets which consisted of standard fetal ultrasound planes acquired during routine clinical practice on the 2nd and 3rd trimester of pregnancy.

On one hand, we considered a large-scale dataset consisting of 12,000 images acquired at Hospital Clínic and Hospital Sant Joan de Déu in Barcelona, Spain [7]. The images were acquired with several curvilinear transducers: Voluson E6, Voluson S8 and Voluson S10 (GE Medical Systems, Zipf, Austria), and Aloka (Aloka CO., LTD.)

On the other hand, we considered a small-scale dataset from five African countries (Algeria, Egypt, Ghana, Malawi and Uganda) containing a total of 450 images. These images were acquired with a diversity of ultrasound devices. The specific details can be found in [5].

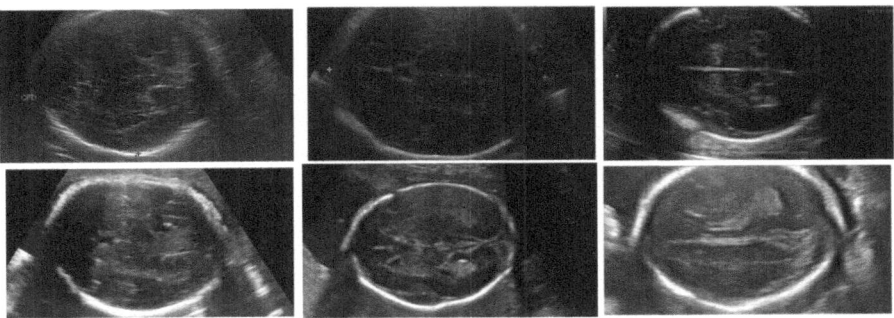

Fig. 1. Fetal standard planes for the brain acquired at different centers. From left to right and from top to bottom, the planes are from Spain, Ghana, Egypt, Malawi, Algeria, Uganda

All images are labeled with five standard plane categories: *abdomen*, *brain*, *femur*, *thorax* and *other* (the latter includes other less common planes such as maternal cervix, umbilical cord or extremities). These labels were assigned by clinicians or technicians with extensive training in ultrasound fetal plane classification. Figure 1 shows some examples of these images. The number of planes per category and country is shown in Table 1. There are two things to highlight about this distribution, the first one and the most evident is that the African datasets have a very few amount of samples in comparison with the Spanish dataset. In order to try to mitigate this unbalance we perform data augmentation to the African datasets by adding randomly rotated as well as randomly cropped images between a factor of 0.8 and 1.0 of its original size. The second point is that only the dataset from the Spanish hospitals include this last category, so we removed it from the study in order to avoid an even more unbalanced distribution. In this proof-of-concept study, we did not incorporate additional balancing techniques, as the primary goal was to test out the deployed platform rather than optimize or fine-tune the model's performance. Our focus was on demonstrating the feasibility of the proposal, leaving the exploration of advanced balancing methods for future research since it is true that in real life we will find unbalanced datasets specially when dealing with African rural health centers that could have less data

All of the images were preprocessed by resizing them to 256×256 pixels, allowing us to create mini-batches and still run given the compute limitations. Intensity values were normalized between 0 and 1.

Since all datasets used in this study are publicly available, no ethical approval was needed to conduct the current study.

Table 1. Number of images per category and country

Country	Abdomen	Brain	Femur	Thorax	Other
Spain	711	1781	1040	1718	4213
Malawi	25	25	25	25	0
Egypt	25	25	25	25	0
Uganda	25	25	25	0	0
Ghana	25	25	25	0	0
Algeria	25	25	25	25	0

2.2 Training

FL consists of a collaborative training in which the model weights are distributed among client nodes, each with its own local data. The models are distributed to and refined at various centers, then returned and aggregated on a central server. This process is repeated for several rounds, with the updated models being redistributed to the original nodes. While there are several aggregation techniques available such as model selection, weighted averaging, median aggregation or voting aggregation, in this proof-of-concept study, we utilize federated averaging, which involves averaging the weights of all the different local models [16]. This aggregation process extracts insights from all datasets without transferring raw data beyond each client. Subsequently, the updated aggregated weights are broadcast back to all clients, who then continue training the aggregated model with their respective private datasets. This iterative process retains data locally and only shares model weights reducing privacy risks associated with data sharing. Additionally, it distributes computational load across clients.

In the current work, we implemented an FL platform using the Flower library (*flwr*) [6], PyTorch [9] and PyTorch Lightning [10]. *flwr* is used as a framework for communicating with the different nodes from the central server through Secure Socket Layer (SSL) protocol (guaranteeing a secure connection) and for performing the aggregation of the client models. Then, each client has its own local model that is trained with its local data, in this part, Lightning was used to obtain optimal speed and reliability by optimizing computation and memory usage, enabling more efficient model training. Additionally, we used the automatic mixed precision feature to accelerate training and reduce memory usage without sacrificing model accuracy. The optimizer used was SGD with a learning rate scheduler that dynamically adjusts the learning rate based on the training progress, avoiding unnecessary stagnation of the training loop.

Our framework includes different models and variants of them, but for this study we use ResNet-50 [11] since it provides a good trade-off between accuracy and resource usage.

In synchronous training, the primary challenge when using a Raspberry Pi in federated frameworks is that the training speed is limited by the slowest node. This is because the framework must wait for all clients to respond, potentially leading to impractically long training times. On the other hand, asynchronous training allows the server to proceed without waiting for all nodes to complete each round. However, this approach carries the risk that slower nodes, such as the Raspberry Pi, may not contribute sufficiently to the training. We decided to train using synchronous steps to have all the nodes contribute equally and to measure how much the whole system slows down when adding an edge device and thus, to determine the feasibility of using it.

To optimize performance across different nodes, we configured each node to train for 10 epochs per round, except for the Raspberry Pi, which was set to 3 epochs per round. We trained for 20 rounds, with each node having a training dataset size of 80%, except for the Raspberry Pi, which had a dataset size of only 40%. The testing, validation and training datasets were divided into non-overlapping partitions.

2.3   Hardware

A variety of hardware components was used in this study, with specific details provided in Table 2. Only one computer had access to a GPU for training, one node was the Raspberry Pi 5 with an ARM processor and the other computers trained only using the CPU.

Table 2. Hardware used for the different nodes during the training process.

Node	Location	Processor	GPU	OS
Central	Spain	Intel Core i7-9700K	RTX 2060 8 GB	Ubuntu 22.04 LTS
1	Argelia	Intel Core i7-8650U	–	Windows 11 Pro
2	Ghana	Intel Core i5-1145G7	–	Windows
3	Spain	Arm Cortex-A76	–	Ubuntu 24.04 Raspberry
4	Spain	Intel Core i7-9700K	RTX 2080 8 GB	Ubuntu 22.04 LTS
5	Uganda	Intel 300	–	Windows 10 Pro
6	Uganda	Intel Core i7-6700	–	Ubuntu 22.04 LTS

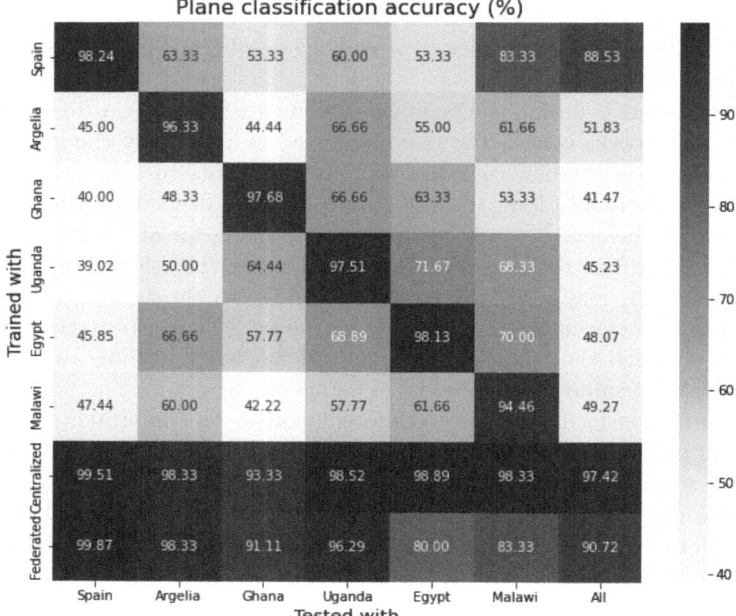

Fig. 2. Accuracy achieved with different datasets for Resnet-50

Table 3. Performance related information

Dataset	Device	Samples	Batch	Epochs/Round	Iter./s	Avg Time/Round (min)
Spain	GPU	6549	8	10	~15.6	10.48
Malawi	Raspberry	240	2	3	~0.3	26.78
Egypt	CPU	480	8	10	~2.5	4.05
Uganda	CPU	360	8	10	~0.57	12.96
Ghana	CPU	360	8	10	~0.67	14.88
Algeria	CPU	480	8	10	~0.79	14.81

3 Results and Discussion

We trained 8 different models. Six local models were trained with the respective dataset from each node that represents each country. One centralized model with all the datasets merged together, that represents the situation in which all the centers manage and agree to transfer all their data to one single center, training node 4 in Table 2. And one final model trained with our federated framework, in which all the datasets were kept in their respective centers. Figure 2 shows the accuracy for each model when tested on the different datasets in order to measure the achieved generalizability.

It is possible to observe that the models trained with individual datasets are not getting enough generalization when tested with external datasets. For example, the model trained with the Ugandan dataset achieves only 39% of accuracy on the Spanish dataset. In contrast, the centralized model gets a high level of generalization across all test datasets. Lastly, our federated trained model also achieves a good level of generalization but without sharing data between the nodes and with important hardware constrains. Figure 3 shows the evolution of the accuracy during training process of centralized and federated approaches. While the centralized training is more stable and achieves higher performance, the accuracy of the federated model is comparable with very few rounds into the training.

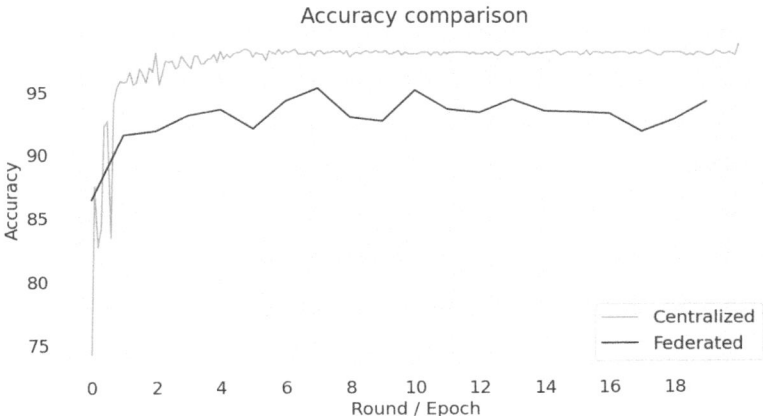

Fig. 3. Accuracy evolution during training of the centralized and federated models.

The Raspberry Pi not only successfully processed and trained a model locally with its dataset but also could participate in the FL training among all of the other nodes. With this, we validate its capability as a low-cost, effective tool for AI deployment in resource-constrained settings. Also, we show the feasibility and potential of training models with a FL approach using edge devices such as Raspberry Pi.

Table 3 shows information related to the performance of every node. In all cases the epochs per round was fixed to 10 except in the case of the Raspberry, that was configured to perform only 3 to compensate for the performance gap with the other nodes. The batch size was also smaller in this case to avoid memory bottlenecks. By looking at the iterations per second and the average time per round, one can notice that the Raspberry Pi node is the slowest. However, the difference can be compensated by reducing the batch size and the epochs per round. While it is slowing the whole process, the accuracy is still high as we can corroborate in Fig. 2.

4   Conclusions

This study aims to be the beginning of a series of research efforts focused on utilizing FL in Africa. We demonstrated that using limited resources devices within a federated project such as Raspberry Pi is feasible. Our developed platform was successfully deployed and tested in a realistic scenario with diverse constrained devices. This deployment marks a significant starting point in our project, demonstrating both the robustness of our solution and that such a framework will be able to handle future real-world applications. The platform is now fully operational and ready to deliver its intended benefits to users encouraging the formation of partnerships between local healthcare providers, international research institutions, and technology developers.

Acknowledgments. This research was funded by a grant from the European Research Council (ERC) under the European Union's Horizon Europe research and innovation programme (AIMIX project - Grant Agreement No. 101044779).

References

1. Martin-Isla, C., et al.: Image-based cardiac diagnosis with machine learning: a review. Front. Cardiovasc. Med. **7**, 1 (2020)
2. Sohan, M.F., Basalamah, A.: A systematic review on federated learning in medical image analysis. IEEE Access **11**, 28628–28644 (2023)
3. Linardos, A., Kushibar, K., Walsh, S., Gkontra, P., Lekadir, K.: Federated learning for multi-center imaging diagnostics: a simulation study in cardiovascular disease. Sci. Rep. **12**(1), 3551 (2022)
4. Rehman, M.H., Hugo Lopez Pinaya, W., Nachev, P., Teo, J.T., Ourselin, S., Cardoso, M.J.: Federated learning for medical imaging radiology. Br. J. Radiol. **96**(1150), 20220890 (2023)
5. Sendra-Balcells, C., et al.: Generalisability of fetal ultrasound deep learning models to low-resource imaging settings in five African countries. Sci. Rep. **13**(1), 2728 (2023)
6. Beutel, D.J., et al.: Flower: a friendly federated learning research framework. arXiv preprint arXiv:2007.14390 (2020)
7. Burgos-Artizzu, X.P., et al.: Evaluation of deep convolutional neural networks for automatic classification of common maternal fetal ultrasound planes. Sci. Rep. **10**(1), 10200 (2020)
8. Soltan, A.A.S., et al.: A scalable federated learning solution for secondary care using low-cost microcomputing: privacy-preserving development and evaluation of a COVID-19 screening test in UK hospitals. Lancet Digit. Health **6**(2), e93–e104 (2024)
9. Ansel, J., et al.: PyTorch 2: faster machine learning through dynamic python byte-code transformation and graph compilation. In: Proceedings of the 29th ACM International Conference on Architectural Support for Programming Languages and Operating Systems (ASPLOS '24), vol. 2. ACM (2024)
10. Falcon, W., The PyTorch Lightning Team: PyTorch Lightning (Version 1.4) [Computer software] (2019). https://github.com/Lightning-AI/lightning

11. He, K., Zhang, X., Ren, S., Sun, J.: Deep residual learning for image recognition. In: Proceedings of the IEEE Conference on Computer Vision and Pattern Recognition, pp. 770–778 (2016)
12. Moshawrab, M., Adda, M., Bouzouane, A., Ibrahim, H., Raad, A.: Reviewing federated learning aggregation algorithms; strategies, contributions, limitations and future perspectives. Electronics **12**(10), 2287 (2023)
13. Mollura, D.J., et al.: Artificial intelligence in low- and middle-income countries: innovating global health radiology. Radiology **297**(3), 513–520 (2020). https://doi.org/10.1148/radiol.2020201434
14. Birur, N.P., et al.: Field validation of deep learning based point-of-care device for early detection of oral malignant and potentially malignant disorders. Sci. Rep. **12**(1), 14283 (2022)
15. Maraci, M.A., et al.: Toward point-of-care ultrasound estimation of fetal gestational age from the trans-cerebellar diameter using CNN-based ultrasound image analysis. J. Med. Imaging **7**(1), 014501 (2020)
16. McMahan, B., Moore, E., Ramage, D., Hampson, S., y Arcas, B.A.: Communication-efficient learning of deep networks from decentralized data. In: Proceedings of Artificial Intelligence and Statistics, pp. 1273–1282. PMLR (2017)
17. Yang, Q., Liu, Y., Chen, T., Tong, Y.: Federated machine learning: concept and applications. ACM Trans. Intell. Syst. Technol. (TIST) **10**(2), 1–19 (2019)
18. Rieke, N., et al.: The future of digital health with federated learning. NPJ Digit. Med. **3**(1), 1–7 (2020)
19. Dayan, I., et al.: Federated learning for medical imaging: our experience and next steps. J. Am. Coll. Radiol. **18**(9), 1213–1221 (2021)

Generative Style Transfer for MR Image Segmentation: A Case of Glioma Segmentation in Sub-Saharan Africa

Rancy Chepchirchir[1,2], Jill Sunday[3], Raymond Confidence[4,5], Dong Zhang[6], Talha Chaudhry[7], Udunna C. Anazodo[4,5,8,9], Kendi Muchungi[10], and Yujing Zou[11(✉)]

[1] Institute of Mathematical Science, Strathmore University, Nairobi, Kenya
[2] Faculty of Science and Engineering, University of Hull, Hull, UK
r.chepchirchir-2023@hull.ac.uk
[3] Department of Medical Engineering, Technical University of Mombasa, Mombasa, Kenya
jillselesa35@gmail.com
[4] Medical Artificial Intelligence Lab, Lagos, Nigeria
raymondconfidence@gmail.com
[5] Lawson Health Research Institute, London, ON, Canada
[6] Department of Electrical and Computer Engineering, University of British Columbia, Vancouver, BC, Canada
donzhang@ece.ubc.ca
[7] University of Nairobi, Nairobi, Kenya
talhahchaudhry99@gmail.com
[8] Department of Clinical and Radiation Oncology, University of Cape Town, Cape Town, South Africa
udunna.anazodo@mcgill.ca
[9] Montreal Neurological Institute, McGill University, Montreal, Canada
[10] Brain Mind Institute, The Aga Khan University, Nairobi, Kenya
kendi.muchungi@aku.edu
[11] Medical Physics Unit, McGill University, Montreal, Canada
yujing.zou@mail.mcgill.ca

Abstract. In Sub-Saharan Africa (SSA), the utilization of lower-quality Magnetic Resonance Imaging (MRI) technology raises questions about the applicability of machine learning (ML) methods for clinical tasks. This study aims to provide a robust deep learning-based brain tumor segmentation (BraTS) method tailored for the SSA population using a threefold approach. Firstly, the impact of domain shift from the SSA training data on model efficacy was examined, revealing no significant effect. Secondly, a comparative analysis of 3D and 2D full-resolution models using the nnU-Net framework indicates similar performance of both the models trained for 300 epochs achieving a five-fold cross-validation score of 0.93. Lastly, addressing the performance gap observed in SSA validation as opposed to the relatively larger BraTS glioma (GLI) validation set, two strategies are proposed: fine-tuning SSA cases using the GLI + SSA best-pretrained 2D fullres model at 300 epochs, and introducing a novel neural style transfer-based data augmentation technique for the SSA cases. This investigation underscores the potential of enhancing brain tumor prediction within SSA's unique healthcare landscape.

© The Author(s), under exclusive license to Springer Nature Switzerland AG 2025
U. Anazodo et al. (Eds.): MImA 2024/EMERGE 2024, CCIS 2240, pp. 110–123, 2025.
https://doi.org/10.1007/978-3-031-79103-1_12

Keywords: Brain Tumor Segmentation · Neural style transfer · nnU-Net

1 Introduction

Brain tumors present a substantial health challenge in Africa. The efforts of research on brain tumors have barely made any positive change in survival rate in low-and middle-income countries (LMICs). The rate of mortalities from glioma is among the highest in the world, with the Sub-Saharan Africa experiencing a rise of 25% [1].

Accurate segmentation of distinct sub-regions within gliomas such as peritu- moral edema, necrotic core, enhancing, and non-enhancing tumor core, based on multimodal MRI scans, hold clinical relevance for the diagnosis, prognosis, and treatment of brain tumors. Accurately delineating the regions of interest within a tumor provides essential insight about its size, location, and shape, enabling the determination of the extent of tumor involvement [2].

However, the segmentation of these sub-regions presents a formidable challenge due to the heterogeneity of brain tumors [3] and in resource-constrained settings such as the SSA, due to the propensity for suboptimal image contrast and resolution [1] from lower quality MRI scanners and lack of availability of advanced imaging techniques [4]. These challenges raise uncertainty about the feasibility of implementing ML methods for clinical purposes [4, 5]. Furthermore, suboptimal image contrast and resolution [6], necessitates advanced image pre-processing to enhance their resolution before employing ML techniques for tasks like tumor segmentation, classification, or outcome prediction [1].

In response to these challenges, this study aims to develop a generalizable deep learning-based brain tumor segmentation method. Our approach addresses the challenge of lower quality MRI scans in the SSA by developing an effective model that handles data variability. Through advanced preprocessing and optimized segmentation, we aim to create a robust model that improves diagnostic outcomes in under-resourced settings.

1.1 Related Work

Neural style transfer (NST) has been employed as a data augmentation technique in various medical domains, including COVID-19 diagnosis classification [7] and 3D cardiovascular MR image segmentation [8]. However, this study marks the first application of NST to brain tumor segmentation within a Sub-Saharan Africa context. Moreover, Tomar et.al. [9] have explored a self-supervised style transfer technique as data augmentation to improve brain tumor segmentation performances. This comprehensive methodology was, however, more computationally expensive than the NST approach. Bouter et al. [10] demonstrated the feasibility of artificially creating super-resolution MR images from low-resolution counterparts, indicating that such an approach could be leveraged for our 2021 BraTS dataset. Lastly, the work conducted by Sendra et al. [11] addressed similar domain-shift challenges using comparable approaches. Their findings suggested that employing transfer learning for domain adaptation could integrate modest-sized African samples into extensive databases of developed nations. Notably, both studies highlighted the imbalance between African and high-resource country cases, with

Sendra et al. studying 25 patients from five African centers, while our study examined 60 African cases.

2 Methods

Implementation
Our solution, which is implemented with PyTorch, is an extension of NVIDIA's nnUNet [12], which is publicly available on GitHub[1]. The baseline model training and inference were done with mixed precision to minimize costs (i.e. time and memory). The experiments were run on Tesla T4 Turing GPUs and NVIDIA's V100 system on Compute Canada cluster. We then stored both the latest and best checkpoint models based on the Dice score on the validation dataset for use during inference.

Datasets
This work was conducted as part of the BraTS 2023 Challenge, where the datasets preselected by the organizers were used. The datasets were multi-center MRI scans of 1251 adult glioma (GLI) cases from the 2021 Continuous Evaluation sub-challenge [13] and 60 adult glioma cases acquired in SSA (SSA) from the BraTS-Africa sub challenge [1] - the largest publicly available African adult glioma MRI data. Thus, a total of 1311 MRI scans of adults with pre-operative glioma including both low-grade glioma (LGG) and high-grade glioma (GBM/HGG) were used to train and validate the proposed model. The MRI scans comprised of routine T1-weighted (T1), post-contrast T1-weighted (T1ce), T2-weighted (T2) and T2-weighted Fluid Attenuated Inversion Recovery (FLAIR) images acquired as part of standard of care [1, 13]. Each case also contained pre-labelled brain tumor sub-region masks, namely, necrotic tumor core (NCR), enhancing tumor (ET), and peritumoral edematous tissue (ED).

The datasets were split into GLI (1251 cases) and GLI + SSA (1311) and used separately to train the model, while a five-fold cross-validation approach was used to evaluate the performance of the model. Four cases from the SSA dataset were excluded as outliers based on image quality inconsistencies. To investigate the impact of the outliers in real-world applications, the GLI + SSA data were also trained excluding the four outlier cases (GLI + SSA2; 1307 cases).

Data Preprocessing
The data were preprocessed following Futrega et.al Optimized U-Net approach [14] and involved foreground cropping operation, intensity normalization, and resampling. These preprocessing steps were taken to establish data coherence among the multi-center data, enhance image quality, and standardize the format for subsequent processing steps. Commencing with loading the dataset, organized in alignment with a designated data path, we extracted crucial metadata from a JSON file. Enhancing image quality and uniformity was realized through the application of a crop foreground operation. This step effectively removed extraneous background regions. Additionally, the process encompassed intensity normalization. Notably, for MRI scans, normalization was confined to non-zero regions. Addressing variations in voxel spacing among diverse scans, resampling played

---

[1] https://github.com/MIC-DKFZ/nnUNet.

a pivotal role. The aforementioned measures are crucial in reducing variability in input data and improving reliability of the model's predictions [14].

2.1 Baseline Model (Optimized U-Net)

The baseline for our model was inspired by the work of Futrega et al. [14], the Optimized U-Net model with deep supervision to improve the gradient flow by calculating the loss function at various decoder levels. Here, each experiment was trained for 2, 5, 10, and 30 epochs using the Adam optimizer with varying learning rates for the three experiments: 1) GLI, 2) GLI + SSA, and 3) GLI + SSA2.

2.2 nnU-Net Model

nnU-Net, an image segmentation method introduced in [12], adapts to specific datasets by autonomously configuring a U-Net-based segmentation pipeline. It simplifies model training by creating multiple U-Net configurations for different datasets, effectively handling diverse input modalities and class imbalances. Notably, nnU-Net version 2 brings enhancements, offering a user-friendly development framework.

Two- and Three-Dimensional (2D & 3D) Configuration
We have employed a 3D full resolution with a batch size of 2, a patch size of [128, 128, 128], 32 U-Net base features, a per-stage encoder and decoder of [2, 2, 2, 2, 2, 2], kernel sizes of [[1, 1, 1], [2, 2, 2], [2, 2, 2], [2, 2, 2], [2, 2, 2], [2, 2, 2]], and convolution kernel sizes of [[3, 3, 3], [3, 3, 3], [3, 3, 3], [3, 3, 3], [3, 3, 3], [3, 3, 3]] on both the GLI dataset and the GLI+SSA dataset. We also employed a 2D full resolution with a batch size of 105, a patch size of [192, 160], 32 U-Net base features, a per-stage encoder and decoder of [2, 2, 2, 2, 2, 2], kernel sizes of [[1, 1], [2, 2], [2,2], [2, 2], [2, 2], [2, 2]], and convolution kernel sizes of [[3, 3], [3, 3], [3, 3], [3, 3], [3, 3], [3, 3]] on the GLI and GLI+SSA dataset. Each experiment was trained for 2, 5, 10, 30, and 300 epochs using the Adam optimizer with varying learning rates

2.3 Proposed Methods: Neural Style Transfer Augmentation and 2D Full-Res Model Finetuning

In the context of resource-constrained settings where limited training data continues to pose a challenge, we employed neural style transfer (NST), first proposed by Gatys et al. [15], as a data augmentation technique to enhance the effectiveness of our model training process. Neural style transfer is a technique rooted in deep learning that enables the separation and combination of content and style aspects within images. Specifically, for our application, we leveraged neural style transfer to enhance the SSA MR images. This technique involved using an SSA MR image as the content (or source) image and performing a random pairing with a GLI MR image as the style (or target) image. The neural style transfer process entailed adapting the stylistic features of the GLI image onto the SSA image, thereby creating new, augmented training samples for SSA cases. Using the Keras functional API, the intermediate layers of a pretrained VGG19 image classification network [16] were used as a feature extractor to obtain the content and style representations of an image. The neural style transfer algorithm computes a loss by evaluating the discrepancies between the stylistic and content features of the generated image and target images. The style loss measures stylistic differences using Gram matrices of feature maps, while the content loss quantifies content dissimilarity through feature map comparisons.

Overall Workflow. The proposed methodology is illustrated below in Fig. 1. At the time of submission for this short paper, we hereby present results for the nnU-Net best 2D and 3D models, as well as the best pre-trained $2D^2$ fullres nnU- Net model trained from GLI + SSA training data fine-tuned on the original and NST-augmented SSA training data. The 5-fold cross validation model evaluation was based on mean Dice Similarity Coefficient (DSC) after each epoch. A Paired Samples t-test was conducted to compare model performances between datasets and as a function of training time/epochs (significance at $p < 0.05$).

3   Results

3.1   Baseline Model (Optimized U-Net)

To establish a robust evaluation, we conducted experiments using the Optimized U-Net [14] on both the GLI and GLI + SSA datasets as part of our foundational benchmarking process. This comparison aimed to elucidate how distributional shifts within training data can influence cross-validated model performance. Table 1 presents outcomes from both datasets. It includes their Dice Similarity Coefficient (DSC), training, and validation

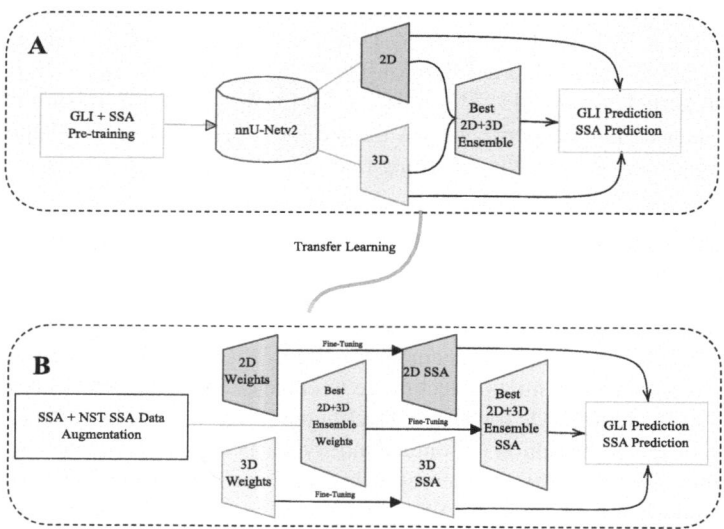

Fig. 1. Fine tuning small-scale SSA data with neural style transfer data augmentation techniques (B) with pretrained weights from large-scale GLI data via transfer learning (A).

---

[2] https://github.com/CAMERA-MRI/SPARK2023/tree/main/SPARK_BTS_KIFARU.

losses for the GLI and GLI + SSA datasets, with or without four outlier cases (00051, 00097, 00041, and 00084), as shown in Fig. 1 in Appendix. We halted training at 30 epochs for comparative purposes with an experiment using the complete dataset for the same epoch count. From the aforementioned results, a decrease in model performance was observed from a Dice score of 0.89 for the GLI trained dataset to 0.88 for the GLI + SSA dataset at 30 epochs (Table 1). There was no statistically significant difference between the Dice scores of the GLI and GLI + SSA models ($p > 0.05$) or between models when outliers were excluded **($p > 0.05$)** (Table 1). However, this exclusion was specific to this experiment, aimed at probing the domain shift issue in multi-institutional data, particularly in the Sub-Saharan Africa context.

3.2 nnU-Net Model (Version 2)

We compared the 3D fullres versus 2D fullres configurations of nnU-Net trained on the GLI + SSA dataset. The best model was considered to have obtained the best 5-fold cross-validation averaged Dice Similarity Coefficient.

Configuration Name: 3D Full Resolution. Here, we employed the 3D configuration models for the GLI and GLI + SSA datasets, as represented in Table 2. It is evident that the 300 epoch model achieved the best pseudo Dice score at 0.95 and 0.93 for the GLI and GLI + SSA datasets respectively; followed by the 30 epoch model at 0.90 for both the datasets. The paired t-test comparisons between models, showed no statistically significant difference ($p > 0.05$) between the pseudo Dice scores (Table 2).

Configuration Name: 2D Full Resolution. Table 3 reveals that the 2D full-resolution nnU-Net model, trained for 300 epochs, achieved five-fold cross- validation scores of 0.91 and 0.93 for the GLI and GLI + SSA training datasets respectively. Despite the pseudo Dice score of the GLI + SSA dataset being slightly higher than that of the GLI

Table 1. Comparison of the GLI, GLI + SSA, and GLI + SSA2(Excl. 4 SSA outliers) Datasets for the Optimized U-Net.

Datas	Results			Paired t-test
	Dice Similarity Coefficient	Train Loss	Val Loss	
GLI	0.79	1.83	0.21	GLI vs. GLI+SSA:
	0.84	1.12	0.15	t-stat=1.82, p=0.17, df=3
	0.86	0.85	0.14	
	0.89	0.59	0.10	
GLI+SSA	0.77	1.90	0.23	GLI+SSA vs. GLI+SSA2:
	0.84	1.01	0.15	t-stat=-0.51, p=0.65, df=3
	0.86	0.76	0.14	
	0.88	0.57	0.12	
GLI+SSA2	0.79	1.58	0.20	
	0.83	1.01	0.17	
	0.86	0.56	0.14	
	0.88	0.55	0.11	

Val loss = validation loss

model, the paired t-test revealed that the difference between the pseudo Dice scores is not statistically significant ($p > 0.05$) (Table 3). Also, at 300 epochs of training, for the GLI + SSA dataset, the five-fold cross-validation performance of 2D fullres nnU-Net is equal to that of 3D (0.93). However, the GLI + SSA dataset has significantly longer training times than that of the GLI dataset, with the 3D fullres nnU-Net training faster. The leision-wise Dice Similarity Coefficients for each fold of the 2D fullres nnU-Net model are presented in Table 4 of the Appendix.

Table 2. 3D fullres nnU-Net configuration: A representation of the average of Pseudo Dice Score per epoch number for the GLI and the GLI + SSA datasets, with Paired t-test results included.

Dataset	Epoch	Learning Rate	Train Loss	Val Loss	Pseudo Dice	Epoch Time (s)	Paired t-test
	2	0.00536	-0.67	-0.71	0.78	169.28	
	5	0.00235	-0.86	-0.87	0.86	165.65	
GLI	10	0.00126	-0.80	-0.78	0.83	195.28	GLI vs GLI+SSA
	30	0.00047	-0.83	-0.85	**0.90**	190.9	Pseudo Dice:
	300	6e-05	-0.88	-0.87	**0.95**	201.65	
	2	0.00536	-0.42	-0.64	0.78	449.16	**t-stat: 1.12**
	5	0.00235	-0.72	-0.72	0.79	467.69	
GLI+SSA	10	0.00126	-0.78	-0.79	0.84	427.12	**p-value: 0.33**
	30	0.00047	-0.84	-0.85	**0.90**	427.3	**df:4**
	300	6e-05	-0.89	-0.86	**0.93**	496.42	

Val loss = validation loss

Table 3. 2D fullres nnU-Net: A representation of the average of prediction Pseudo Dice Scores per epoch number for the GLI and the GLI + SSA datasets, with t-test results.

Dataset	Epoch	Learning Rate	Train Loss	Val Loss	Pseudo Dice	Time (s)	Paired t-test
	2	0.00536	-0.79	-0.82	0.85	193.98	
	5	0.00235	-0.86	-0.87	0.85	190.19	
GLI	10	0.00126	-0.89	-0.89	0.88	192.5	GLI vs GLI+SSA
	30	0.00047	-0.83	-0.83	0.89	294.5	Pseudo Dice:
	300	6e-05	-0.89	-0.90	**0.91**	330.0	
	2	0.00536	-0.80	-0.82	0.84	337.17	**t-stat: -1.43**
	5	0.00235	-0.86	-0.87	0.86	488.99	
GLI+SSA	10	0.00126	-0.89	-0.88	0.88	300.19	**p-value: 0.22**
	30	0.00047	-0.92	-0.91	0.89	230.32	**df: 4**
	300	6e-05	-0.91	-0.90	**0.93**	501.19	

Val loss = validation loss

3.3 Fine-Tuning and Neural Style Transfer on SSA Training Data as a Data Augmentation Technique Together Improves SSA Validation Results

Despite the 2D fullres nnU-Net trained using GLI + SSA data displaying good performance in five-fold cross-validation and generating satisfactory predictions for previously unseen GLI validation data as illustrated in Fig. 3 in the Appendix, its performance on SSA validation data was weaker. It particularly performed badly on the SSA MR images that appeared to be incomplete. For example in the left image of Fig. 4 in the Appendix, an empty mask was predicted. The neural style transfer technique used as a data augmentation method helped solve this problem. Furthermore, the best 2D fullres nnU-Net trained model at 300 epochs was used as a pretrained model to fine-tune the original 60 cases of SSA training data as well as its stylized augmented SSA data. This directly resulted in an improvement in prediction on the same unseen SSA validation example shown in the right image of Fig. 4 of the Appendix whereas originally an empty mask was predicted before the NST data augmentation and fine-tuning step.

4 Discussion

In summary, at the time of this short paper submission, we have demonstrated the following results. *Firstly*, we highlighted the similarity in the performance of the model with or without the SSA datasets. The results of the Paired Samples t-test between datasets revealed that there is no significant statistical difference between the performance of the three models as shown in Table 1. The results showed that the model performs equally on both the GLI and GLI + SSA datasets, demonstrating the its generalizability. *Secondly*, we conducted a comparative analysis between the performance of 3D and 2D full-resolution models, utilizing the latest iteration (version 2) of nnU-Net [12]. Our investigation revealed that both the 2D and 3D full-res nnU-Net models trained for 300 epochs yielded an average pseudo Dice score of 0.93 for the GLI + SSA training data via a 5-fold cross-validation strategy. *Thirdly*, our validation process on the provided GLI and SSA cases - without ground truth annotations - revealed a significant performance discrepancy between the GLI and SSA validation sets. This can be visually inspected by the relatively good prediction for the GLI validation data shown in Fig. 3 as compared to Fig. 4A in the Appendix. The fusion of the neural style transfer data augmentation with subsequent fine-tuning targeted specifically at SSA cases have demonstrated significant improvements in results for the SSA validation set (Fig. 4B).

An important limitation of our study was the scarcity of African datasets. Future work should extend this approach to a larger African dataset to enhance its applicability. Thus, as proposed in the overall methodology workflow in Fig. 1, in the near future, we will: ensemble the best 2D fullres and 3D full res nnU-Net model trained from the combined GLI and SSA training data, before repeating the fine-tuning experiment with the neural style transfer data augmented SSA plus the original SSA training data. Moreover, we will capitalize on the availability of higher-quality GLI data that could result in more neural style transfer random pairing with the limited SSA training data. We will then find an optimal number of data augmentation pairs while examining model performances. Further post- processing methods will also be investigated.

5 Conclusion

In this investigation, we have established the viability of enhancing brain tumor prediction within the limited-resource context of Sub-Saharan Africa (SSA). By utilizing a pretrained and high-performing 2D fullres nnU-Net model, we achieved refinement through fine-tuning using SSA training data augmented via neural style transfer. This methodology underscores the potential for notable performance improvements within SSA's unique healthcare setting.

Acknowledgments. The authors would like to thank the following instructors of the Sprint AI Training for African Medical Imaging Knowledge Translation (SPARK) Academy 2023 summer school on deep learning in medical imaging for providing insightful background knowledge on brain tumors that informed the research presented here; Craig Jones, Eranga Ukwatta, Esin Ozturk-Isik, Evan Calabrese, Jeff Rudie,

Konstantinos Gousias, MacLean Nasrallah, Malhar Patel, Mueez Waqar, Nicole Levy, Peizhi Yan, Piotr Pater, Ujjwal Baid, and Yihao Liu. The authors would also like to thank Talha Chaudhry for his clinical input with regard to the Sub- saharan Africa dataset and Linshan Liu for administrative assistance in supporting the SPARK Academy training and capacity-building activities. The authors acknowledge the computational infrastructure support from the Digital Research Alliance of Canada (The Alliance) and knowledge translation support from the McGill University Doctoral Internship program through student exchange program for the SPARK Academy. The authors are grateful to McMedHacks for providing foundational information on python programming for medical image analysis as part of the 2023 SPARK Academy program. This research was funded by the Lacuna Fund for Health and Equity (PI: Udunna Anazodo, grant number 0508-S-001) and National Science and Engineering Research Council of Canada (NSERC) Discovery Launch Supplement (PI: Udunna Anazodo, grant number DGECR-2022-00136).

Appendix

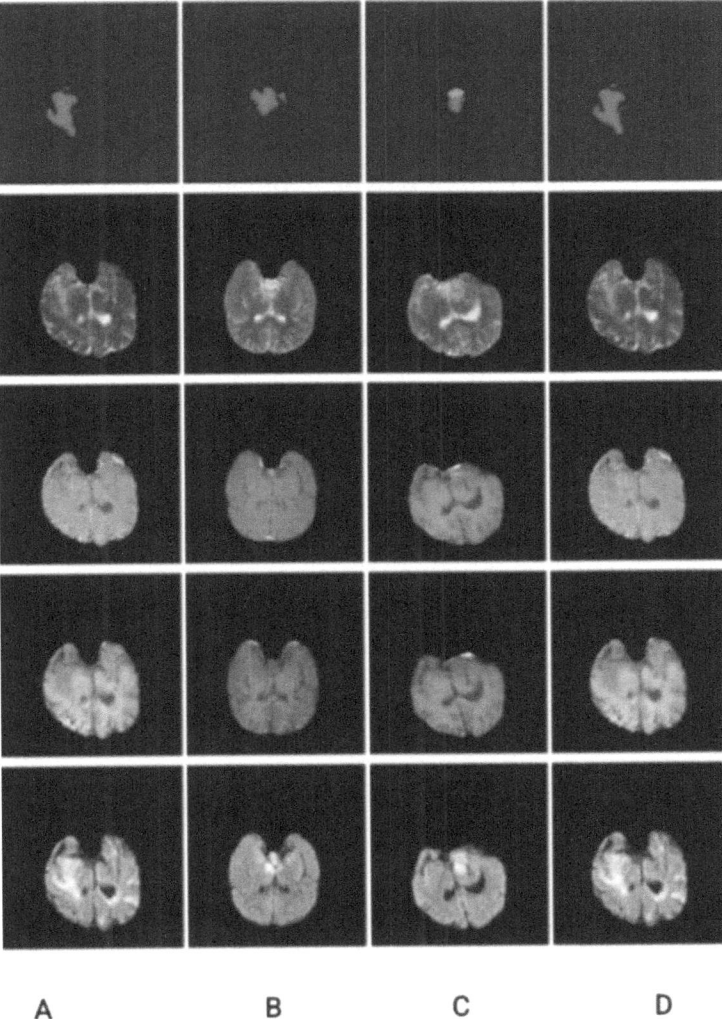

Fig. 2. Excluded SSA training Cases A) 00051, B) 00097, C) 00041, & D) 00084 only for the optimized UNet baseline experiment [Predicted masks (top row), T2 (second row), T1ce (third row),T1 (fourth row), and T2-FLAIR (bottom row)]. This was not employed for the rest of the experiments. The predicted masks are shown with color coding as follows: background: purple, necrotic tumor core (NCR): blue, enhancing tumor (ET): yellow, peritumoral edematous tissue (ED): turquoise. (color figure online)

Table 4. Lesion-wise dice score from five-fold cross-validation for the best 2D fullrest nnU-Net trained mdoel.

Model	Dice Score			Epochs
	Dice_ET	**Dice_TC**	**Dice_WT**	
Fold 0	0.8689	0.8205	0.8082	2
	0.9131	0.8534	0.8304	5
	0.8963	0.8231	0.8415	10
	0.9388	0.9031	0.8991	30
	0.9471	**0.9179**	**0.9179**	300
Fold 1	0.8745	0.7932	0.776	2
	0.9234	0.866	0.8509	5
	0.9327	0.8839	0.8922	10
	0.9294	0.8949	0.8958	30
	0.9488	**0.9051**	**0.8890**	300
Fold 2	0.8805	0.8024	0.8212	2
	0.9178	0.8568	0.8352	5
	0.9388	0.9062	0.8977	10
	0.9469	0.9145	0.902	30
	0.9369	**0.8955**	**0.8989**	300
Fold 3	0.8197	0.7734	0.7608	2
	0.9129	0.837	0.8247	5
	0.9303	0.8709	0.8414	10
	0.9395	0.9096	0.9009	30
	0.9441	**0.9101**	**0.9011**	300
Fold 4	0.8706	0.8028	0.7922	2
	0.8914	0.8247	0.8004	5
	0.9248	0.8637	0.8592	10
	0.9294	0.8949	0.8958	30
	0.9459	**0.9222**	**0.9211**	300

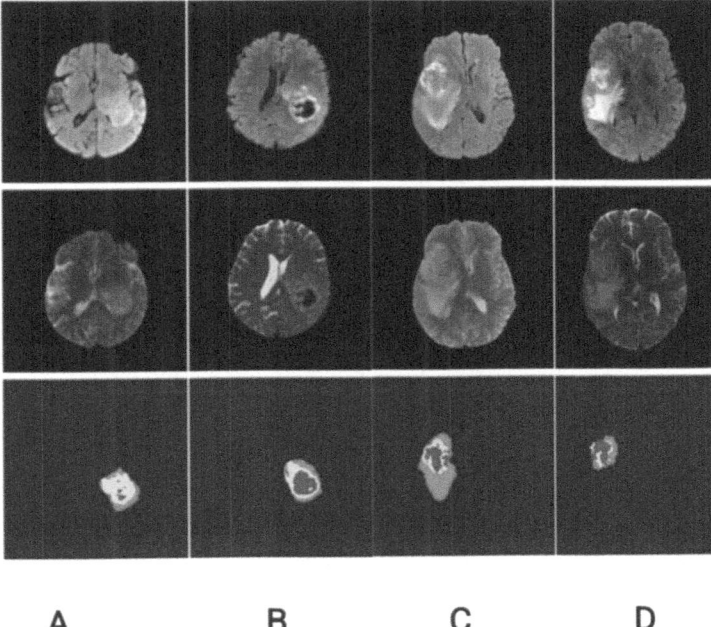

A   **B**   **C**   **D**

Fig. 3. Predicted Masks for validation data BraTS-GLI-00001-000 (A), BraTS-GLI- 00001-001 (B), BraTS-GLI-00013-000 (C), and BraTS-GLI-00013-001 (D), [T1 (top row), T2 (bottom row)] cases using the well-performing best 2D fullres nnUNet model without fine-tuning. The predicted masks (bottom row) are shown with color coding as follows: background: purple, necrotic tumor core (NCR): blue, enhancing tumor (ET): yellow, peritumoral edematous tissue (ED): turquoise. (color figure online)

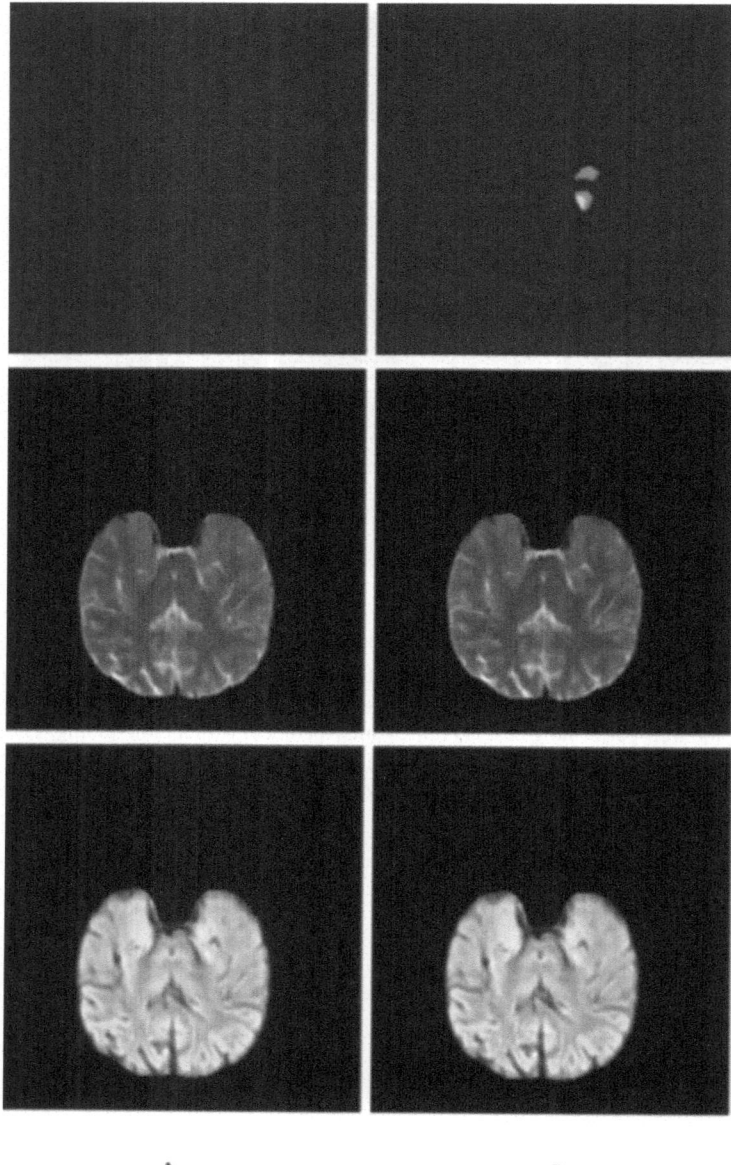

A   B

Fig. 4. Tumor segmentation improvement for SSA validation case BraTS-SSA-00192-000 before (A) and after (B) neural style transfer data augmentation. [T1 (bottom row), T2(middle row)]. This was also after fine-tuning on SSA training data only from the best GLI pretrained model at the best 2D fullres nn-Unet. The color coding is the following: background: purple, necrotic tumor core (NCR): blue, enhancing tumor (ET): yellow, peritumoral edematous tissue (ED): turquoise. (color figure online)

References

1. Adewole, M., et al.: The brain tumor segmentation (brats) challenge 2023: Glioma segmentation in sub-saharan africa patient population (brats-africa). *arXiv preprint* arXiv:2305.19369 (2023)
2. Liu, J., Li, M., Wang, J., Fangxiang, W., Liu, T., Pan, Y.: A survey of mri-based brain tumor segmentation methods. Tsinghua Sci. Technol. **19**(6), 578–595 (2014)
3. Feng, X., Tustison, N.J., Patel, S.H., Meyer, C.H.: Brain tumor segmentation using an ensemble of 3d u-nets and overall survival prediction using radiomic features. Front. Comput. Neurosci. **14**, 25 (2020)
4. Anazodo, U.C., et al.: A framework for advancing sustainable magnetic resonance imaging access in Africa. NMR Biomed. **36**(3), e4846 (2023)
5. Zhang, D., Confidence, R., Anazodo, U.: Stroke lesion segmentation from low-quality and few-shot mris via similarity-weighted self-ensembling framework. In: International Conference on Medical Image Computing and Computer-Assisted Intervention, pp. 87–96. Springer, Cham (2022)
6. Lin, H., et al.: Deep learning for low-field to high- field MR: image quality transfer with probabilistic decimation simulator. In: Machine Learning for Medical Image Reconstruction: Second International Workshop, MLMIR 2019, Held in Conjunction with MICCAI 2019, Shenzhen, China, October 17, 2019, Proceedings, vol. 2, pp. 58–70. Springer, Cham (2019)
7. Hernandez, A.G., et al.: Improving image quality in low-field MRI with deep learning. In: International Conference on Information Photonics (2021)
8. Ma, C., Ji, Z., Gao, M.: Neural style transfer improves 3d cardiovascular MR image segmentation on inconsistent data (2019). arXiv:1909.09716 arXiv:1909.09716
9. Tomar, D., Bozorgtabar, B., Lortkipanidze, M., Vray, G., Rad, M.S., Thiran, J.P.:Self- supervised generative style transfer for one-shot medical image seg- mentation. In: 2022 IEEE/CVF Winter Conference on Applications of Computer Vision (WACV), pp. 1737–1747. IEEE. Waikoloa (2022)
10. de Leeuw, M.L., den Bouter, G., Ippolito, T.P.A.O., Remis, R.F., van Gijzen, M.B., Webb, A.G.: Deep learning-based single image super-resolution for low-field mr brain images. Sci. Rep. **12**(1), 6362 (2022)
11. Sendra-Balcells, C., et al.: Generalisability of fetal ultrasound deep learning models to low-resource imaging settings in five African countries. Sci. Rep. **13**(1), 2728 (2023)
12. Isensee, F., Jaeger, P.F., Kohl, S.A.A., Petersen, J., Maier-Hein, K.H.: nnu-net: a self-configuring method for deep learning-based biomedical image segmentation. Nat. Methods, 18(2), 203–211 (2021)
13. BraTS Challenge Organizers. Brats continuous evaluation sub-challenge 2021 (2021). https://www.synapse.org/Synapse:syn27046444/files/. Accessed 19 Aug 2024
14. Futrega, M., Milesi, A., Marcinkiewicz, M., Ribalta, P.: Optimized u-net for brain tumor segmentation. In: International MICCAI Brainlesion Workshop, pp. 15–29. Springer (2021)
15. Gatys, L.A., Ecker, A.S., Bethge, M.: A Neural Algo- rithm of Artistic Style (2015). arXiv: 1508.06576
16. Simonyan, K., Zisserman, A.: Very deep convolutional networks for large-scale image recognition. In: International Conference on Learning Representations (ICLR) (2015)

Impact of Skin Tone Diversity on Out-of-Distribution Detection Methods in Dermatology

Assala Benmalek[(✉)], Celia Cintas, and Girmaw Abebe Tadesse

IBM Research,Nairobi, Kenya
a.benmalek@univ-boumerdes.dz

Abstract. Addressing representation issues in dermatological settings is crucial due to variations in how skin conditions manifest across skin tones, thereby providing competitive quality of care across different population segments. Although bias and fairness assessment in skin lesion classification has been an active research area, substantially less exploration has been done of the implications of skin tone representations on Out-of-Distribution (OOD) detectors' performance. Most dermatology datasets are reported to suffer from bias in skin tone distribution, which could lead to skewed model performance across skin tones. This paper explores the impact of variations of representation rates across skin tones during the training of OOD detectors and their downstream implications on performance. We review and compare state-of-the-art OOD detectors across two categories of skin tones, FST I-IV (lighter tones) and FST V-VI (brown and darker tones), over samples collected from different clinical protocols. Our experiments conducted using multiple skin image datasets reveal that increasing the representation of FST V-VI during training reduces the representation gap by $\approx 5\%$. We also observe an increase in the overall performance metrics for FST V-VI when more representation is shown during training. Furthermore, the group fairness metrics evaluation yields that increasing the FST V-VI representation leads to improved group fairness. The code is publicly available at the repository (https://github.com/assalaabnk/OOD-in-Dermatology/tree/proportions).

Keywords: Algorithmic fairness · Skin tone representation · Out-of-distribution detection · Dermatology

---

G. A. Tadesse—Girmaw contributed to the foundation of this work while he was at IBM Research, and he is now at Microsoft AI for Good Lab.

Supplementary Information The online version contains supplementary material available at https://doi.org/10.1007/978-3-031-79103-1_13.

U. Anazodo et al. (Eds.): MImA 2024/EMERGE 2024, CCIS 2240, pp. 124–133, 2025.
https://doi.org/10.1007/978-3-031-79103-1_13

1   Introduction

Skin disease remains a global health challenge, with skin cancer being the most common cancer worldwide. Following the recent success of Deep Learning (DL) in various computer vision problems, Convolutional Neural Networks (CNNs) have been employed for skin disease classification with improved performance. However, DL models are prone to and exacerbate existing societal biases [5]. Imbalance in the representation of different skin tones in data used for training models is one of the concerning biases. Academic materials in dermatology are also reported to suffer from representation bias [21,24]. These challenges will then adversely affect the quality of care for patients of color [21]. As we observe a growing interest in DL for dermatology [7,10,29], it is imperative to address the robustness and fairness of these solutions [1,14,16,24] to build trustworthy solutions for positive societal impact.

Many existing DL techniques [2,29] achieve high performance on publicly available dermatology datasets [23]. The study of the fairness of these models has attracted attention recently, e.g., particularly the bias of skin disease classification models [14,16]. Still, many studies did not provide information on the representation of skin tones in their datasets used for modelling [18]. When we look at robustness, we are interested in the ability of the models to identify out-of-distribution (OOD) samples that differ from the training distribution. For example, OOD samples may come from new skin conditions, new collection protocols [15], or new patient sub-populations. Benmalek et al. reported the need to understand OOD methods' performance beyond average metrics to develop more fair solutions [4].

In this paper, we want to explore further the implications of different proportions of representations during the training of these OOD methods and their downstream evaluation. Specifically, we are interested in addressing the following two questions - **Q1**: How much does the skin tone representation of the In-Distribution Dataset (ID) impact the OOD overall performance? **Q2**: Do we observe changes in group-fairness assessment across different proportions? Our contributions are three-fold. First, we assess OOD detectors under different skin tones and the impact of representation bias at training time. Second, we create manually labeled categories of skin images from the public benchmark dataset SD-198 [23]. The categories include binary splits of the Fitzpatrick Skin Types (FST), i.e., FST I-IV and FST V-VI. The code and label resources from this work are available for public use[1]. Lastly, our experiments yield that increasing the representation of FST V-VI during training improves the fairness between the two groups and reduces the performance gap by $\approx 5\%$.

---

[1] https://github.com/assalaabnk/OOD-in-Dermatology/tree/proportions.

2   Related Work

2.1   Algorithmic Fairness Studies

Multiple approaches used Individual Typology Angle (ITA) computed from pixel intensity values [12,16] in analyzing skin tones in the literature. The ITA values were then mapped to FST indices [27]. This information is key to stratifying further studies regarding the algorithm fairness of classifiers. Rezk et al., [22] proposed data augmentation techniques to improve the diversity of skin tones at the training time of DL models. Moreover, the proximity of skin tones between train and test sets was found to play a significant impact on the classification performance [11,12]. Lastly, state-of-the-art OOD methods for skin images were found to significantly differ in their performance, impacting FST I-IV and FST V-VI differently [4]. Nonetheless, this study analyzed the datasets as a whole without understanding the impact of FST I-IV and FST V-VI sample sizes during training.

2.2   Out-of-Distribution Detection Methods

Existing OOD detection methods could be grouped into *ensemble methods*, such as Isolation Forest (IF) [19], OneClassSVM [6] and *deep learning approaches* [13, 15,17] based on the type of models employed. Xuan Li [28] proposed DeepIF, which used the IF approach on the features computed by a pre-trained CNN to detect OOD images of skin lesions. ODIN [17] and NN-Softmax methods [13] used CNNs trained for classification to build robust OOD detectors. [13] used temperature scaling in the last layer, and [17] extended this approach, adding small perturbations to the input to separate the softmax score distributions between in- and out-of-distribution images, allowing for more effective detection. Lastly, autoencoders (AE) are common alternatives for OOD detection due to their capability to learn training data distribution. In this work, we aim to analyze the impact of rarely observed skin tones (FST V-VI) on the performance of OOD methods by varying their proportion during training.

3   Materials and Methods

Our framework to assess the impact of skin tone representation on OOD detection models is shown in Fig. 1. Particularly, we aim to understand the impact of skin tone representations in ID used to train OOD detection models. Given the ID and OOD datasets, we first employ stratification of each dataset across skin tones and categorize them into two groups: FST I-IV and FST V-VI. Then, a train-test split is performed on the ID dataset, and the OOD detection model is trained. Both the test set of the ID dataset and the OOD dataset are used to test the model. Our evaluation of the framework comprises two main categories: OOD detection performance and group fairness metrics. The details of each step in the pipeline are described below.

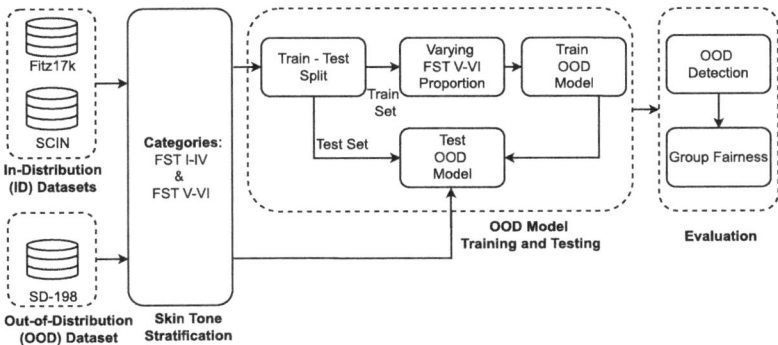

Fig. 1. Overview of our pipeline that takes both In-distribution and Out-of-distribution datasets. We first stratify the datasets between the two skin tone categories: FST I-IV and FST V-VI. The OOD model is then trained with varying proportions of FST V-VI samples. The evaluation step quantifies both the performance of the OOD detection and the fairness across these two categories.

3.1 Datasets

In this work, we explored three existing datasets: SD-198 [23], SCIN [25], and Fitzpatrick 17k [9] for clinical samples from different collection protocols. See Fig. 2 for reference examples for each category across these datasets.

Fitzpatrick 17k Dataset [11,12], referenced as Fitz17k afterward, contains 16,577 clinical images with skin type labels based on the Fitzpatrick scoring system [9]. The images are sourced from two online open-source dermatology atlases and annotated with Fitzpatrick skin type labels by a team of human annotators. The Fitzpatrick labeling system is a six-point scale originally developed for classifying sun reactivity of skin and adjusting clinical medicine according to skin phenotype. Fitz17k dataset is then grouped into FST I-IV and FST V-VI categories - consisting of 13844 and 2168 images, respectively. Given that the totality of FST V-VI samples is 11% with respect to the overall dataset. We created three different proportions of the FST V-VI category: 11% (12222 images), 8% (11879 images), and 5% (11510 images).

SCIN [25] dataset consists of 5032 images from clinical resources among various clinical skin conditions and population groups. The Dataset is broadly distributed across Fitzpatrick skin types. Dermatologists estimated the FST indices of the dataset's samples based. After carefully curating the labels by taking the most occurring label out of the three skin tone labels of each sample, we have 3931 samples categorized as FST I-IV and 438 as FST V-VI. We used SCIN dataset an ID dataset.

SD-198 [23] dataset consists of 6,473 clinical images among diagnostic categories. We manually annotated the skin tone for this dataset, as this information was missing; the labels are available at the repository[2]. This results 6214 samples

---

[2] https://github.com/assalaabnk/OOD-in-Dermatology/tree/proportions.

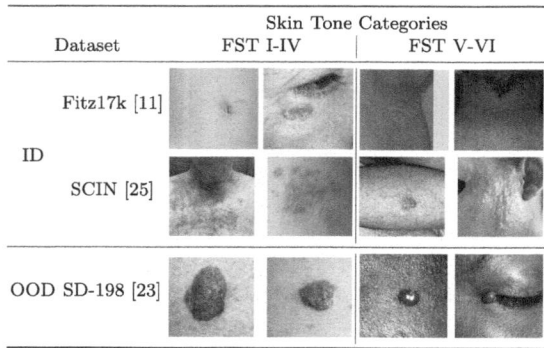

Fig. 2. Examples for FST I-IV and FST V-VI skin types across all datasets. We follow the labeling proposed in Benmalek et al. [4] for Fitz17k [11]. Lastly, SD-198 was labeled manually by the authors.

categorized as FST I-IV and 210 as FST V-VI. We use this dataset for as OOD dataset for our experiments.

3.2   OOD Models and Training Setup

We adopt Isolation Forest [19] and OneClassSVM [6] as baselines, and AutoEncoders [20] as state-of-the-art OOD methods.

The OOD methods were trained on three different proportions consisting of 11%, 8%, and 5% of FST V-VI samples relative to the overall Fitz17k dataset. The same hyperparameters were used for each OOD method across all the proportions. The IF was configured with 300 estimators and a contamination of 0.1; the OneSVM was configured with $\nu = 0.01$, and $\gamma = 0.0001$ during training with Fitz17k. The parameters for both traditional methods were obtained via grid search. The AE [20] which consists of 12 layers in total, (3 convolutional layers + 3 ReLU layers) in the encoder and (3 transposed convolutional layers + 2 ReLU layers + 1 Sigmoid layer) in the decoder, was trained on the Fitz17k dataset by applying Early Stopping with a patience of 5 steps to get the best-performing model and avoid overfitting; the training halted after 33, 49, and 45 epochs for Fitz17k of proportions 11%, 8%, and 5% respectively. This approach ensured we obtained the best-performing models for each dataset proportion. The threshold for OOD detection was calculated using Brent's method to find the root of the reconstruction error distributions for ID and OD samples.

3.3   Evaluation: OOD Detection Performance and Fairness Metrics

To measure the OOD detection performance across all models, we employ Area Under Receiver Operating Characteristic Curve ($AUROC$), F_1-score (F_1), and a Representation Gap (RG) score [4]. RG is the difference in the performance of an OOD detector under different skin tones compared to overall performance. We hypothesize that fair OOD detectors with small $\mathcal{RG}$ score will get an improvement in representation gap in different proportions ($\mathcal{P}_i$); this means that the model's performance and fairness are affected by the skin tone representation in the training data. Finally, based on these scores and proportions, we evaluate the fairness of all OOD models across different skin tones and propose a ranking based on the aforementioned score.

To measure the fairness of our OOD detectors across different proportions, we employ known group fairness metrics: Statistical parity difference SPD [26] and Disparate Impact DI [8] using AIF3360 toolkit [3]. SPD measures the difference in the proportion of positive outcomes between the FST I-IV and FST V-VI categories, and DI evaluates the ratio of the positive prediction rates between the two skin tone categories. In accordance with the fairness literature, we define the FST I-IV skin category as the privileged group and the FST V-VI group as the unprivileged group, given the disproportionate representation ratios between the two skin categories. Thus, $SPD = 0$: an equal rate, $SPD < 0$: the unprivileged group is at a disadvantage, and $SPD > 0$: the privileged group is at a disadvantage. A value of $DI \approx 1$ indicates equal fairness between the two groups, where the rate of receiving the favorable outcome is the same for both the unprivileged and privileged groups. $DI < 1$ indicates that the unprivileged group is at a higher disadvantage, and a value $DI > 1$ indicates that the privileged group is at a disadvantage. Note that $SPD = 0$ and $DI = 1$ suggest a fair OOD detector across the two skin categories.

4   Results

Table 1. OOD detection performance when Fitz17k and SD-198 datasets were used as In-Distribution Dataset (ID) and Out-of-distribution Dataset (OD), respectively. Methods include IF: Isolation Forest, OneSVM: One-class SVM, and AE: Autoencoder. Bold represents the best performance in each column. FST V-VI (%): percentage of FST V-VI samples in the ID.

		AUROC ↑			F_1 ↑			
FST V-VI (%)	Methods	FST I-IV	FST V-VI	All	FST I-IV	FST V-VI	All	$\mathcal{RG}$ ↓
11%	OneSVM [6]	0.50±0.00	0.50±0.01	0.49±0.01	**0.88±0.01**	**0.99±0.01**	**0.87±0.01**	0.11
	IF [19]	0.50±0.01	0.63±0.03	0.52±0.02	0.84±0.01	0.94±0.01	0.84±0.02	0.11
	AE [20]	**0.88±0.03**	**0.93±0.03**	**0.94±0.01**	0.82±0.03	0.78±0.01	0.86±0.04	**0.03**
8%	OneSVM [6]	0.49±0.01	0.50±0.01	0.50±0.00	**0.87±0.00**	**0.99±0.00**	**0.87±0.01**	0.12
	IF [19]	0.51±0.01	0.64±0.05	0.51±0.01	0.84±0.01	0.94±0.01	0.83±0.01	0.11
	AE [20]	**0.90±0.03**	**0.92±0.03**	**0.95±0.00**	0.85±0.02	0.76±0.01	**0.87±0.00**	**0.08**
5%	OneSVM [6]	0.50±0.01	0.50±0.01	0.50±0.00	0.87±0.01	**0.99±0.02**	0.86±0.01	0.12
	IF [19]	0.50±0.01	0.63±0.04	0.51±0.01	0.83±0.01	0.94±0.01	0.83±0.01	0.11
	AE [20]	**0.94±0.00**	**0.95±0.00**	**0.95±0.00**	**0.89±0.00**	0.82±0.00	**0.88±0.00**	**0.07**

4.1 Q1: Impact of FST V-VI Proportion on OOD Performance

Table 1 shows the OOD detection performance across different proportions of the FST V-VI skin type in the Fitz17k dataset used to train the models. We can observe that low performing models such as IF (AUROC $\approx$ 0.5), the representation gap measured between the two skin categories is wider across the different proportions (0.1). AE ([20]) achieved the best detection performance across different proportions of FST V-VI samples with $\approx$ 0.95 AUROC and the lowest RG (0.03) between the two categories. We notice an increase in the representation gap across the proportions in both datasets for each reduction in the FST V-VI proportion. As for the Fitz17k dataset, the RG score increased from 0.03 (11% FST V-VI) to 0.08 (8% FST V-VI), and 0.07 (5% proportion). Similar results are observed with the same OOD detectors trained on the SCIN dataset over different FST V-VI proportions, as detailed in the supporting sections of this paper (See Supplementary Material) (Fig. 3).

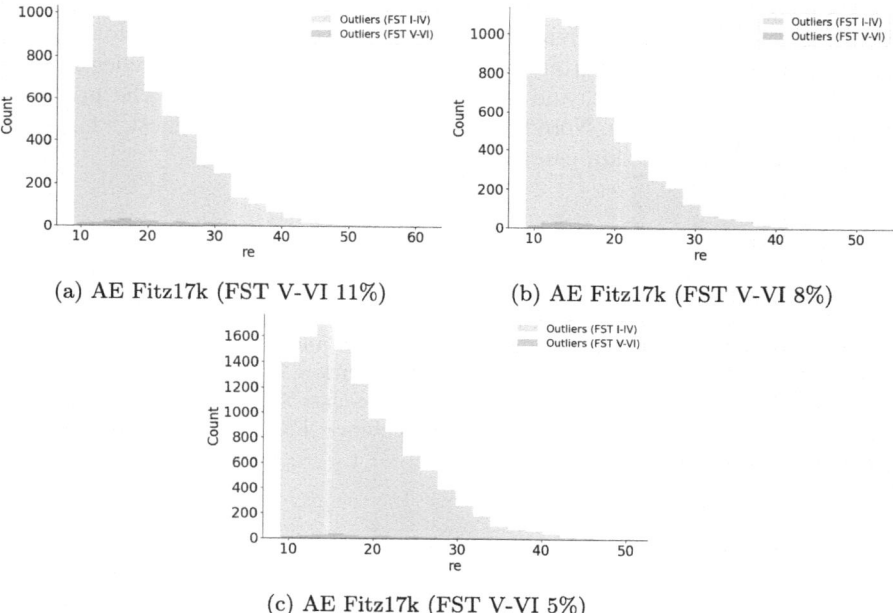

(a) AE Fitz17k (FST V-VI 11%)   (b) AE Fitz17k (FST V-VI 8%)

(c) AE Fitz17k (FST V-VI 5%)

Fig. 3. Abnormal scores distributions for the AE trained on Fitz17k with different proportions of FST V-VI samples. In (a) with $p = 11\%$, AE assigned $\approx$ 96% above threshold for FST I-IV and $\approx$ 99% for FST V-VI; in (b) with $p = 8\%$, $\approx$ 94% for FST I-IV and $\approx$ 97% for FST V-VI; in (c) with $p = 5\%$, $\approx$ 92% for FST I-IV and $\approx$ 96% for FST V-VI. Indicating that reducing the proportion of FST V-VI in the training data slightly decreases the OOD performance.

4.2   Q2: Impact of Different FST V-VI Proportions on OOD Group Fairness Metrics

Table 2 shows the group fairness metrics results of OOD detection methods trained on different proportions of the Fitz17k dataset. We observe that for the best-performing OOD model, AE trained on Fitz17k with different proportions, the group fairness metrics SPD and DI are improved by 0.13 and 0.37, respectively when proportions increase from 5% to 11%, which means samples from FST V-VI (*unprivileged* group) are being at a greater advantage. Additionally, there is a correlation between the group fairness metrics and the $\mathcal{RG}$ score across the OOD detectors trained on the different proportions of both datasets. This indicates that a smaller RG enhances the fairness of the OOD detectors between the *privileged* (FST I-IV) and *unprivileged* (FST V-VI) groups, even with different representation rates of the *unprivileged* group in the training data.

Table 2. Results from evaluation of group fairness using SPD: Statistical Parity Difference and DI: Disparate Impact metrics. Fitz17k and SD-198 datasets were used as In-Distribution and Out-of-Distribution datasets, respectively.

FST V-VI% in ID	Methods	Group Fairness Metrics	
		SPD	DI
11%	OneSVM [6]	−0.069	0.863
	IF [19]	0.267	3.991
	AE [20]	**0.477**	**1.929**
8%	OneSVM [6]	0.033	1.065
	IF [19]	0.228	3.468
	AE [20]	**0.464**	**1.916**
5%	OneSVM [6]	0.009	1.018
	IF [19]	0.307	4.157
	AE [20]	**0.341**	**1.551**

5   Conclusion

In this study, we assess the impact of different proportions of FST V-VI skin types during the training of OOD detectors on model performance and fairness metrics. To this end, we employed multiple skin image datasets (Fitz17k, SCIN, SD-198) for our validation using three OOD models (OneSVM, IF, and AE). Our results confirm that the representation of skin tones significantly affects both the OOD detection performance and fairness metrics. Thereby, increasing the proportion of FST V-VI samples retained overall model performance while the gap between group-based performance was reduced. We also identified that models behaved differently across different proportions of FST V-VI samples, where Autoencoders achieved the best performance while OneSVM performed the least. This work has the potential to serve as a benchmark for the broader research community in addressing skin tone bias within skin datasets. Despite its contributions, this study has two limitations. First, we simplified the skin tone labels to a binary problem. As we know, the diversity of skin tones is vast, and more comprehensive categories are needed to capture accurate variations of the skin categories. Second, more state-of-the-art OOD DL-based models need to be incorporated into our study. Future work will incorporate other alternative skin tone estimation scales with higher granularity as well as explore other fairness metrics.

Compliance with Ethical Standards. This research uses human subject data made available in open access by the corresponding authors (SD-198 [23], and Fitzpatrick 17k [11], and SCIN [25] datasets) licensed under a Creative Commons Attribution - Non-Commercial 4.0 International License, and ethical approval was not required. No funding was received for this study.

References

1. Adamson, A.S., Smith, A.: Machine learning and health care disparities in dermatology. JAMA Derm. **154**(11), 1247–1248 (2018)
2. Ahmed, S.A.A., Yanikoglu, B., Aptoula, E., Goksu, O.: Skin lesion classification with deep learning ensembles in ISIC 2019 (2019)
3. Bellamy, R.K., et al.: Ai fairness 360: an extensible toolkit for detecting, understanding, and mitigating unwanted algorithmic bias. arXiv preprint arXiv:1810.01943 (2018)
4. Benmalek, A., Cintas, C., Tadesse, G.A., Daneshjou, R., Varshney, K., Dalila, C.: Evaluating the impact of skin tone representation on out-of-distribution detection performance in dermatology. In: IEEE International Symposium on Biomedical Imaging (2024)
5. Birhane, A., Prabhu, V., Han, S., Boddeti, V.N.: On hate scaling laws for data-swamps. arXiv preprint arXiv:2306.13141 (2023)
6. Dreiseitl, S., Osl, M., Scheibböck, C., Binder, M.: Outlier detection with one-class svms: an application to melanoma prognosis. In: AMIA Annual Symposium Proceedings, vol. 2010, p. 172. American Medical Informatics Association (2010)
7. Esteva, A., et al.: Dermatologist-level classification of skin cancer with deep neural networks. Nature **542**(7639), 115–118 (2017)
8. Feldman, M., Friedler, S.A., Moeller, J., Scheidegger, C., Venkatasubramanian, S.: Certifying and removing disparate impact. In: Proceedings of the 21th ACM SIGKDD International Conference on Knowledge Discovery and Data Mining, pp. 259–268 (2015)
9. Fitzpatrick, T.B.: The validity and practicality of sun-reactive skin types i through vi. Arch. Dermatol. **124**(6), 869–871 (1988)
10. Gomolin, A., Netchiporouk, E., Gniadecki, R., Litvinov, I.V.: Artificial intelligence applications in dermatology: where do we stand? Front. Med. **7** (2020)
11. Groh, M., Harris, C., Daneshjou, R., Badri, O., Koochek, A.: Towards transparency in dermatology image datasets with skin tone annotations by experts, crowds, and an algorithm. arXiv preprint arXiv:2207.02942 (2022)

12. Groh, M., et al.: Evaluating deep neural networks trained on clinical images in dermatology with the fitzpatrick 17k dataset. In: Proceedings of the IEEE/CVF Conference on Computer Vision and Pattern Recognition, pp. 1820–1828 (2021)

13. Hendrycks, D., Gimpel, K.: A baseline for detecting misclassified and out-of-distribution examples in neural networks. arXiv preprint arXiv:1610.02136 (2016)

14. Kalb, T., Kushibar, K., Cintas, C., Lekadir, K., Diaz, O., Osuala, R.: Revisiting skin tone fairness in dermatological lesion classification. In: Workshop on Clinical Image-Based Procedures, pp. 246–255. Springer, Heidelberg (2023)

15. Kim, H., Tadesse, G.A., Cintas, C., Speakman, S., Varshney, K.: Out-of-distribution detection in dermatology using input perturbation and subset scanning. In: 2022 IEEE 19th International Symposium on Biomedical Imaging (ISBI), pp. 1–4. IEEE (2022)

16. Kinyanjui, N.M., Odonga, T., Cintas, C., Codella, N.C.F., Panda, R., Sattigeri, P., Varshney, K.R.: Fairness of classifiers across skin tones in dermatology. In: Martel, A.L., et al. (eds.) MICCAI 2020. LNCS, vol. 12266, pp. 320–329. Springer, Cham (2020). https://doi.org/10.1007/978-3-030-59725-2_31

17. Liang, S., Li, Y., Srikant, R.: Principled detection of out-of-distribution examples in neural networks. arXiv:1706.02690 (2017)

18. Liang, W., Tadesse, G.A., Ho, D., Fei-Fei, L., Zaharia, M., Zhang, C., Zou, J.: Advances, challenges and opportunities in creating data for trustworthy ai. Nat. Mach. Intell. **4**(8), 669–677 (2022)

19. Liu, F.T., Ting, K.M., Zhou, Z.H.: Isolation forest. In: 2008 Eighth IEEE International Conference on Data Mining, pp. 413–422. IEEE (2008)

20. Masci, J., Meier, U., Cireşan, D., Schmidhuber, J.: Stacked Convolutional auto-encoders for hierarchical feature extraction. In: Honkela, T., Duch, W., Girolami, M., Kaski, S. (eds.) ICANN 2011. LNCS, vol. 6791, pp. 52–59. Springer, Heidelberg (2011). https://doi.org/10.1007/978-3-642-21735-7_7

21. Mcfarling, U.L.: Dermatology faces a reckoning: lack of darker skin in textbooks and journals harms care for patients of color (2020)

22. Rezk, E., Eltorki, M., El-Dakhakhni, W., et al.: Improving skin color diversity in cancer detection: deep learning approach. JMIR Derm. **5**(3), e39143 (2022)

23. Sun, X., Yang, J., Sun, M., Wang, K.: A benchmark for automatic visual classification of clinical skin disease images. In: Leibe, B., Matas, J., Sebe, N., Welling, M. (eds.) ECCV 2016. LNCS, vol. 9910, pp. 206–222. Springer, Cham (2016). https://doi.org/10.1007/978-3-319-46466-4_13

24. Tadesse, G.A., et al.: Skin tone analysis for representation in educational materials using machine learning. npj Digit. Med. **6**(1), 151 (2023)

25. Ward, A., et al.: Crowdsourcing dermatology images with google search ads: creating a real-world skin condition dataset (2024)

26. Wick, M., Mokhtari, G., Tristan, J.B.: Fairness metrics: a comparative analysis. arXiv preprint arXiv:2004.04870 (2020)

27. Wilkes, M., Wright, C.Y., du Plessis, J.L., Reeder, A.: Fitzpatrick skin type, individual typology angle, and melanin index in an African population. JAMA Derm. **151**(8), 902–903 (2015)

28. Li, X., Lu, Y., Desrosiers, C., Liu, X.: Out-of-distribution detection for skin lesion images with deep isolation forest. In: Liu, M., Yan, P., Lian, C., Cao, X. (eds.) MLMI 2020. LNCS, vol. 12436, pp. 91–100. Springer, Cham (2020). https://doi.org/10.1007/978-3-030-59861-7_10

29. Zhang, P., Zhong, Y., Li, X.: Melanet: a deep dense attention network for melanoma detection in dermoscopy images (2019)

Deployment and Evaluation of Intelligent DICOM Viewers in Low-Resource Settings: Orthanc Plugin for Semi-automated Interpretation of Medical Images

Andrew Shawa[1]([✉]) [iD] and Lighton Phiri[2] [iD]

[1] Department Computer Science, University of Zambia, Lusaka, Zambia
andrew.shawa@cs.unza.zm
[2] Department Library and Information Science, University of Zambia, Lusaka, Zambia
lighton.phiri@unza.zm

Abstract. Medical image interpretation is a crucial part of radiological workflows, providing valuable input to the final output of the process: medical image interpretation reports. Radiologists typically employ Digital Imaging and Communication in Medicine (DICOM) Viewers. While numerous types of DICOM Viewers have been implemented, there has been arguably little focus on how such software tools can be made more effective by integrating them with Artificial Intelligence (AI) services. This paper presents a study conducted to design and implement an Orthanc Web-based Picture Archiving and Communication System (PACS) plugin DICOM Viewer for facilitating the semi-automated interpretation of medical images using AI. An Orthanc DICOM Viewer plugin, interoperable with two (2) models—Pneumonia Classification and Detection models—was implemented using the Python programming language. The plugin was evaluated with Radiologist Residents at a large University Teaching Hospital in a controlled setting. TAM 2 instrument was used to assess its perceived usefulness and ease of use. The TAM 2 constructs were rated positively, with study participants expressing a desire to incorporate other pathologies to the plugin and, additionally, integrating similar tools in more widely used DICOM Viewers such as RadiAnt and Weasis. The integration of AI models has the potential to reduce the turnaround time and workload involved in interpreting medical images. More significantly, the positive responses related to perceived usefulness of the tool suggest a potential that such tools have in improving delivery of services at the point of clinical care.

Keywords: Artificial Intelligence · DICOM Viewer · Pneumonia

1 Introduction

Medical image interpretation is an essential part of radiological workflows and generally involves visualisation and analysis of medical images that are generally encoded using the Digital Imaging and Communication in Medicine (DICOM) standard [1]. The DICOM

© The Author(s), under exclusive license to Springer Nature Switzerland AG 2025
U. Anazodo et al. (Eds.): MImA 2024/EMERGE 2024, CCIS 2240, pp. 134–143, 2025.
https://doi.org/10.1007/978-3-031-79103-1_14

standard ensures that medical image data and corresponding metadata (Data Elements) are encoded in a standardised way, to facilitate efficient and effective interoperability and transmission of the files.

Experts such as Radiologists use specialised software applications referred to as DICOM Viewers [2] to visualise and analyse medical images encoded using the DICOM standard. In addition to visualisation and analysis, DICOM Viewers generally offer additional multi-modality functionalities to enable annotation and marking, measurement and windowing, among other features.

While numerous DICOM Viewers exist [2], there has been limited exploration into enhancing their functionality through integration with Artificial Intelligence (AI) services, to support Radiologists and other exports by facilitating efficient and effective radiological workflow activities.

This paper presents work conducted at the University Teaching Hospitals (UTHs) in order to demonstrate how DICOM Viewers can potentially be extended in order to interface them with AI-centric services. In particular, the work was focused on determining how best to integrate AI models for automatically interpreting medical images could be integrated within DICOM Viewers and, more significantly, perceptions of Radiologists on integrating AI-centric tools within DICOM Viewers. Our most recent work set the stage for the significance of DICOM Viewers and also highlighted how DICOM Viewers are used by Radiologists at UTHs.

The main contributions of this paper are as follows:

- Experimental results outlining the perceived usefulness of integrating AI services within DICOM Viewers by Radiologists with little experience working with AI tools in low-resource settings
- Technique for integrating AI models within PACS DICOM Viewer as PACS is already define

The remainder of this paper is organised as follows: Sect. 2 is a review of existing literature aligned with this work. Section 3 describes the methodology employed to undertake this study. In Sect. 4, the study results are presented and discussed. Finally, Sect. 5 provides concluding remarks.

2   Related Work

2.1   DICOM Viewers

Medical images are generally encoded using the DICOM standard/format [4] in which medical imaging information—bitstreams—are stored in the same file as the auxiliary metadata—DICOM Data Elements. Radiologists use software tools called DICOM Viewers to view and analyse DICOM files. DICOM viewers play a crucial role in radiological workflows by providing radiologists with the necessary tools to visualise and interpret medical images effectively.

RadiAnt DICOM Viewer. RadiAnt offers an intuitive interface designed for ease of use, facilitating quick access to all features. Its multi-slide view allows users to view up to 20 slides simultaneously, enhancing efficiency in image analysis. Additionally, RadiAnt provides advanced measurement tools, including tools for measuring length, ellipse, pencil, angle, and Cobb angle [5]. It also offers image adjustment tools for adjusting and enhancing the quality of the images.

Weasis DICOM Viewer. Weasis is an open-source DICOM viewer for medical imaging professionals. It offers a customizable user interface and comprehensive toolset for image analysis and interpretation. Weasis supports various DICOM modalities, including radiography, CT, MRI, and ultrasound, and provides advanced features such as region-of-interest (ROI) analysis, image fusion, and measurement tools [6]. Its flexibility and extensibility make it suitable for both research and clinical applications.

OsiriX DICOM Viewer. OsiriX is one of the world's most popular DICOM viewers, boasting an intuitive user interface combined with high-performance capabilities. Its advanced post-processing techniques in both 2D and 3D, coupled with innovative 3D and 4D navigation, make it a standout. Furthermore, it integrates seamlessly with any PACS system. OsiriX is optimised for speed, ensuring swift image loading and processing. Its intuitive interface allows medical professionals to navigate and access tools effortlessly. OsiriX offers sophisticated techniques in 2D and 3D image processing, along with 3D and 4D navigation, enhancing diagnostic capabilities [7].

OHIF Viewer. OHIF Viewer is a free, open-source DICOM viewer that combines an intuitive user interface with robust functionality, making it a strong alternative to commercial viewers like OsiriX. Supporting DICOMweb standards such as WADO (Web Access to DICOM Object) for image retrieval and QIDO (Query Retrieve Service) for metadata queries, OHIF Viewer facilitates seamless integration with various PACS systems. The open-source framework allows extensive customization, enabling institutions to tailor the viewer to specific workflows without licensing restrictions. Additionally, OHIF's modular architecture supports the creation of custom extensions and modules, allowing users to enhance and extend the viewer's capabilities to meet specific needs [8].

3D Slicer. 3D Slicer is a free, open-source software platform for medical image processing, visualisation, and analysis, supporting multiple modalities such as DICOM (X-ray, CT, MRI), ultrasound, and microscopy data. It provides advanced visualisation techniques, including volume rendering and segmentation editing, facilitating detailed anatomical and pathological analyses. The platform offers robust tools for quantitative image analysis, including measurements, land-marking, and region-of-interest calculations, aiding in objective assessment and clinical decision-making. Notably, 3D Slicer integrates with various AI frame-works, enabling automated segmentation, lesion detection, and image registration to enhance diagnostic accuracy and streamline workflows. Its extensibility through a rich ecosystem of plugins and modules, coupled with a large user community, makes it a versatile tool for both clinical and research applications [9].

Summary. These DICOM viewers serve as essential tools for Radiologists, allowing them to visualise and analyse medical images with precision and efficiency. However, while these viewers offer robust features for manual interpretation, there has been a growing interest in enhancing their capabilities through integration with AI technologies. This integration aims to automate certain aspects of image analysis, improve diagnostic accuracy, and streamline radiological workflows, as discussed in the following sections. More significantly, while this study uses Orthanc [10] to demonstrate how intelligent DICOM Viewers can be designed and implemented, the idea can be extended to other DICOM Viewers, such as Weasis [6], that are implemented using extensive architectures.

2.2 Artificial Intelligence in Radiological Workflows

Automated Image Analysis. AI technologies have gained significant traction in the field of radiology, offering promising solutions to various challenges encountered in medical image interpretation and diagnosis. Machine learning and deep learning algorithms, in particular, have shown remarkable potential in automating and augmenting radiological workflows [11].

AI algorithms have been increasingly employed to automate the analysis of medical images, aiding radiologists in tasks such as detection, characterization, and segmentation of abnormalities. These algorithms leverage large datasets to learn patterns and features associated with different pathologies, thereby enhancing the efficiency and accuracy of image interpretation.

Improving Diagnostic Accuracy and Efficiency. The integration of AI tools in radiology holds promise for improving diagnostic accuracy by providing radiologists with advanced decision support systems. Numerous studies have demonstrated the efficacy of AI algorithms in detecting and diagnosing various medical conditions, often achieving performance levels comparable to or even surpassing human experts in specific tasks.

In addition to enhancing diagnostic accuracy, AI technologies offer opportunities to streamline radiological workflows and improve operational efficiency [12]. By automating routine tasks, prioritising critical cases, and facilitating image interpretation, AI-driven solutions can help radiologists manage increasing workloads and reduce turnaround times for diagnosis and reporting.

Challenges and Limitations. Despite the potential benefits, integrating AI into radiological workflows presents several challenges and limitations. These include issues related to data quality and quantity, algorithm robustness and generalizability, interpretability of AI-driven results, regulatory compliance, and ethical considerations surrounding patient privacy and consent [13]. One other challenge lies in seamlessly incorporating AI capabilities into widely used DICOM viewers for semi-automated interpretation of medical images. While AI technologies hold promise for enhancing efficiency and accuracy in image interpretation, the limited availability of AI enabled viewers hinders the adoption of AI-driven tools in radiological workflows. This paper addresses this challenge by presenting a study on the design and implementation of an Orthanc DICOM Viewer plugin, aimed at facilitating semi-automated interpretation of medical images using AI.

3 Methodology

A mixed-method approach was used to execute this study, involving the design and implementation of an Orthanc plugin and subsequent usability evaluation of the toolkit. The study setting for this study was UTHs, as outlined in Sect. 1.

Ethical clearance was granted by The University of Zambia Biomedical Research Ethics Committee (Reference Number: 2731-2022) and The National Health Research Authority (Reference Number: NRHA000024/10/05/2022), to conduct this study. In addition, formal permission was granted from The University Teaching Hospitals.

3.1 Designing and Implementing Intelligent DICOM Viewers

The main objective of this work was to demonstrate the feasibility of integrating AI services within DICOM Viewers. UTHs is in the process of deploying a PACS platform that uses the Orthanc platform [14] as the base framework.

An Orthanc plugin [15], was developed in Python[1] by extending the existing Orthanc Web Viewer [16], one of the many available plugins for Orthanc. In our most recent work, Pneumonia classification[2] and detection[3] models were implemented and, additionally, a Web service for facilitating interoperability with the models [17].

This plugin enhances the DICOM Viewer by extracting pixel data from medical images, pre-processing it for transmission to AI models, and retrieving pre- dictions from an AI service. Figure 1 illustrates the plugin's interaction with the Orthanc Web Viewer and the AI models web service.

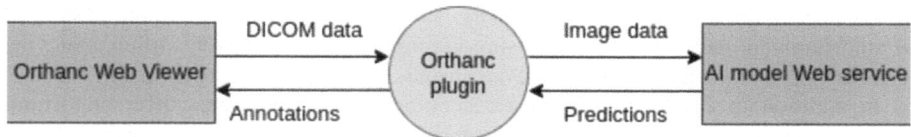

Fig. 1. Dataflow Level 0 diagram showing the interaction between the Orthanc plugin, Orthanc Web Viewer and the AI model Web service.

3.2 Evaluating Usefulness of Intelligence DICOM Viewers

To evaluate the perceived usefulness of the plugin, a controlled experiment was conducted that involved participants performing a predefined task and, subsequently responding to a Technology Acceptance Model (TAM 2) questionnaire [18].

The study setting for the experiment was UTHs and participants— Radiologists, Radiology Residents and General Practitioners—were recruited from UTHs using convenience sampling. UTHs has a shortage of Radiologists, as outlined in our prior work [14] and as a result, Radiology Residents and General Practitioners are compelled to interpret medical images.

---

[1] https://github.com/Enterprise-Medical-Imaging-in-Zambia/emiz_project-wp6-orthanc_plugin.

[2] https://github.com/Enterprise-Medical-Imaging-in-Zambia/emiz_project-wp6-binary-classifier-code.

[3] https://github.com/Enterprise-Medical-Imaging-in-Zambia/emiz_project-wp6-simple-bounding-box-model.

Participants were required to use the Orthanc plugin to interpret an image associated with pneumonia. They subsequently responded to the modified version of the TAM 2 instrument.

4 Results and Discussion

4.1 Orthanc Intelligent DICOM Viewer

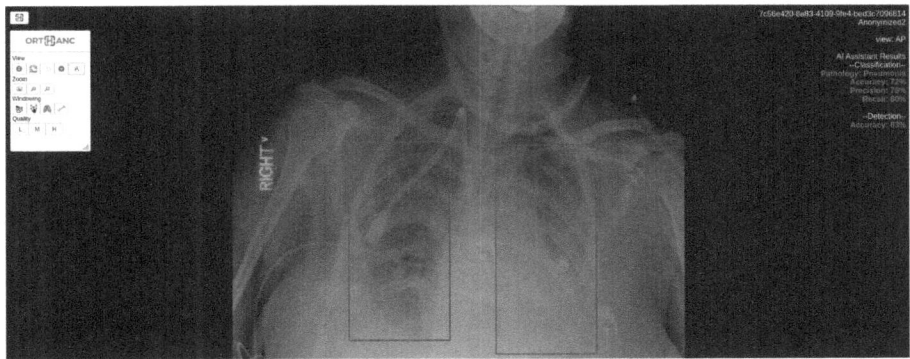

Fig. 2. Screenshot of Intelligent DICOM Viewer Orthanc plugin integrated with an external Pneumonia classification and detection Artificial Intelligence model.

Figure 2 illustrates how the plugin renders the responses from the AI models. Within the figure, the data from the AI model is highlighted in red—the red text (Top Right) are important metrics from the models enabling end users to have an understanding of the reliability of the AI models, while the red bounding boxes (Centre) highlights the potential problematic areas identified by the detection model.

The integration of the plugin with the DICOM Viewer and the AI service aims to facilitate a streamlined workflow, fostering enhanced integration between the viewer and the automatic image interpretation system. This seamless communication ensures that radiologists receive real-time feedback on their diagnoses, empowering them to make more accurate assessments and identify potential problems sooner.

Post-processing Using DICOM Data Elements. A crucial design consideration made during the implementation of the plugin was deciding how the AI model response results would be stored; more specifically, the proof of concept described in this paper is associated with two different types of model responses: the binary classification results from the classification model, indicating whether a DICOM file is positive or negative for pneumonia and the results of the detection model specifying the coordinates of the potential areas with anomalies in the medical image.

The DICOM standard enables the creation of additional private tags [19], and this feature makes it possible for information from the AI models to be added as Data Elements within the DICOM input file. Figure 3 illustrates the standard DICOM Data Element structure.

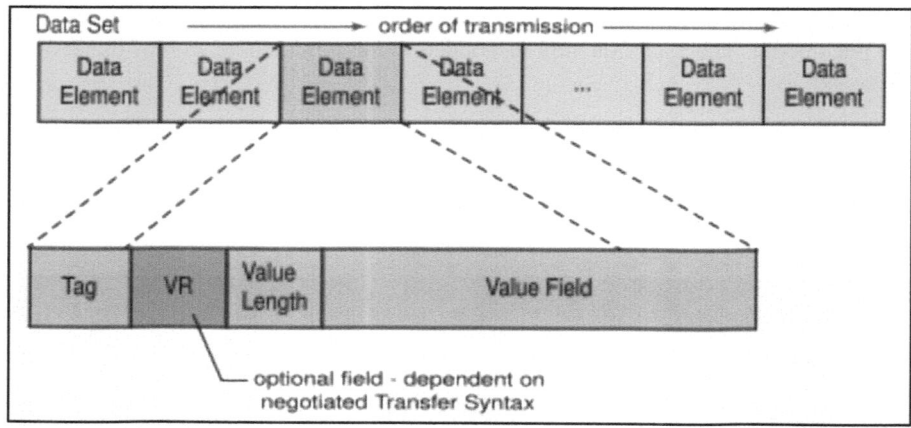

Fig. 3. DICOM Data Set and Data Element Structures [20].

4.2 Usability Evaluation

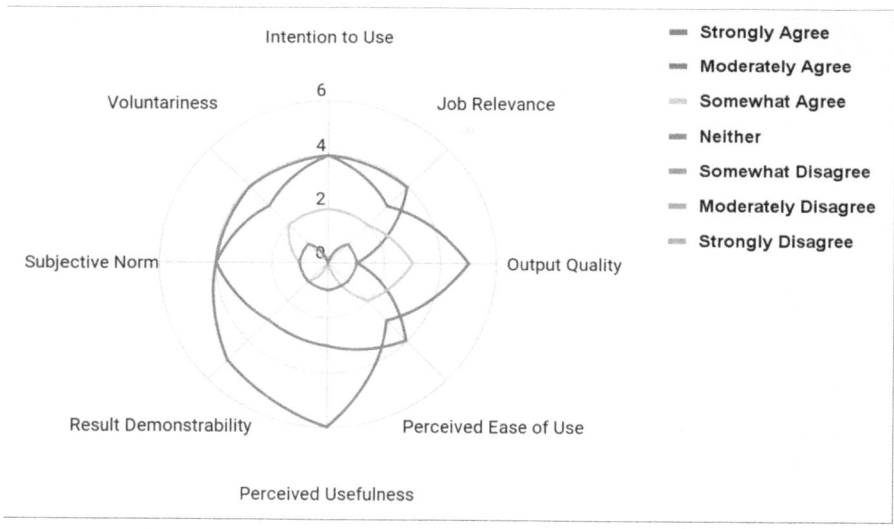

Fig. 4. Radar Chart illustrating the summary of the usability evaluation results.

The results from the TAM 2 questionnaires were analysed using the eight (8) TAM 2 constructs using the standard approach for analysing TAM 2 responses. The results of the analysis are shown in Fig. 4. The plot displays multivariate data in a two-dimensional format, with each axis representing one of the eight TAM 2 constructs. The distance from the centre of the chart to a point on an axis indicates the score for that construct, and the different coloured lines or markers correspond to different ratings: Strongly Agree, Moderately Agree, Somewhat Agree, Neither, Somewhat Disagree, Moderately Disagree, and Strongly Disagree.

Most of the participants had positive responses for all the eight TAM 2 construction, with participants overwhelmingly agreeing with "Intention to Use" and "Perceived Usefulness". The positive responses across all eight (8) constructs, indicated by the outermost layers—Strongly Agree, Moderately Agree—, shows strong agreement with constructs such as Intention to Use and Perceived Usefulness.

A few participants had neutral responses, particularly visible in the 'Neither Agree nor Disagree' category. This is represented by points closer to the center of the chart for most constructs.

Some construct-specific insight from the study are as follows:

- "Intention to Use" received completely positive responses, with no participants falling into neutral or negative categories
- Other constructs such as Perceived Ease of Use and Job Relevance also had high levels of agreement but included a small number of neutral responses.
- Constructs like Output Quality and Result Demonstrability showed some variation in agreement levels, reflecting differing perceptions among participants.

Participants' Comments. Participants had the option to offer supplementary remarks relating to the AI intervention. The comments were mostly suggestions on what the participants would prefer the interface and underlying AI should perform. We did have a few cautions as well.

"Search tool should allow incomplete word searches" [Radiology Resident 1] [sic].

"Is there a possibility that a tool like this would result in bias? i.e. only focusing on abnormalities identified by the tool"[Radiologist 1][sic].

"Yes, I have the following suggestions. 1. The icon for accessing the tool is too small in the top left corner of the screen. Consideration should be made to enlarge it. 2. The steps taken to finally submit an image for interpretation (i.e. accessing the "A" button) should be reduced in order to expedite the process. Should"[Radiologist 2][sic].

"The interface should be more clear and user friendly" [Medical Doctor 1][sic].

"Access should be restricted to medically trained professionals" [Radiology Resident 2][sic].

"Can we find a way to further improve the detection by narrowing the bounding boxes, they seem to cover majority of the lung field on the sample images provided?"[Radiology Resident 3][sic].

5 Conclusion

This paper describes the implementation of an Intelligent DICOM Viewer toolkit that integrates results from AI models within DICOM Viewers in order to facilitate the efficient and effective interpretation of medical images using AI techniques.

The work presented in this paper demonstrates how an existing DICOM Viewer—Orthanc—can be extended in order to integrate AI techniques. In addition, the paper highlights how the existing DICOM standard could complement such integration. More importantly, the work provides insight into potential practical AI perceptions of radiologists in low-resource settings.

With the rapid adoption of AI in the healthcare sector, it becomes necessary to explore avenues of integrating AI within tools and technologies used by experts in the field.

Future work could explore how AI models could potentially be deployed locally to improve the efficiency of the Web service response time.

References

1. DICOM Standards Committee: PS3.1. https://dicom.nema.org/medical/dicom/current/out put/html/part01.html. Accessed 08 Mar 2024
2. Haak, D., Page, C.-E., Deserno, T.M.: A survey of DICOM viewer software to integrate clinical research and medical imaging. J. Digit. Imaging **29**, 206–215 (2015). https://doi.org/10.1007/s10278-015-9833-1
3. Shawa, A., Zulu, E.O., Phiri, L.: Software tools for supporting automatic interpretation of medical images. In: Proceedings of the Pan African Conference on Science, Computing and Telecommunications, pp. 49–52. ICICT (2023)
4. Current Edition. https://www.dicomstandard.org/current. Accessed 30 Aug 2024
5. DICOM Viewer – RadiAnt. https://www.radiantviewer.com/products/radiant-dicom-viewer-standard/. Accessed 07 Mar 2024
6. Roduit, N.: Weasis DICOM medical viewer. https://weasis.org/en/index.html. Accessed 07 Mar 2024
7. OsiriX DICOM Viewer, https://www.osirix-viewer.com/, last accessed 2024/03/07
8. Website, Ohif Viewer. https://docs.ohif.org/
9. Website, 3D Slicer. https://download.slicer.org/
10. Jodogne, S.: The orthanc ecosystem for medical imaging. J. Digit. Imaging **31**, 341–352 (2018). https://doi.org/10.1007/s10278-018-0082-y
11. Ranschaert, E.R., Morozov, S., Algra, P.R.: Artificial Intelligence in Medical Imaging: Opportunities, Applications and Risks. Springer (2019)
12. Sim, J.Z.T., Bhanu Prakash, K.N., Huang, W.M., Tan, C.H.: Harnessing artificial intelligence in radiology to augment population health. Front. Med. Technol. **5**, 1281500 (2023). https://doi.org/10.3389/fmedt.2023.1281500
13. Warren, B.E., et al.: An introductory guide to artificial intelligence in interventional radiology: part 2: implementation considerations and harms. Can. Assoc. Radiol. J. 8465371241236377 (2024). https://doi.org/10.1177/08465371241236377
14. Zulu, E.O., Phiri, L.: Enterprise medical imaging in the global south: challenges and opportunities. In: Proceedings of the 2022 IST-Africa Conference. IEEE (2022). https://doi.org/10.23919/IST-Africa56635.2022.9845508
15. Creating new plugins — Orthanc Book documentation. https://orthanc.uclouvain.be/book/developers/creating-plugins.html. Accessed 30 Aug 2024
16. Orthanc - DICOM Server. https://www.orthanc-server.com/static.php?page=web-viewer. Accessed 30 Aug 2024
17. Muzumala, M., Zulu, E.O., Chibuta, P., Nyirenda, M., Phiri, L.: Evaluating perceived workload, usability and usefulness of artificial intelligence systems in low-resource settings: semi-automated classification and detection of community acquired pneumonia. In: Marrakesh, M. (ed.) Third Workshop on Applications of Medical Artificial Intelligence. Springer, Cham (2024)
18. Venkatesh, V., Davis, F.D.: A theoretical extension of the technology acceptance model: four longitudinal field studies. Manage. Sci. **46**, 186–204 (2000). https://doi.org/10.1287/mnsc.46.2.186.11926

19. 7.8 Private Data Elements. https://dicom.nema.org/dicom/2013/output/chtml/part05/sect_7.8.html. Accessed 08 Mar 2024
20. 7 The Data Set. https://dicom.nema.org/dicom/2013/output/chtml/part05/chapter_7.html. Accessed 08 Mar 2024

Enhancing Soil-Transmitted Helminths Diagnosis Through AI: A Self-supervised Learning Approach with Smartphone-Based Digital Microscopy

Lin Lin[1,2(✉)] (iD), Daniel Cuadrado[1], Roberto Mancebo-Martín[1] (iD), Stella Kepha[3], Paul Gichuki[3], Charles Mwandawiro[3], María Jesús Ledesma-Carbayo[2] (iD), Miguel Luengo-Oroz[1] (iD), Elena Dacal[1] (iD), and David Bermejo-Peláez[1]

[1] Spotlab, R&D, Madrid, Spain
lin@spotlab.ai
[2] ETSI Telecomunicación, Universidad Politécnica de Madrid & CIBER-BBN, Madrid, Spain
[3] ESACIPAC, Kenya Medical Research Institution (KEMRI), Nairobi, Kenya

Abstract. Soil-transmitted helminths (STH), including hookworm, *Ascaris lumbricoides*, and *Trichuris trichiura*, impose significant health burdens in low- and middle-income tropical and subtropical regions, infecting over 1.5 billion people globally. Traditional diagnostic methods like the Kato-Katz technique are time consuming. This study introduces an innovative AI-driven system utilizing affordable 3D-printed adapters and smartphones to digitize Kato-Katz microscopy samples, capturing high-resolution images for subsequent analysis. These digitized images can be uploaded to a telemedicine platform for remote diagnosis and expert consultation.

Central to our system is the development of a foundational AI model for parasite detection and classification. The model operates in two stages: First, an object detection algorithm identifies all parasites in the image, achieving a mean average precision (mAP) of 97.90% on the validation set using the YOLOv8 architecture. Second, a classification algorithm categorizes each detected parasite by species. The classification model is initially trained on a large, unannotated dataset of parasite images using a self-supervised learning (SSL) approach to learn domain-specific visual features, which are often missed while using generic pre-training datasets. Subsequently, it is fine-tuned on a labeled dataset, significantly improving performance. The model initialized with SSL on STH images achieved an F1 score of 91.70%, outperforming those initialized with random weights (F1 score of 55%) and those trained on DINO-Imagenet weights (F1 score of 53%).

By integrating AI with low-cost digital imaging, our approach aims to revolutionize STH diagnosis in resource-constrained settings, aligning with the WHO's 2030 Roadmap for the elimination of neglected tropical diseases.

Keywords: STH · NTD · AI · foundation model · object detection · self-supervised learning

U. Anazodo et al. (Eds.): MImA 2024/EMERGE 2024, CCIS 2240, pp. 144–152, 2025.
https://doi.org/10.1007/978-3-031-79103-1_15

1   Introduction

Soil-transmitted helminths (STH), including hookworm, *Ascaris lumbricoides*, and *Trichuris trichiura*, are common in low- and middle-income tropical and subtropical countries. Over 1.5 billion people are infected, suffering from anemia, gastrointestinal distress, and chronic fatigue [1]. The World Health Organization (WHO) estimates STH infections cause over 3 million disability-adjusted life years (DALYs) lost annually. The WHO 2030 Roadmap for NTDs aims to eliminate these parasites through mass drug administration (MDA) with albendazole and mebendazole [2, 3].

The Kato-Katz technique, used to diagnose STH infections, involves preparing stool samples on microscope slides for visual inspection [4]. While simple and cost-effective, its sensitivity drops if samples are not examined within 30 to 60 min due to egg degradation or hatching, especially with hookworms. Subjective visual assessment also leads to variability and errors in diagnosis.

AI integration in medical imaging has transformed fields like radiology and cardiology in high-income countries. However, its use in low- and middle-income countries (LMICs) is limited. Developing accessible and reliable AI solutions for these regions is vital for global healthcare. AI can enhance the diagnosis and management of diseases like STH infections, common in LMICs [5–8].

AI algorithms for medical imaging need abundant labeled data for training. Self-supervised learning (SSL) has emerged as an alternative, learning features from large unlabeled datasets to reduce the need for labeled data [9].

This work proposes a system that digitizes Kato-Katz samples using a 3D-printed adapter and smartphones. This method captures and stores high-resolution images of stool samples shortly after preparation, preserving their quality for later analysis. These images can be uploaded to a telemedicine platform for remote diagnosis and expert opinions. We also developed a foundation model for stool parasites using a SSL approach, allowing visual representation acquisition without labeled images. This promising foundation AI algorithm is capable of analyzing a single stool sample to detect and classify multiple types of parasites simultaneously.

2   Material and Methods

2.1   Dataset

We collected an extensive dataset composed of 1,380 stool samples from children aged between 5–15 years in Kwale, Kenya. Each stool sample was prepared using the Kato-Katz thick smear method and visually analyzed by conventional microscopy. In parallel, each stool sample was digitized by taking pictures of the field of view (FoV) and the images were transferred to a telemedicine platform. Both processes were made at 100x magnification (~0.08 μm/pixel).

The proposed digitization system is based on a 3D-printed adapter that allows coupling a smartphone to a conventional microscope by aligning the smartphone camera with the objective of the microscope to acquire the images. This adapter can convert any conventional microscope into a digital one and enables the digitization of microscope samples without the need for expensive scanners. Additionally, the system has been designed to be universal, working with any microscope model and any smartphone model.

From this dataset, a total of 163 stool samples (3,075 FoV images) were further analyzed, with all visible parasites labeled by species, including *Ascaris lumbricoides*, *Trichuris trichiura* and hookworm. This annotated dataset was split at patient level into training (65%), validating (20%) and testing (15%). This split was used for both the parasite detection algorithm and the species classification. All images from the remaining unannotated stool samples (1,217 stool samples with 14,680 FoV images) were used for training the self-supervised phase of the pipeline.

For detection, the entire FoV is used, while classification relies on patches cropped from areas with identified parasites. Instead of random cropping for the unsupervised dataset, we used a trained object detector to identify possible parasites, significantly increasing the amount of relevant data. Table 1 illustrates the dataset distribution, which is maintained consistently for both the detection and classification tasks.

Table 1. Data Distribution: Breakdown of unannotated (SSL) and annotated (Train; Validation; Test) data sets. Expert labels include *Ascaris*, *Trichuris*, and hookworm classes, along with artifacts (false positives generated by the detection algorithm). "Images" represents the number of FoV images, while "Total" indicates the number of parasites, which corresponds to the number of cropped patches.

Data set	#Patients	#Images	*Ascaris*	*Trichuris*	Hookworm	Artifact	Total
SSL	1,217	14,680	-	–	–	–	104,885
Train	84	1,637	6,294	1,283	586	2,253	10,416
Validation	43	817	1,351	698	325	937	3,311
Test	36	621	899	811	201	932	2,843

2.2 Overview of the Proposed Foundational Method

The proposed foundational model for parasite detection and classification operates as follows: First, an object detection algorithm identifies all parasites present in an image, regardless of their species. Then, a classification algorithm categorizes each parasite into its specific species. This classification algorithm is based on a foundational model for species differentiation. Initially, it is trained on a large unannotated dataset of parasite images to learn visual representations of stool parasites using a self-supervised approach. After learning these domain specific image-based features, the algorithm is further fine-tuned on a labeled dataset to accurately discern the species of each parasite. Figure 1 illustrates the proposed approach.

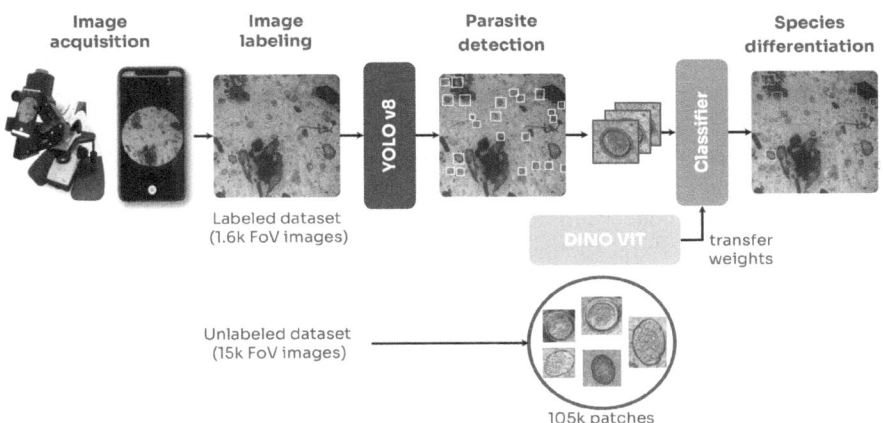

Fig. 1. Overview of the proposed pipeline comprising: (1) image acquisition using a 3D-printed adapter; (2) Image labeling; (3) parasite detection using YOLOv8; (4) self-supervised pre-training with DINO; (5) species differentiation with our DINO pretrained ViT classifier.

2.3 Slide-Level Parasite Detection

Each field of view image is processed through an object detection algorithm to detect all possible parasites regardless of the species. The proposed algorithm for parasite detection in stool sample images was based on a YOLO architecture (YOLOv8) [10], which is a single-stage object detection algorithm that uses a convolutional neural network (CNN) as backbone. Unlike two-stage algorithms, single-stage detection models like YOLO offer enhanced processing speed, making them highly suitable for mobile deployment. For each detected parasite, an image patch is then extracted and further processed by the patch-level parasite classification to determine the specie.

2.4 Patch-Level Parasite Classification

In this study, we aimed to enhance stool parasite differentiation by utilizing SSL to train a feature extractor (backbone) for better data generalization using unlabeled datasets. Specifically, we propose the use of a vision transformer (ViT) [11] (ViT-S/16) as the backbone, which has achieved significant success in various domains and whose application in the medical field is rapidly expanding. During the pre-training (self-supervised phase), we adopted the DINO technique [12], a knowledge distillation method that does not require labeled data.

ViT operates by dividing an image of fixed-size patches (NxN), each patch is passed to a linear operator to obtain patch embedding. To preserve spatial information of each patch, their position is encoded and is added later to each patch embedding. The resulting sequence is fed to the transformer encoder. In order to perform classification, an extra learnable classification token (CLS) is added to the patch embedding, and the result is passed to a linear classification head to classify the image. This classification token and head is not required for SSL.

The general concept of DINO is summarized in Fig. 2. It has two models that follow the same architecture, teacher and student, parameterized by t and s. For an input image I, patches of different sizes were generated: large patches (global crops) and small patches (local crops). All crops are followed by extensive augmentation. For a crop x, pair of view (x1, x2) were generated with random augmentation, both models produce output probabilities Pt and Ps, obtained by normalizing the output of the model with a softmax function. The student model learns to match distribution by minimizing the cross-entropy loss mins CE(Pt(x), Ps(x)). The parameters of the student s are updated by stochastic gradient descent, and the parameters of the teacher t are updated using the exponential moving average (EMA) of the students.

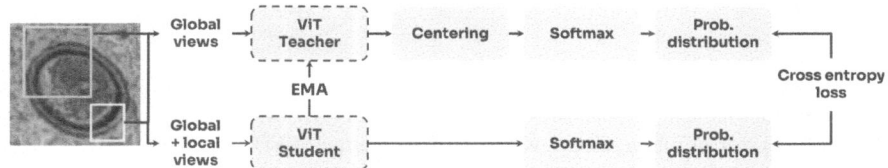

Fig. 2. Overview of the proposed pipeline for training a ViT architecture with DINO SSL strategy.

After pre-training, we conducted supervised training. In this phase, we add a multilayer perceptron (MLP) layer on top of the ViT encoder to perform classification. To ensure the need for SSL pretraining we tried two different approaches: first, by freezing all weights of the architecture except those from the classification layer (linear probing), and second, by enabling fine-tuning of all weights in the architecture, including the ViT encoder.

2.5 Experimental Setup

For comparative purposes, we compared different versions of the YOLOv8 architecture for parasite detection. This comparison aimed primarily to identify the optimal architecture for deployment on edge devices, such as smartphones, enabling real-time detections. Testing multiple YOLOv8 variants allowed us to determine their performance and suitability for this specific application.

Additionally, we trained the ViT model without pre-training to assess the improvement conferred by SSL. The comparison of these models with and without pre-training provided insights into the benefits of SSL in improving model performance.

All comparisons, including those for parasite detection and classification, were conducted on a validation set to ensure consistency and reliability in the performance assessments. The optimal configurations, determined through these validation tests, were then evaluated on an independent test set to confirm their efficacy and generalizability. This comprehensive approach allowed us to identify the best-performing models for both tasks and ensure their readiness for practical deployment.

The metrics used for evaluating the object detection algorithm included mean average precision (mAP), the precision and recall. The performance of the classification algorithm was assessed by measuring the balanced accuracy (BACC) and F1-score to account for data imbalance and to provide a more comprehensive evaluation of the classifier's effectiveness across all classes.

3 Experiments and Results

3.1 Slide-Level Parasite Detection

We conducted a set of experiments to evaluate the performance of various YOLOv8 models: YOLOv8-n (nano), YOLOv8-s (small), YOLOv8-m (medium), and YOLOv8-l (large). Instead of training the model from scratch, we used the weights pretrained with COCO image dataset and fine-tuned on our dataset. Our experiments utilized an input size of 640 × 640 pixels, a learning rate set to 0.01, and we trained each model for 100 epochs with early stopping implemented after 20 epochs without improvement by monitoring the loss in the validation set. We evaluated the mAP, precision, recall, and inference time of each one.

Table 2 illustrates the performance of each YOLOv8 model. Our results indicate that all four models achieved a comparable mAP of 97%. However, YOLOv8-n demonstrated significantly faster inference times, being approximately three times quicker than YOLOv8-l. This makes YOLOv8-n particularly well-suited for deployment on edge devices where computational resources are limited.

Table 2. Parasite detection performance on the validation set. Note: the inference time was calculated on an Intel i5 CPU processor.

Architecture	mAP	Precision	Recall	Inference time/image (ms)
YOLOv8-n	97.07	**91.27**	87.19	**372**
YOLOv8-s	**97.79**	90.72	91.36	465
YOLOv8-m	97.47	89.18	92.33	696
YOLOv8-l	97.04	87.02	**94.36**	1060

3.2 Patch-Level Parasite Classification

During the self-supervised training phase, we employed the following setup: a ViT-S/16 model with a patch size of 16. The model underwent training for 200 epochs with a batch size of 96 and an input size of 224 × 224, utilizing a cosine learning rate scheduler (initial LR: $5e^{-4}$, minimum LR: $1e^{-6}$). The backbone was trained using patches generated by the object detection algorithms (N = 104,885). Figure 3 illustrates the efficacy of the pretrained model in our domain-specific dataset, in extracting relevant features from stool parasites, demonstrating that features belonging to the same class are clustered together while those from different classes are distinctly separated.

To assess the benefits of using SSL on a domain-specific dataset, we conducted an experiment comparing three different pre-training approaches for the ViT architecture: SSL on a domain-specific dataset (DINO-STH), SSL on a domain-agnostic dataset

Fig. 3. Uniform manifold approximation and projection (UMAP) visualization of the features extracted from our pre-trained ViT architecture.

(DINO-ImageNet) as it demonstrated strong performance on various computer vision tasks, and no pre-training. In all three cases, we utilized the same ViT-S/16 architecture to ensure a fair comparison. This experiment was designed to determine whether our model, trained on domain-specific data, extracts more relevant features compared to a model trained on domain-agnostic data.

All experiments were conducted under the same hyperparameters: a batch size of 128, learning rate of 0.001, training for 100 epochs with early stopping patience of 10 epochs, and images resized to 224 × 224 pixels. During the supervised training phase, we executed two types of training strategies: fine-tuning, where both the backbone and the linear classifier (MLP) were trained together, and linear probing, where only the linear classifier was trained while keeping the backbone fixed. All these experiments used all available data of the training set ($N = 10,416$). Table 3 presents the performance of all models on the validation set.

Table 3. Parasite classification performance on the validation set, obtained from full fine-tuning and linear probing.

Pre-training	Fine tuning		Linear probing	
	BACC	F1-score	BACC	F1-score
None	57.48	55.13	47.90	46.61
DINO-ImageNet	55.45	53.48	85.10	85.15
DINO-STH	**91.48**	**91.70**	90.51	90.86

In addition, and for comparison purposes, we performed supervised training with only 200 images per parasite class, instead of using all available dataset, to assess the models' capacity when only a limited labeled dataset is available. When evaluated on the validation set, the model pre-trained on our domain-specific dataset (DINO-STH) achieved a BACC of 86.57% and a F1-Score of 86.24%, whereas when it was pre-trained on the domain-agnostic dataset (DINO-ImageNet) achieved 74.76% and 74.05% of BACC and F1-Score respectively.

3.3 Evaluation of the Proposed System on an Independent Test Set

To evaluate the generalization capabilities of our proposed system, we set aside a subset of labeled images, independent from those used for training and validation. After determining the best configurations from the validation set, we applied these configurations to the test set.

The evaluation revealed that the optimal configuration for parasite detection was YOLOv8-n, which achieved a mean Average Precision (mAP) of 94.93%, precision of 89.63%, and recall of 88.50% on the test set. For parasite species classification, the best performance was achieved using the ViT-S/16 architecture pre-trained on our domain-specific unlabeled dataset through SSL. All detected boxes (those with a probability greater than 0.05) were then processed by the classification algorithm, and the predictions were compared to the annotations made by experts. The performance of the parasite classification algorithm was 80.32% BACC) and 80.17% F1-score. The whole system achieved a mAP of 82.5%, precision of 89.63% and recall of 82.14%.

This assessment on an independent test set underscores the robustness and generalizability of our approach, demonstrating its potential for accurate parasite detection and classification in practical applications.

4 Conclusions

In this work, we presented a comprehensive approach for the detection and classification of soil-transmitted helminth (STH) parasites using YOLOv8 and Vision Transformers (ViT). For the classification task, we created a foundation model based on SSL techniques to leverage a large amount of unannotated data, enabling the model to learn meaningful features. The proposed classification system trained on our domain-specific data achieved better results compared to the ViT backbone pre-trained on domain-agnostic dataset (ImageNet). This improvement is particularly noticeable when the available annotated training set is small. With only 200 training images per parasite class, the performance improved by 12% when comparing our SSL-pretrained model to the one trained on a domain-agnostic dataset. This work is highly relevant because, in the field of medical imaging, the availability of labeled data is often limited. By incorporating SSL and leveraging unannotated data, we have shown that it is possible to enhance model performance, especially in data-scarce environments. This approach, which also leverages 3D-printing technologies and smartphones to enable data digitization without the need for expensive hardware, holds great promise for improving diagnostic accuracy and accessibility in low-resource settings, ultimately aiding in the fight against parasitic diseases. This marks a step toward meeting the World Health Organization's performance benchmarks for in-vitro diagnostic devices, leveraging AI to combat parasitic diseases.

References

1. Soil-transmitted helminth fact sheet, https://www.who.int/news-room/fact-sheets/detail/soil-transmitted-helminth-infections, Accessed 25 May 2023

2. Pullan, R.L., Freeman, M.C., Gething, P.W., Brooker, S.J.: Geographical inequalities in use of improved drinking water supply and sanitation across -Saharan Africa: mapping and spatial analysis of cross-sectional survey data. PLoS Med. **11**, e1001626 (2014). https://doi.org/10.1371/journal.pmed.1001626

3. World Health Organization ed: Ending the neglect to attain the Sustainable Development Goals: a road map for neglected tropical diseases 2021–2030. World Health Organization (2021)

4. Katz, N., Chaves, A., Pellegrino, J.: A simple device for quantitative stool thick-smear technique in Schistosomiasis mansoni. Rev. Inst. Med. Trop. Sao Paulo **14**, 397–400 (1972)

5. Ward, P., et al.: Affordable artificial intelligence-based digital pathology for neglected tropical diseases: a proof-of-concept for the detection of soil-transmitted helminths and Schistosoma mansoni eggs in Kato-Katz stool thick smears. PLoS Negl. Trop. Dis. **16**, e0010500 (2022). https://doi.org/10.1371/journal.pntd.0010500

6. Meulah, B., et al.: A review on innovative optical devices for the diagnosis of human soil-transmitted helminthiasis and schistosomiasis: from research and development to commercialization. Parasitology **150**, 137–149 (2023). https://doi.org/10.1017/S003118202200 01664

7. Yang, A., et al.: Kankanet: an artificial neural network-based object detection smartphone application and mobile microscope as a point-of-care diagnostic aid for soil-transmitted helminthiases. PLoS Negl. Trop. Dis. **13**, e0007577 (2019). https://doi.org/10.1371/journal.pntd.0007577

8. Li, Q., et al.: FecalNet: automated detection of visible components in human feces using deep learning. Med. Phys. **47**, 4212–4222 (2020). https://doi.org/10.1002/mp.14352

9. Huang, S.-C., Pareek, A., Jensen, M., Lungren, M.P., Yeung, S., Chaudhari, A.S.: Self-supervised learning for medical image classification: a systematic review and implementation guidelines. npj Digital Med. **6**, 74 (2023). https://doi.org/10.1038/s41746-023-00811-0

10. Jocher, G., Chaurasia, A., Qiu, J.: Ultralytics YOLOv8 (2023)

11. Dosovitskiy, A., et al.: An Image is Worth 16 × 16 Words: Transformers for Image Recognition at Scale. arXiv. (2020). https://doi.org/10.48550/arxiv.2010.11929

12. Caron, M., et al.: Emerging properties in self-supervised vision transformers. arXiv. (2021)

Capturing Complexity of the Foot Arch Bones: Evaluation of a Statistical Modelling Framework for Learning Shape, Pose and Intensity Features in a Continuous Domain

Catherine Namayega[1](✉) ⓘ, Bhushan Borotikar[1,2,7] ⓘ, Martin Menten[3,4] ⓘ,
Victoria Gibbon[1] ⓘ, Xolisile Thusini[1] ⓘ, Bernhard Egger[5] ⓘ, Arne Burssens[6] ⓘ,
Emmanuel Audenaert[6] ⓘ, and Tinashe E. M. Mutsvangwa[1,7] ⓘ

[1] University of Cape Town, Cape Town, South Africa
nmycat001@myuct.ac.za
[2] Symbiosis International University, Pune, India
[3] Technical University Munich, Munich, Germany
[4] Imperial College London, London, UK
[5] Friedrich-Alexander-University of Erlangen-Nuremberg, Erlangen, Germany
[6] Ghent University Hospital, Ghent, Belgium
[7] IMT Atlantique, Brest, France

Abstract. Advances in medical imaging have enabled detailed digitisation and representation of human anatomy, but challenges remain when modelling complex structures. Statistical models have been developed to capture variations in shape, pose, and intensity features of anatomical structures. However, these models often embed a single feature. This paper investigates how a novel dynamic multi-feature-class Gaussian process modelling framework (DMFC-GPM) designed to learn shape, pose, and intensity features in continuous domains, facilitates the modelling of complex anatomical structures. The work evaluates the framework's ability to capture multi-feature variations within complex anatomy. Computed tomography image data was processed to build and validate a statistical shape, pose and intensity neutral-arched foot model (12 bones). Framework evaluation was done by validation of the model using specificity, and generality. Fitting the model globally to all objects resulted in specificity and generality reported as average root mean square (RMS) of 0.61 ± 0.11 mm and 1.02 ± 0.21 mm, and average Hausdorff distance (Hd) of 3.47 ± 0.98 mm and 7.56 ± 1.33 mm, respectively. Further validation of the model marginalised to the talus bone resulted in specificity and generality of 0.81 ± 0.25 mm and 1.20 ± 0.48 mm average RMS and 3.24 ± 1.10 mm and 6.50 ± 2.34 mm average Hd, respectively. The talus model variations were consistent with literature. Thus, the novel DMFC-GPM framework can model complex anatomies such as the foot arch.

U. Anazodo et al. (Eds.): MImA 2024/EMERGE 2024, CCIS 2240, pp. 153–163, 2025.
https://doi.org/10.1007/978-3-031-79103-1_16

Keywords: Statistical Shape Pose and Intensity Models · Foot Arch model · Weight Bearing Cone-Beam CT

1   Introduction

X-ray imaging is widely used in resource-limited settings for its cost-effectiveness and lower radiation exposure, but its two-dimensional (2D) images complicate the interpretation of three-dimensional (3D) anatomy [17,27]. Statistical modelling methods have been used to learn the natural variation of shape, intensity and pose features in biological objects towards improvement of 3D reconstruction of X-ray images [6,10,15,28]. However, most statistical methods describe features of a single or multiple structures without embedding the shape, pose, and intensity correlations in a continuous domain. Modelling these features individually limits the effectiveness of statistical models in 3D image reconstruction tasks.

A novel statistical modelling framework called the dynamic multi-feature class Gaussian process modelling (DMFC-GPM) framework is capable of probabilistic modelling of a latent space embedding shape, pose, and intensity features in a joint statistical space within a continuous domain [9]. This framework also permits on-demand marginalisation to a region or feature of interest improving the performance in 3D reconstruction tasks. However, despite the advantages of this method, it has only been applied to a few structures such as the shoulder and hip joints with limited objects [8,9]. Hence, this research investigates the efficacy of the DMFC-GPM framework to model the complex anatomy such as the human foot arch bones. The human foot arch is a structure formed by 7 tarsal and 5 metatarsal bones, ensuring optimized weight distribution during movement [4]. A neutral-arched foot is characterised by a moderate medial longitudinal arch, a feature that can be confirmed through medical imaging [1,17]. To understand the effect of weight on the foot bone structure, imaging of the feet is typically done using 2D X-ray modalities due to their low-cost and lower effective radiation exposure [17]. However, specialised computed tomography (CT) scans called weight-bearing cone-beam CT (WB-CBCT) are used to obtain images for 3D visualisation, model building and analysis of bone under weight-bearing [5].

This study aimed to evaluate the efficacy of the DMFC-GPM framework in modelling complex structures through development, evaluation, and validation of multi-object statistical shape, pose, and intensity model (SSPIM) of a neutral-arched foot [9]. The foot model consisted of 7 tarsal and 5 metatarsal bones which was evaluated and validated using qualitative (sample visualisation) and quantitative (specificity and generality) [7] methods.

2   Methodology

2.1   Imaging Dataset

The study sample consisted of a dataset of 11 WB-CBCT patient feet images (7 females and 4 males) obtained from Ghent University hospital, imaged in a standing anatomical position and clinically classified as neutral-arched. Ethics approval was obtained from the University of Cape Town Human Research Ethics Committee (HREC:476/2022). The segmented images of individual bones (or objects, as they will be referred to in this study) were converted into different forms, which include the triangle surface meshes (output of segmentation), tetrahedral volume meshes (defining the interior structure of each triangle surface mesh), scalar volume meshes (tetrahedral volume meshes with embedded CT intensities), and scalar volume mesh joint (anatomically correct combination of scalar volume meshes). Each plays a role in the development of the model as described below.

As adopted from [8], the scalar volume mesh, V is represented as a triplet of features consisting of shape class (S), pose class (P), and intensity class (I) succinctly described as $V = (S, P, I)$. The object shape class is a volumetric mesh defined by a set of 3D geometric points joined to make vertices. The pose class is a set of points capturing the spatial orientation of 3D points. Finally, the intensity class is a set of scalar values of each 3D point. The next sections detail the processes followed to build and evaluate a validated multi-object SSPIM.

2.2   Data Processing Procedure

The WB-CBCT images of feet, obtained as digital imaging and communications in medicine (DICOM) files, were cropped, and due to normative/asymmetrical differences the right feet were mirrored [23] to the left foot, resulting in total 22 feet sample datasets. All the CT images were semi-automatically segmented in 3D slicer-v5.4.0 for the 7 tarsals (talus and calcaneus, navicular, cuboid and 3 cuneiforms), and 5 metatarsals bones to obtain surface triangle meshes. However, building a multi-object SSPIM necessitated an initial establishment of dense correspondence for shape, pose, and intensity features of the objects. Dense correspondence entails identifying a one-to-one relationship between surface points that define an object's geometry and the intensity distribution. Achieving dense correspondence across the individual objects was a two-step process implemented using the Scalismo library: firstly, rigid registration that employed landmarks selected on the objects, and secondly, non-rigid registration that utilised Gaussian process morphable model (GPMM) based priors.

Prior to the rigid registration step, a reference image was selected, and all its segmented objects were converted to mesh grids using Amira-v5.4.3. This was followed by conversion of the object grids to a readable tetrahedral mesh format compatible with Scalismo using the AVS UCD reader in ParaView-v5.10.1. The process was repeated for each object of the reference. To achieve rigid registration, a set of anatomical landmarks was manually selected by a trained personnel three independent times on the reference objects. Following this, the average of these landmarks was used as reference landmarks to increase intra-observer precision [20]. These reference landmarks were used to visually aid the location of the target object landmarks. The reference and target landmarks were used to calculate the rigid transform between the reference and the target objects to align the entire dataset to the reference coordinate.

Non-rigid transformation was performed using the GPMM-based freeform model prior (FFM) approach [15]. A GPMM is a generalisation of statistical shape modelling where the shape, $s = x + u(x)|x \in \Gamma_R$, is modelled as a deformation, u, from the reference shape, $\Gamma_R \subset \mathbb{R}^3$. The Gaussian process, $u \sim \mathcal{GP}(\mu, k)$ is completely defined using μ the mean deformation field ($\mu : \Omega \to \mathbb{R}^3$) and k the covariance function or kernel ($k : \Omega \times \Omega \to \mathbb{R}^{3 \times 3}$). To build a data representative FFM template the iterative median closest point Gaussian mixture model (IMCP-GMM) was applied [19]. The FFM for each template object shape (N models) was built for both the surface and tetrahedral volume meshes. A sampling-based computation using Markov chain Monte Carlo (MCMC) and Metropolis-Hastings algorithms was used to fit the FFM to the target object and the mean sample of the posterior model selected as the new target [16,18,24]. To get tetrahedral volume meshes in correspondence, a continuous correspondence transformation over the volumetric domain of the reference tetrahedral mesh and its neighbourhood points on the outer surface was defined. The target surface mesh points of the neighbourhood of the reference were then used to define the tetrahedral mesh in correspondence with the reference mesh. The results were objects in shape correspondence.

After registration, intensity correspondence was obtained by aligning the CT image and the tetrahedral volume mesh, then assigning the voxel intensity value of the closest voxel to a mesh point. The process was repeated for all the registered objects to generate the scalar volume meshes. This was followed by aligning the volumes to the calcaneus (selected as a pivot object) to obtain pose due to anatomical movement or position. The calcaneus was selected because it is the largest of the tarsal and foot bones and provides a strong lever [12,26]. This was achieved by calculating the transform between the reference calcaneus and the target calcanei in the dataset then applying the transform to the rest of the objects. The result was scalar volume mesh joints with embedded pose.

2.3   Multi-object SSPIM Building

Once shape, pose, and intensity correspondence across individual objects was established, the subsequent phases involved discovering a concise representation that captures the data variability across joints (multi-objects). This process unfolded in three main stages as detailed in [8,9] with code[1]: 1) linearizing rigid transformations, 2) homogenising shape, pose, and intensity attributes, 3) computing a latent space based on Gaussian processes and versatile kernel.

Linearisation of Rigid Transformation. This involved obtaining the pose as a deformation field, which entailed selecting a center of rotation (CoR) on each object as a point around the rotation center of each object. This was followed by calculating the relative pose as a deformation field between the CoR of the reference and target objects. The method of selecting the CoR worked on the assumption that the center of mass was close to the CoR; otherwise, a less

---

[1] https://github.com/rassaire/Dmfc-gpm.

subjective and more computationally expensive method like sphere fitting might be used to compute the CoR [14].

Homogenizing Shape, Pose, and Intensity Attributes. After obtaining shape, pose, and intensity deformation fields, the next step was to establish shared deformation fields across the training dataset. A reference multi-object scalar volume mesh was selected, and deformation fields were defined for each example. This was followed by obtaining the unified space of the three feature classes, and the pose transformation was projected into a linear space using exp/log bijective mapping.

Computing a Latent Space Based on Gaussian Processes and Kernel. In this step a Gaussian process was used to define a latent space as a set of orthogonal functions, given the training dataset of n with N scalar volume mesh objects with shape, pose and intensity deformation fields, $\{F_i\}_{i=1}^n$ defined on the continuous domain by $F_i; \Omega \to \mathbb{R}^7$ where $\Omega = U_{j=1}^N \Omega^j$. The Gaussian process is defined by the estimated mean of the deformation fields $(\bar{F})$ and kernel $(\mathcal{K})$ to increase variations in the model and together define the multi feature-class deformation fields.

$$\mathcal{F} = \mathcal{GP}\left(\bar{F}, \mathcal{K}\right) \tag{1}$$

The multi feature-class deformation fields $\mathcal{F}$ are represented by an orthogonal set of basis functions. After combining the latent space functions, the $\mathcal{GP}$ was made continuous through interpolation and the could now be marginalised by making any of the deformation fields for a feature or object constant.

2.4   Model Marginalisation

One of the inherent attributes of the DMFC-GPM framework is marginalisation, which allows exclusion of any feature in the model. For example, the multi-object SSPIM of the foot can be marginalised to shape and/or intensity, shape and/or pose of multiple or single objects. In this study, the probabilistic foot model defined on a continuous domain, $\mathcal{P}(\Omega)$, was marginalised to a talus, (T) shape to give a probabilistic model $\mathcal{P}(\Omega_T)$ defined by $\mathcal{GP}(\bar{F}, \mathcal{K})\mid_{T \subset \Omega}$ where pose and intensity are marginalised from the deformation field $\bar{F}$ and only the first j^{th} object in the matrix (talus) is varied.

2.5   Validation of the Multi-object SSPIM

Validation of the global multi-object SSPIM poses significant challenges due to the unavailability of ground-truth data and the high computational cost of processing the entire CT. Ideally, the model would be fitted to a CT image, and the best-fit prediction would be compared to the ground-truth segmentation for evaluating shape and pose accuracy. Following this, one could compute voxel-wise intensity differences, although this approach is computationally

demanding. Alternatively, projecting the volumes to 2D for simpler comparison could be considered, but this method was beyond the scope of the current
study. Model validation in this paper was achieved through qualitative visual
inspection of model PCs and quantitative evaluation using MCMC fitting [24].
Quantitative validation included determining the specificity and generality of
the model marginalised into two models; 1) the shape and pose of 12 objects
and 2) the shape of a single object (talus). To determine model specificity, all
patient data that were included in the training dataset were predicted using
the model and the average error calculated. While, the K-Fold cross-validation
method was used to measure generality of both models to unseen data. This was
done by building a model without the samples from the identity (left and right)
that would be used as test data. The process was repeated for entire dataset
and the average error calculated. The MCMC fitting process was run for 1000
iterations, and the predictions were compared to the ground-truth segmented
data using average root mean square (RMS) and Hausdorff distance (Hd).

2.6   Model Evaluation

The multi-object mean model surface mesh was analysed following the protocol in [2] to confirm whether the objects are representative of a neutral-arched
foot. Further evaluation on the talus shape model was carried out to determine
whether the multi-object SSPIM is capable of modelling the morphology of a
single object within the modelled complex anatomy. To facilitate a morphometric comparison of this bone with previous research, the model was sampled for
the first three principal components (PCs) for samples from -2 to 2 standard
deviation (STD) [3,11]. Each sample was compared to the mean using the signed
closest point metric in the model to model distance module in 3D slicer-v5.4.0
to create distance maps.

3   Results

A multi-object SSPIM of neutral-arched foot structure and its PCs are illustrated
in Fig. 1. The multi-object mean mesh has a talonavicular coverage of 100%,
angle between the mid-axes of the 1st and 2nd metatarsals of 9°, calcaneal pitch
of 21.6° and the 1st and 2nd metatarsal length was approximately 62.36 mm
and 74.18 mm (effective 1st metatarsal length of 11.82 mm), respectively. The
first PC described variations due to size and intensity increase (2 STD) and
decrease (-2 STD). The second PC described variations due to abduction and
dorsiflexion (2 STD), and adduction and plantarflexion of the mid and forefoot
(-2 STD). The third PC described variations due to widening (-2 STD) and
narrowing (2 STD) of the foot.

Model specificity and generality obtained from fitting the SSPIM model to
segmented feet volumes returned an RMS error of 0.61 $\pm$ 0.11 mm and 1.02 $\pm$
0.21 mm, and Hd of 3.47 $\pm$ 0.98 mm and 7.56 $\pm$ 1.33 mm, respectively. The best
and worst reconstruction results are shown in Fig. 2A (i) and (ii). The specificity

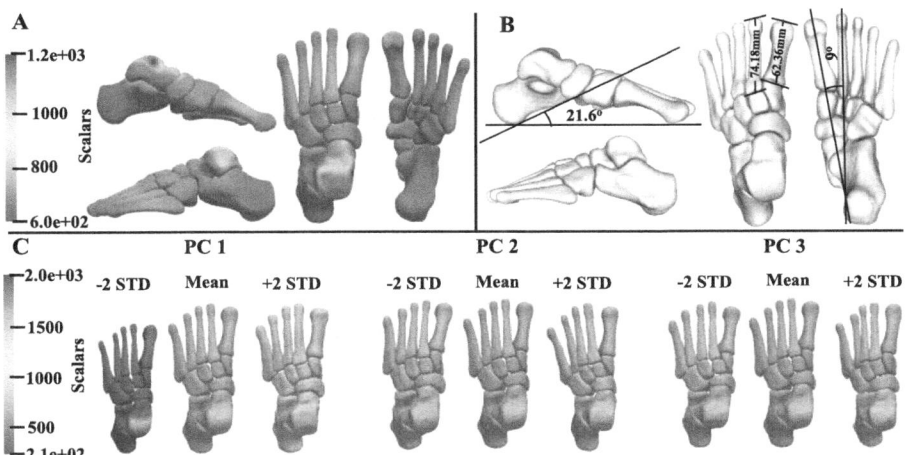

Fig. 1. Illustration of the multi-object SSPIM mean model (A) showed a higher intensity distribution (density) at the superior medial talar body than the rest of the foot. The mean model marginalised to shape and pose (B) with the measurement to confirm the neutral-arched foot. The first three PCs from −2 to 2 STD of the model (C) showed shape, pose and intensity variations. The colour scale indicated higher intensity values in red and lower intensity values in blue, measured as scalars values in Hounsfield units (HU). (Color figure online)

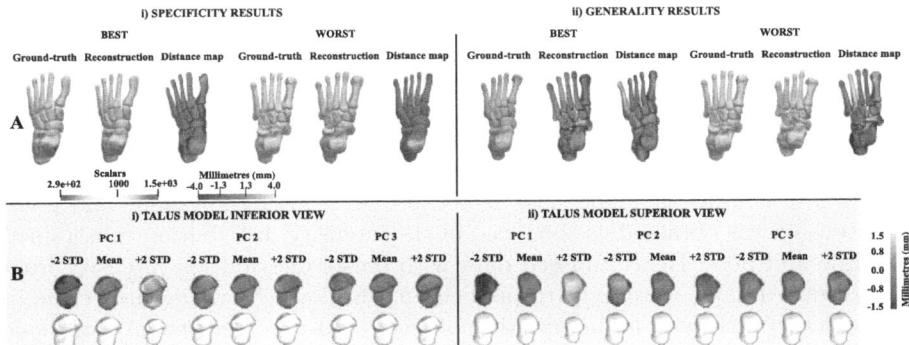

Fig. 2. Illustration A shows the best and worse reconstructions for both specificity (i) and generality (ii) showing the ground-truth scale volume mesh, the reconstruction and distance map (reconstructed shape error) obtained from fitting CT segmented volumes to the multi-object SSPIM model. The scalar volume scale illustrates intensity values in red and lower intensity values in blue, measured as scalars values in HU. Figure B shows the inferior (i) and superior (ii) views of the first three PCs of the neutral-arched model marginalised to the shape of the talus. The PCs variations (distance colour map - first row) are indicated as distance maps between the mean and the sample and PC shape instances (second row). The scale on the distance maps indicated areas with negative if the source surface (mean) at a given point was inside the target surface (blue) and vice versa (yellow) measured in millimetres (mm). (Color figure online)

and generality values obtained from fitting the SSPIM marginalised segmented feet volumes were 0.81 ± 25 mm and 1.20 ± 0.48 mm RMS with the Hd of 3.24 ± 1.10 and 6.50 ± 2.34 mm, respectively.

In Fig. 2, first PC described the highest variation due to general talar size increase (-2 STD). The second PC (-2 STD) described variations due to general size increase on the talar anterior body, posterior subtalar facet laterally, head, neck, and a size decrease of the talar posterior body, sulcus tali and posterior subtalar facet medially. The third PC (-2 STD) described size decrease on the sulcus tali, medial subtalar facet, and size increase of the posterior subtalar facet, anterior talar head, and medial malleolar surface. The opposite variations for all the PCs were observed in the 2STD.

4   Discussion

The intensity distribution represented as HU obtained from CT images can directly be interpreted as the relative density of bone tissues [21]. In neutral-arched feet, the talus bone, especially at the dome, head and medial malleolar articular surface, has a higher density to allow weight transmission to the rest of the foot bones, which correlates with our observations in the mean model [13, 25]. The mean shape and pose model anatomical measurements Fig. 1B) confirm the mean model as a neutral-arched foot [2]. The multi-object SSPIM model outperformed the marginalised model based on the fitting results due to the available pose in the SSPIM that accounts for the correlation with the surrounding bones. The RMS errors of the talus model using standard model validation metrics in Scalismo are similar to [3, 11] which suggests that the model can accurately fit surface mesh data despite the training set being low. However, higher Hd values indicate a need to increase the data to improve model robustness. The predicted intensity values from the samples within the training dataset are visually similar to the ground-truth volumes, while the intensity of unseen data is quite different. This is because after fitting for shape and pose, the intensity was obtained as the mean of the intensity distribution constrained by shape and pose. Hence, unseen data with out-of-distribution intensity from the training dataset were hard to predict. In addition, the visual inspection of the model PCs showed no intersecting objects, and observed model variations expressed the variation observed in the human feet samples. The observed PC variations in the talus marginalised model showed high variations in the sulcus tali, talar anterior body, neck, head, and posterior subtalar facet analogous to the results reported in the morphological analysis of the talus in previous studies [3, 22]. These observations further confirmed the ability of the novel statistical method to model multi-object SSPIM to also capture morphological variations of single bone.

5   Conclusion

The study evaluated a DMFC-GPM framework by building the first multi-object SSPIM of 12 bones of the foot. Despite the limited data the results demonstrated the framework's ability to capture the complexity and variability of anatomical structures in the data sample. The framework enabled on-demand marginalisation of the model to features of interest, allowing for multiple downstream applications in analysing medical images of the human foot. In addition, although global validation of such models is challenging, it is still possible to understand and evaluate the entire foot architecture globally based on single object models. Such evaluations are useful in clinical procedures and in 3D reconstruction from 2D X-rays.

Acknowledgments. Funding was provided by the Swedish International Development Cooperation Agency under the Organization for Women in Science for the Developing World (OWSD); and from the South African National Research Foundation (NRF) and University of Cape Town postgraduate funding office (PGFO). Mentorship was facilitated by the AFRICAI/MICCAI Summer School 2023.

Disclosure of Interests. No potential conflict of interest was reported by the author(s).

References

1. Akoh, C.C., Phisitkul, P.: Clinical examination and radiographic assessment of the cavus foot. Foot Ankle Clin. **24**(2), 183–193 (2019). https://doi.org/10.1016/j.fcl.2019.02.002
2. Anazor, F., Sibanda, V., Abubakar, A., Dhinsa, B.S.: Computed tomography scan architectural measurements in adult foot and ankle surgery: a narrative review for orthopaedic trainees. Cureus **14**(11) (2022). https://doi.org/10.7759/cureus.32039
3. Arbabi, S., et al.: Statistical shape model of the talus bone morphology: a comparison between impinged and nonimpinged ankles. J. Orthop. Res.® **41**(1), 183–195 (2023). https://doi.org/10.1002/jor.25328
4. Asghar, A., Naaz, S.: The transverse arch in the human feet: a narrative review of its evolution, anatomy, biomechanics and clinical implications. Morphologie **106**(355), 225–234 (2022). https://doi.org/10.1016/j.morpho.2021.07.005
5. Barg, A., et al.: Weightbearing computed tomography of the foot and ankle: emerging technology topical review. Foot Ankle Int. **39**(3), 376–386 (2018). https://doi.org/10.1177/1071100717740330
6. Bossa, M., Olmos, S.: Statistical linear models in procrustes shape space (2006). https://inria.hal.science/inria-00635901/
7. Davies, R., Twining, C., Cootes, T., Waterton, J., Taylor, C.: A minimum description length approach to statistical shape modeling. IEEE Trans. Med. Imaging **21**(5), 525–537 (2002). https://doi.org/10.1109/TMI.2002.1009388
8. Fouefack, J.R.: Towards a framework for multi class statistical modelling of shape, intensity, and kinematics in medical images. Ph.D. thesis, Ecole nationale supérieure Mines-Télécom Atlantique, University of Cape Town (2021). https://open.uct.ac.za. http://hdl.handle.net/11427/35730
9. Fouefack, J.R., Borotikar, B., Lüthi, M., Douglas, T.S., Burdin, V., Mutsvangwa, T.E.: Dynamic multi feature-class Gaussian process models. Med. Image Anal. **85**, 102730 (2023). https://doi.org/10.1016/j.media.2022.102730
10. Goswami, B., Misra, S.K.: 3D modeling of X-ray images: a review. Int. J. Comput. Appl. **132**(7), 40–46 (2015). https://doi.org/10.5120/ijca2015907566

11. Grant, T.M., et al.: Development and validation of statistical shape models of the primary functional bone segments of the foot. PeerJ **8**, e8397 (2020). https://doi.org/10.7717/peerj.8397

12. Keener, B.J., Sizensky, J.A.: The anatomy of the calcaneus and surrounding structures. Foot Ankle Clin. **10**(3), 413–424 (2005). https://doi.org/10.1016/j.fcl.2005.04.003. iD: 273322

13. Kelikian, A.S., Sarrafian, S.K.: Sarrafian's Anatomy of the Foot and Ankle: Descriptive, Topographic, Functional. Lippincott Williams & Wilkins (2011)

14. Lempereur, M., Leboeuf, F., Brochard, S., Rousset, J., Burdin, V., Rémy-Néris, O.: In vivo estimation of the glenohumeral joint centre by functional methods: accuracy and repeatability assessment. J. Biomech. **43**(2), 370–374 (2010). https://doi.org/10.1016/j.jbiomech.2009.09.029

15. Lüthi, M., Gerig, T., Jud, C., Vetter, T.: Gaussian process morphable models. IEEE Trans. Pattern Anal. Mach. Intell. **40**(8), 1860–1873 (2018). https://doi.org/10.1109/TPAMI.2017.2739743. iD: 1

16. Madsen, D., Morel-Forster, A., Kahr, P., Rahbani, D., Vetter, T., Lüthi, M.: A closest point proposal for MCMC-based probabilistic surface registration. In: Vedaldi, A., Bischof, H., Brox, T., Frahm, J.-M. (eds.) ECCV 2020. LNCS, vol. 12362, pp. 281–296. Springer, Cham (2020). https://doi.org/10.1007/978-3-030-58520-4_17

17. Mirbagheri, H.R.: Radiologic evaluation of chronic foot pain. Iran. J. Radiol. **14**(5) (2017). https://doi.org/10.5812/iranjradiol.48199

18. Morel-Forster, A., Gerig, T., Lüthi, M., Vetter, T.: Probabilistic fitting of active shape models. In: Reuter, M., Wachinger, C., Lombaert, H., Paniagua, B., Lüthi, M., Egger, B. (eds.) ShapeMI 2018. LNCS, vol. 11167, pp. 137–146. Springer, Cham (2018). https://doi.org/10.1007/978-3-030-04747-4_13

19. Mutsvangwa, T., Burdin, V., Schwartz, C., Roux, C.: An automated statistical shape model developmental pipeline: application to the human scapula and humerus. IEEE Trans. Biomed. Eng. **62**(4), 1098–1107 (2015). https://doi.org/10.1109/TBME.2014.2368362

20. Mutsvangwa, T., Veeraragoo, M., Douglas, T.S.: Precision assessment of stereo-photogrammetrically derived facial landmarks in infants. Ann. Anatomy-Anatomischer Anzeiger **193**(2), 100–105 (2011). https://doi.org/10.1016/j.aanat.2010.10.008

21. Nackaerts, O., Maes, F., Yan, H., Souza, P.C., Pauwels, R., Jacobs, R.: Analysis of intensity variability in multislice and cone beam computed tomography. Clin. Oral Implant Res. **22**(8), 873–879 (2011). https://doi.org/10.1111/j.1600-0501.2010.02076.x

22. Peterson, A.C., et al.: Multi-level multi-domain statistical shape model of the subtalar, talonavicular, and calcaneocuboid joints. Front. Bioeng. Biotechnol. **10** (2022). https://doi.org/10.3389/fbioe.2022.1056536

23. Rokkedal-Lausch, T., Lykke, M., Hansen, M.S., Nielsen, R.O.: Normative values for the foot posture index between right and left foot: a descriptive study. Gait Posture **38**(4), 843–846 (2013). https://doi.org/10.1016/j.gaitpost.2013.04.006

24. Schönborn, S., Egger, B., Morel-Forster, A., Vetter, T.: Markov chain monte carlo for automated face image analysis. Int. J. Comput. Vis. **123**, 160–183 (2017). https://doi.org/10.1007/s11263-016-0967-5

25. Stolle, J., et al.: A statistical analysis of human talar shape and bone density distribution. J. Orthop. Res. 30 (2024). https://doi.org/10.1002/jor.25835

26. White, T.D., Black, M.T., Folkens, P.A.: Human Osteology. Academic Press (2011)

27. Wijk, M.V., Barnard, M.M., Fernandez, A., Cloete, K., Mukosi, M., Pitcher, R.D.: Trends in public sector radiological usage in the Western Cape Province, South Africa: 2009–2019. SA J. Radiol. **25**(1) (2021). https://doi.org/10.4102/sajr.v25i1. 2251
28. Zheng, G., Yu, W.: Statistical shape and deformation models based 2D–3D reconstruction. In: Statistical Shape and Deformation Analysis, pp. 329–349. Elsevier (2017). https://doi.org/10.1016/B978-0-12-810493-4.00015-8

Explainability-Guided Deep Learning Models For COVID-19 Detection Using Chest X-Ray Images

Houda El Mohamadi[1]([✉]) and Mohammed El Hassouni[2]

[1] LRIT, FS, Mohammed V University in Rabat, Rabat, Morocco
houda.elmohamadi@um5r.ac.ma
[2] FLSH, Mohammed V University in Rabat, Rabat, Morocco
mohamed.elhassouni@flsh.um5.ac.ma

Abstract. The classification and detection of COVID-19 from chest X-ray images using deep learning models is challenging due to the limited amount of training data and the potential for overfitting. Data augmentation remains one of the most effective solutions to address these issues. However, traditional augmentation techniques often generate images that lack the visual characteristics that are most informative for the model. In this paper, we propose a data augmentation approach based on explainable deep learning for COVID-19 diagnosis using chest X-ray images. Our method leverages explainable deep learning techniques to identify the most significant regions in chest X-ray (CXR) images, ensuring that the model focuses on critical areas for disease detection. We extract saliency maps, which act as an importance filter derived from explainability methods. These saliency maps are then used to guide the data augmentation process, ensuring that the generated images emphasize the features most relevant for accurate diagnosis. This method enhances the learning process of complex models by providing a more informative training dataset. It can result in a significant increase in accuracy (approximately 3–20%), a reduction in false-positive images, and a minimization of loss functions and overfitting issues.

Keywords: Explainable method · saliency map · Deep learning · X-ray images · retrain models · minimize Overfitting

1 Introduction

Deep learning (DL) has revolutionized image classification, offering high efficiency and accuracy. This advancement has been particularly impactful in medical imaging, where DL models have significantly enhanced the precision and accuracy of disease diagnosis. During the COVID-19 pandemic, for instance, DL models were used to analyze chest X-rays and CT scans, facilitating diagnosis and treatment of the virus [1]. These models using CXR images can provide valuable information about the presence and severity in detecting lung diseases

© The Author(s), under exclusive license to Springer Nature Switzerland AG 2025
U. Anazodo et al. (Eds.): MImA 2024/EMERGE 2024, CCIS 2240, pp. 164–173, 2025.
https://doi.org/10.1007/978-3-031-79103-1_17

associated with COVID-19, including pneumonia and acute respiratory distress syndrome [2].

Despite their impressive capabilities, deep learning (DL) models can sometimes fall short of providing comprehensive and reliable results, leading to bias and inaccuracies in the outputs, particularly when confronted with limited data. This often results in overfitting and an increased false positive rate [3]. Traditional data augmentation, a widely used solution in the literature to figure out these issues by employing techniques such as rotation, scaling, and flipping [4]. Artificially they increase the size of the training dataset, thereby contributing to reduce overfitting and improve model robustness. However, while these traditional augmentation methods increase the number of images, they do not necessarily enhance the visual information that is crucial for the models to focus on the most pertinent regions for accurate prediction. To address these limitations, there is a growing interest in explainable deep learning methods (XDL) [5]. These methods serve a dual purpose, which aim not only to provide greater transparency and interpretability to the outputs of deep learning models but also by extracting maps as saliency maps that could guide models. Saliency maps highlight the most significant regions in images, enabling the model to focus on these areas for better predictions, and address common challenges faced by DL models, such as overfitting Leading to poor performance on unseen data, and the need of large datasets. In the literature, several explainability methods have been proposed, namely CAM (Class Activation Mapping) [6]. GradCam (Gradient-weighted Class Activation Mapping) refines CAM by considering the gradients of the target class to generate more accurate heatmaps [6]. FEM (Feature Extraction Map) visualizes the activated regions in intermediate layers of the model to understand its hierarchical feature extraction process [7]. One of the key techniques in XDL is the extraction of features not only from the last layers(about shape) but also about textures from 1st layers of models as in TS-EXM (texture-shape explainable method) [8]. These features can be used to generate an explanation maps with more details, which visually represent the areas of the input image that are most relevant in the model to improve their prediction. This approach enhances our understanding of the model's behavior, making its predictions more interpretable and reliable.

To mitigate these issues and drawing inspiration from various works [9–11], our main objective is to propose a novel strategy utilizing an explainable method known as the Texture Shape Explainable Method (TS-EXM), alongside other explainable methods such as Grad-CAM (Gradient-weighted Class Activation Mapping) and FEM (Feature Extraction Mapping). By increasing the dataset using saliency map extracted from EXMs to reconstruct images and strategically retraining the model under the guidance of these EXMs, we aim to enhance the model's performance and generalizability. This approach helps the model better understand the true classification, thereby minimizing the problem of overfitting.

The rest of the paper is organized as follows. In Sect. 2 describes our proposed strategy in detail. Section 3 presents the experiments result with ablation study. Finally, Sect. 4 concludes our work and identifies potential improvements in the future.

2   Proposed Method

The proposed method, as illustrated in Fig. 1, operates in three phases: initial training, explainability pipeline, and model refinement using updated data.

2.1   Initial Training Phase

In the initial phase, we collect and input data into the system to train deep learning models. For this purpose, we utilize pre-trained models through transfer learning for their proven effectiveness in extracting complex features from images. The pre-trained models are fine-tuned on our specific data to generate initial prediction and features maps. This step ensures that our models are well-initialized with a robust features extraction capabilities, which crucial for subsequent steps.

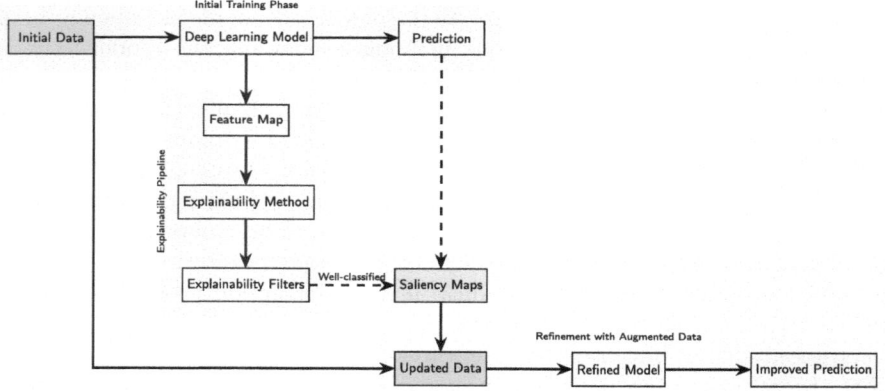

Fig. 1. Deep Learning Scheme with Explainability-Based Data Augmentation and Direct Data Flow

2.2   Explainability Pipeline

The proposed method integrates advanced explainable deep learning (DL) techniques to enhance both model training and interpretability. We leverage feature maps extracted from the model using three explainability methods: Grad-CAM, FEM, and Texture-Shape Explainable Method (TS-EXM). These methods offer complementary strengths: Grad-CAM focuses on capturing the spatial importance of features in the final convolutional layers, while FEM prioritizes extracting crucial texture information from the initial convolutional layers. Finally, TS-EXM specifically targets both shape and Textural information from the final and the initial layers, respectively. These different types of information are then fused to create comprehensive explanation maps. Furthermore, when the model provides correct predictions, particularly those with high true positive rates, these explanation maps are used to generate saliency maps. Saliency maps emphasize the significant features identified during the classification process, ensuring that only well-classified instances, with strong supporting evidence, are considered for further steps. By combining the insights from these diverse explainability methods, we aim to achieve a more comprehensive understanding of the model's decision-making process.

2.3 Data Augmentation Using Saliency Maps

This phase involves a data augmentation process that leverages saliency maps to act as an "explainable filter," improving the model performance and generalization capabilities. Here, saliency maps are first generated using explainability methods. These maps highlight the image regions most influential in the model predictions. The saliency maps are then combined with the original images. This is achieved by weighting the pixel values from both sources, resulting in enhanced images that emphasize the critical features identified by the saliency maps while preserving the original image context. This augmentation enriches the dataset by providing the model with both raw data and its interpretively enhanced versions. Figure 2 exemplifies this process: (a) shows the final result where the saliency map is superposed on the original chest X-ray image to create a reconstructed image, (b) depicts the generated saliency map highlighting the significant regions, and (c) presents the original chest X-ray image

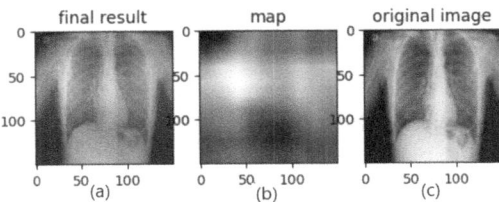

Fig. 2. Saliency-Driven X-ray Image Enhancement: (a): image reconstructed, (b): saliency map generated and (c): the original CXR image

2.4 Model Refinement

In the final phase, the models fine-tuned using transfer learning in the initial phase are retrained with augmented data, which includes both the original images and saliency-enhanced (reconstructed) images. The retraining process involves combining the original and saliency-enhanced images to provide a richer and more comprehensive dataset. This enriched dataset allows the models to learn from a broader range of patterns and features, improving their accuracy and generalization capabilities. By focusing on both the original and highlighted regions, the models can better capture the underlying structures and variations in the data. As a result, the retrained models produce more accurate and robust predictions.

3   Experimental Results

3.1   Data Description

The dataset comprises a total of 9951 medical chest X-ray (CXR) images, categorized into three groups based on their associated medical conditions: Covid-19, Normal, and Viral Pneumonia. The Covid-19 category includes 3616 images, the Normal category consists of 5000 images, and the Viral Pneumonia category encompasses 1345 images. All images are stored in PNG format with a resolution of 299×299 pixels. These images have been collected from various hospitals globally and meticulously curated by a team of radiologists and medical professionals. The dataset divided into two sets: training set and testing set. We used an 80–20 ratio; 80% of the images were used for training, and 20% were used for testing.

3.2   Implementation Details

The dataset used in this study are resized to a resolution of 150×150 pixels. We employed VGG-19, VGG-16, EfficientNetV2S, ResNet-50 and DenseNet121 architectures, with specific adjustments to incorporate (EXM and TS-EXM). Models were trained on an NVIDIA Tesla T4 GPU, with hyperparameters such as learning rate at (0.001), batch size at (64), and number of epochs at (20).

3.3   Neural Network Models

In this study we employed a diverse of convolutional neural network architectures in transfer learning, categorized into less complex and more complex models. The less complex include VGG-16 and VGG-19, both characterized by a straightforward architecture with five blocks of convolutional and max pooling layers. VGG-16 comprises 15,239,235 trainable parameters, while VGG-19, with additional layers, consists of 20,548,931 trainable parameters. The more complex models are ResNet-50, EfficientNetV2S, and DenseNet121. ResNet-50 employs a bottleneck design with 1×1 convolutions to manage computational complexity, featuring 25,688,963 parameters, of which 2,101,251 are trainable. EfficientNetV2S utilizes MBConv and Fused-MBConv blocks for enhanced feature extraction, including 20,259,667 trainable parameters out of a total of 20,413,539 parameters. DenseNet121, with its 16 densely connected blocks, improves feature propagation and mitigates the vanishing gradient problem, comprising 7,103,299 parameters, with 7,019,651 being trainable.

Table 1. Initial Model Performance

Model	Classes	Precision	Recall	F-Score	Accuracy	Loss
VGG-19	Covid	0.964	0.975	0.973	0.977	0.081
	Normal	0.961	0.970	0.954		
	Pneumonia	0.953	0.952	0.925		
VGG-16	Covid	0.937	0.986	0.961	0.961	0.112
	Normal	0.979	0.949	0.964		
	Pneumonia	0.967	0.938	0.952		
ResNet-50	Covid	0.669	0.8797	0.767	0.763	0.352
	Normal	0.890	0.663	0.740		
	Pneumonia	0.766	0.883	0.821		
DenseNet121	Covid	0.938	0.988	0.962	0.953	0.1667
	Normal	0.989	0.944	0.966		
	Pneumonia	0.949	0.978	0.963		
EfficientNetV2S	Covid	0.647	0.994	0.784	0.753	0.688
	Normal	0.993	0.556	0.713		
	Pneumonia	0.705	0.844	0.768		

To assess the classification performance of these transfer learning models, we employed metrics such as accuracy, Precision, recall, and F-score shown in Table 1. VGG-19 and VGG-16 exhibit superior performance compared to DenseNet121, ResNet-50, and EfficientNetV2S. VGG-19 demonstrates excellent accuracy and maintains a very low loss, while VGG-16 also achieves high accuracy with slightly higher loss, and both models show high precision, recall, and F-scores across all classes. DenseNet121 performs well but has a higher loss compared to VGG-19 and VGG-16.

On the other hand, ResNet-50 and EfficientNetV2S perform poorly. ResNet-50 has a significantly higher loss, and EfficientNetV2S suffers from an even greater loss. These models show low precision, recall, and F-scores, particularly for the Covid class, indicating frequent false classifications compared to true classifications.

The high loss and low accuracy in DenseNet121, ResNet-50, and EfficientNetV2S are likely due to insufficient training data and overfitting, as shown in Fig. 3. Overfitting means that the models excel on training data but fail to generalize to test data, resulting in poor performance. When the available data is insufficient such as our data, these lead models cannot fully leverage their capacity, leading to low performance. To resolve those issues we will retrain all models using our proposed data augmentation technique.

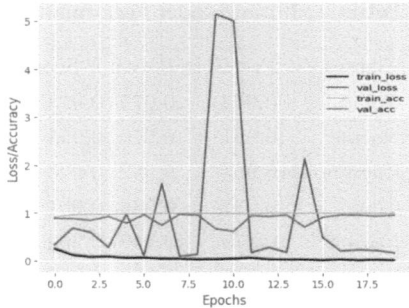

Fig. 3. DenseNet121 models across initial training

3.4   Explainability-Guided Performance

An ablation study was conducted to evaluate the impact of different components on the overall performance and guidance of the deep learning models used. In this study, we systematically added or removed specific elements to understand their contributions to the final outcomes. Additionally, to demonstrate the effectiveness of our approach in enhancing model performance, we compared our method with classical data augmentation techniques, including traditional image generators with early stopping mechanisms ensuring a fair comparison by matching the amount of generated data. Moreover, our methodology does not rely on early stopping mechanisms, as the augmented data inherently enhances model performance by capturing crucial features necessary for growth.

Table 2. Explainability guided model performance

Models	VGG-16		VGG-19		ResNet-50		DenseNet121		EfficientNetV2S	
Methods/Metrics	Acc	Loss	Acc	Loss	Acc	Loss	Acc	Loss	Acc	Loss
Without augmentation	0.961	0.112	0.977	0.813	0.763	0.352	0.953	0.166	0.753	0.688
Classical Augmentation	0.948	0.120	0.963	0.101	0.698	0.643	0.971	0.099	0.935	0.191
EXM	0.956	0.138	0.970	0.095	0.835	0.337	0.976	0.090	0.969	0.119
TS-EXM	0.977	0.096	0.973	0.099	0.838	0.312	0.978	0.075	0.949	0.182
TS-EXM + EXM	0.975	0.090	0.976	0.075	0.861	0.317	0.983	0.049	0.983	0.059

After retraining the models we used previously as shown in Table 2. Our experiments demonstrate a significant improvement in model accuracy, particularly for EfficientNetV2S, DenseNet121, and ResNet-50. We observed that they effectively learned from the explanation techniques employed to augment our data, especially when combining the novel method (TS-EXM) with EXM (Grad-CAM and FEM). For EfficientNetV2S, our approach significantly outperformed the baseline, reducing the loss from 0.688 to 0.095 and increasing the accuracy from 0.753 to 0.983. DenseNet121 also showed substantial improvements, with accuracy rising from 0.953 to 0.983 and loss decreasing from 0.166 to 0.049. Similarly, ResNet-50 exhibited a reduction in loss. The use of TS-EXM alone also demonstrated performance enhancements for complex networks, as shown in Fig. 4a. Applying TS-EXM guide the model learn effectively from saliency maps, leading to a significant reduction and almost elimination of the overfitting problem encountered when using only the original data. However, for VGG-16 and VGG-19, the performance improvements were less pronounced. Despite these smaller gains, it is important to note that our approach did not result in adverse effects. Although there was a slight increase in loss for these models, the number of false classifications decreased, and true classifications increased, indicating an overall positive impact on model performance. Compared to classical data augmentation. Our approach enables the model to learn from the most influential regions, resulting in more accurate predictions and reducing the issue of limited data (overfitting) as shown in Fig. 4b.

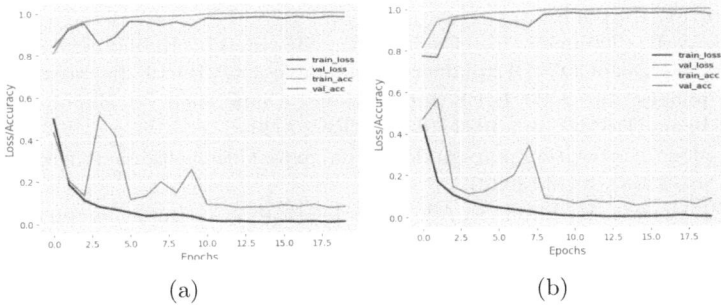

(a)   (b)

Fig. 4. DenseNet21 model: (a) Training using TS-EXM and (b): Training using EXM(GradCam+FEM) and TS-EXM

4 Conclusion

In this study, we introduced an explainability-guided approach for enhancing COVID-19 detection from chest X-ray images. By integrating Texture-Shape Explainable Method (TS-EXM) with Grad-CAM and FEM to augment our original data, our method significantly improved the performance of models, including EfficientNetV2S, DenseNet121, and ResNet-50. The approach effectively increased model accuracy by approximately 3–20%, reduced the loss by about 3–60%, and mitigated overfitting, particularly in complex architectures. Our strategy not only enhanced the models' learning process but also provided better interpretability, making it a valuable tool for reliable COVID-19 diagnosis. While this study focuses on classification, the principles of explainability-guided data augmentation could be adapted for other tasks such as segmentation, where emphasizing critical regions identified by explainability methods could enhance boundary detection and overall accuracy. Future studies could explore the adaptation of our explainability-guided methods to other medical imaging tasks to assess their versatility and effectiveness.

Acknowledgment. This research is supported by the project "Plateforme logicielle d'intégration de stratégies d'immunisation contre la pandémie COVID-19" funded by the grant of the Hassan II Academy of Sciences and Technology of Morocco. Additionally was conducted with the support of the National Center for Scientific and Technical Research (CNRST) under the PhD-Associate Scholarship Program (PASS).

References

1. Toğaçar, M., Muzoğlu, N., Ergen, B., Yarman, B.S.B., Halefoğlu, A.M.: Detection of COVID-19 findings by the local interpretable model-agnostic explanations method of types-based activations extracted from CNNs. Biomed. Signal Process. Control **71**, 103128 (2022)
2. Karim, M.R., Döhmen, T., Cochez, M., Beyan, O., Rebholz-Schuhmann, D., Decker, S.: DeepCOVIDExplainer: explainable COVID-19 diagnosis from chest X-ray images. In: 2020 IEEE International Conference on Bioinformatics and Biomedicine (BIBM), pp. 1034–1037. IEEE (2020)
3. An, J., Joe, I.: Attention map-guided visual explanations for deep neural networks. Appl. Sci. **12**(8), 3846 (2022)
4. Oza, P., Sharma, P., Patel, S., Adedoyin, F., Bruno, A.: Image augmentation techniques for mammogram analysis. J. Imaging **8**(5), 141 (2022)
5. Ribeiro, M.T., Singh, S., Guestrin, C.: "Why should i trust you?" Explaining the predictions of any classifier. In: Proceedings of the 22nd ACM SIGKDD International Conference on Knowledge Discovery and Data Mining, pp. 1135–1144 (2016)
6. Mahmoudi, S.A., Stassin, S., El Habib Daho, M., Lessage, X., Mahmoudi, S.: Explainable deep learning for COVID-19 detection using chest X-ray and CT-scan images. In: Healthcare Informatics for Fighting COVID-19 and Future Epidemics, pp. 311–336 (2022)
7. Zhukov, A., Benois-Pineau, J., Giot, R.: Evaluation of explanation methods of AI-CNNs in image classification tasks with reference-based and no-reference metrics. Adv. Artif. Intell. Mach. Learn. **3**(1), 620–646 (2023)
8. El Mohamadi, H., El Hassouni, M.: Enhanced deep learning explainability for COVID-19 diagnosis from chest X-ray images by fusing texture and shape features. In: 2023 10th International Conference on Wireless Networks and Mobile Communications (WINCOM), pp. 1–6. IEEE (2023)
9. Gebreamlak, M.E., Ayyar, M.P., Benois-Pineau, J., Salmon, J.-P., Zemmari, A.: Leveraging explainability methods in spectral domain for data augmentation and efficient training of CNN classifiers for COVID-19 detection. In: 2023 Twelfth International Conference on Image Processing Theory, Tools and Applications (IPTA), pp. 1–6. IEEE (2023)

10. Watson, M., Awwad Shiekh Hasan, B., Al Moubayed, N.: Using model explanations to guide deep learning models towards consistent explanations for EHR data. Sci. Rep. **12**(1), 19899 (2022)

11. Sun, H., Servadei, L., Feng, H., Stephan, M., Santra, A., Wille, R.: Utilizing explainable AI for improving the performance of neural networks. In: 2022 21st IEEE International Conference on Machine Learning and Applications (ICMLA), pp. 1775–1782. IEEE (2022)

Feasibility of Open-Source Tracking-Based Metrics in Evaluating Ultrasound-Guided Needle Placement Skills in Senegal

Fatou Bintou Ndiaye[2]($\boxtimes$) , Rebecca Hisey[1]($\boxtimes$) , Kyle Sunderland[1]($\boxtimes$),
Idrissa Seck[3]($\boxtimes$), Idy Diop[2]($\boxtimes$), Ron Kikinis[4]($\boxtimes$), Babacar Diao[3]($\boxtimes$),
Gabor Fichtinger[1]($\boxtimes$) , and Mamadou Samba Camara[2]($\boxtimes$)

[1] Laboratory for Percutaneous Surgery, Queen's University, Kingston, Canada
[2] Laboratoire d'Imagerie Médicale et Bio-Informatique, Ecole Supérieure Polytechnique, Dakar, Senegal
fatoubintoundiaye@esp.sn
[3] Université Cheikh Anta Diop, Dakar, Senegal
[4] Surgical Planning Laboratory, Harvard Medical School, Boston, USA

Abstract. Background: Obstructive renal disease is a critical health issue in sub-Saharan countries such as Senegal, necessitating interventions like percutaneous nephrostomy. Unfortunately training for ultrasound-guided procedures such as nephrostomy is limited in Senegal due to the limited availability of physicians to provide feedback to trainees. This work investigates the feasibility of using the open-source platform 3D Slicer to evaluate technical skill in percutaneous nephrostomy in Senegal. Methods: Six novice urology residents and six expert urologists from Ouakam Military Hospital performed four trials of percutaneous nephrostomy each on a low-cost phantom. An electromagnetic tracking was used to track the motion of the ultrasound probe and needle during each trial. Finally, the Perk Tutor extension of 3D Slicer was used to calculate skill assessment metrics that have been previously validated for other ultrasound-guided procedures. Results: Skill assessment metrics were successfully computed using Perk Tutor. The computed metrics showed a significant difference between novice and expert physicians and the novice physicians showing marked improvements across the four trials. Conclusions: The tracking-based metrics computed using Perk Tutor effectively distinguished skill levels and measured performance enhancements between novice and expert physicians practicing percutaneous nephrostomy. This open-source platform provides an accessible option for automating skill assessment in low-resource countries such as Senegal.

Keywords: Tracking-based metrics · percutaneous nephrostomy · skill assessment · medical training · sub-Saharan Africa

1 Introduction

Image-guided medical interventions have been largely unavailable in low-income countries like Senegal, but emerging point-of-care ultrasound (POCUS) imaging now stands to change this status quo. POCUS enables physicians to perform interventions at the

© The Author(s), under exclusive license to Springer Nature Switzerland AG 2025
U. Anazodo et al. (Eds.): MImA 2024/EMERGE 2024, CCIS 2240, pp. 174–180, 2025.
https://doi.org/10.1007/978-3-031-79103-1_18

bedside, whether that bedside is in an urban hospital, rural clinic, or mobile care unit. There are a multitude of applications for ultrasound in medical interventions. One critical application of this technology in sub-Saharan countries, like Senegal, is in the treatment of obstructive renal disease. Obstructive renal disease is highly prevalent in sub-Saharan countries, posing significant health challenges [1]. Effective medical interventions, such as percutaneous nephrostomy, are critical due to their potential to improve patient outcomes and quality of life in regions with limited access to advanced medical facilities [2]. Percutaneous nephrostomy in particular is crucial for treating renal obstruction in Senegal. Its significance lies in providing a direct solution to renal obstruction, enhancing patient outcomes and quality of life [3]. Despite its importance, percutaneous nephrostomy faces challenges that hinder its widespread adoption and effectiveness, particularly due to inadequate skill assessment and training methods for medical professionals performing the procedure [4].

Computer-assisted skill assessment has long been suggested as a method to improve feedback to medical trainees in developed countries; however, its feasibility has yet to be explored in a low-resource setting. Various methods of computer-assisted skill assessment have been proposed such as those using energy and force measurements [5, 6] or computer vision [7] but perhaps the most widely validated are methods that rely on motion tracking [8, 9]. Many researchers have validated the use of tracking-based metrics to assess skill in a variety of ultrasound guided needle interventions [10, 11], though none have explored its application to percutaneous nephrostomy.

In this work, we explore the potential for evaluating skill using the open-source tools provided by 3D Slicer (www.slicer.org) [12]. 3D Slicer is an open-source platform for medical imaging informatics that provides robust functionality to interface easily with hardware and perform many tasks relating to visualization and analysis of medical images and related information. Relevant to this work, 3D Slicer provides an extension called Perk Tutor [13] which provides several modules dedicated to analyzing skill in ultrasound-guided needle interventions. One functionality that Perk Tutor provides is the computation of tracking-based metrics for skill assessment. These metrics have undergone rigorous validation for other ultrasound-guided needle interventions [14–16] but remain untested for percutaneous nephrostomy.

The goal of this work is to demonstrate the feasibility of assessing technical skill in percutaneous nephrostomy in a low-resource setting using the open-source tools provided by 3D Slicer and Perk Tutor.

2 Methods

2.1 Experimental Setup

The experimental setup for this work consisted of a consumer-grade computer, a Telemed C5–2-60 curvilinear ultrasound probe, and an Ascension Trakstar electromagnetic (EM) tracker. Tracking sensors were placed on the ultrasound probe and on the needle, with an additional sensor embedded into the phantom to serve as a reference. A low-cost renal phantom was used to allow participants to practice the procedure. The phantom was constructed from plastisol with added cellulose to simulate tissue properties under ultrasound.

The user interface was implemented within 3D Slicer and provided participants with the ultrasound video along with a list of tasks in the procedure. PLUS Toolkit (www. PlusToolkit.org) [17] was used to provide an interface between the ultrasound, tracking system and 3D Slicer. PLUS Toolkit provides a layer of abstraction that makes it possible to easily swap out different hardware components without changing the underlying application. The synchronized ultrasound video and tracking information were recorded in 3D Slicer, where the Perk Tutor extension was used to perform the analysis. The hardware configuration is shown in Fig. 1.A. Subfigure 1.B provides a schematic visual overview of the setup.

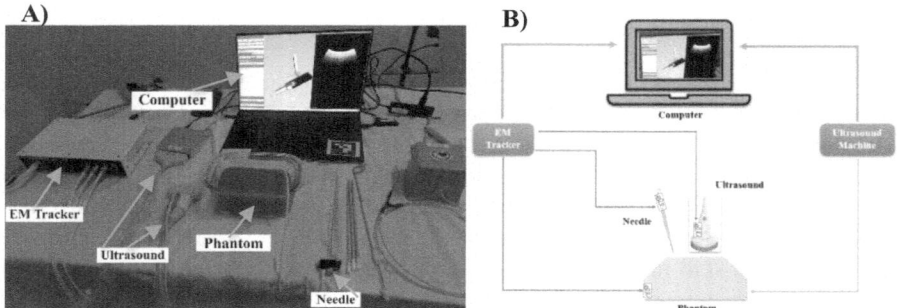

Fig. 1. Nephrostomy tutor physical setup. A) Hardware configuration. B) Schematic visual Overview

2.2   Skill Assessment Metric Calculation

The tracking information was analyzed using the Perk Evaluator, which is a module within Perk Tutor [13]. The Perk Evaluator computes skill-assessment metrics using tracking information, thereby offering a detailed analysis of practitioners' performance. Metrics were calculated to evaluate the practitioner's performance in using both the needle and the ultrasound probe in procedural tasks. Though many metrics are available for use within Perk Tutor, we selected three metrics that have been widely validated for ultrasound-guided interventions [14–16].

Elapsed time: The first metric that we compute is the total elapsed time. This metric measures the entire procedure duration in seconds. This is the only metric not directly related to tracking but is rather the entire duration of the recording.

Path length: This metric measures the total distance travelled by a single instrument relative to the reference sensor. We calculate this metric for both the ultrasound probe and the needle.

Time in tissue: The time in tissue is defined as the total time that the tip of a tracked instrument spends within a defined tissue region. In this case the tracked instrument was the needle tip, and the tissue was defined by a model of the renal cavity within the phantom. The cavity was registered to the reference coordinate system by performing a landmark registration.

Path in tissue: Similarly to time in tissue, the path in tissue is defined as the total path length of a tracked instrument within a defined tissue region. Once again, the tracked instrument in this case was the needle tip and the tissue region was the renal cavity within the phantom.

2.3 Experiments

Six novice urology residents and six expert urologists from Ouakam Military Hospital in Dakar, Senegal were recruited to participate in this study. Each participant performed four trials of simulated percutaneous nephrostomy each using the experimental setup described in Sect. 2.1. For each trial both the ultrasound video and tracking information for the needle and ultrasound probe was recorded. To tracking accuracy was maintained, a pivot and spin calibration was performed prior to each new participant. Skill assessment metrics were computed for each trial as described in Sect. 2.2. Metrics were compared between novice and expert groups using an independent t-test.

3 Results

3.1 Metrics Values

The results for each metric are summarized in Table 1, which includes the t-statistic, p-value, and the average values for each metric for both novice and expert groups. This table provides a clear comparison of performance differences. Additionally, Fig. 2 illustrates the value distributions for all metrics in both novice and expert groups, offering a visual representation of their performance.

Significant differences were observed in three of the five metric values between trials performed by experts and novices. The path length of the ultrasound probe showed the greatest difference between novice and expert participants with novices having an average path length of 5097 ± 2498 mm while the average expert path length was only 2498 ± 1722 mm. Significant differences were also seen for the time in tissue and path length in tissue metrics. The metric that yielded the least difference between the novice and expert participants was the needle path length. The difference between novice and expert participants was not significant for the needle path length nor the total elapsed time.

Table 1. Results for Each Metric for the novice group vs. expert Group

Metric	t-statistic	p-value	Novice Average	Expert Average
Elapsed Time	1.39	0.17	460 ± 194 s	308 ± 164 s
Path Length (Needle)	1.08	0.28	15179 ± 8296 mm	10623 ± 4238 mm
Path Length (Ultrasound probe)	2.76	0.01	5097 ± 2498 mm	2498 ± 1722 mm
Time in Tissue	2.30	0.03	40 ± 40 s	13 ± 21 s
Path in Tissue	2.35	0.02	809 ± 970 mm	216 ± 333 mm

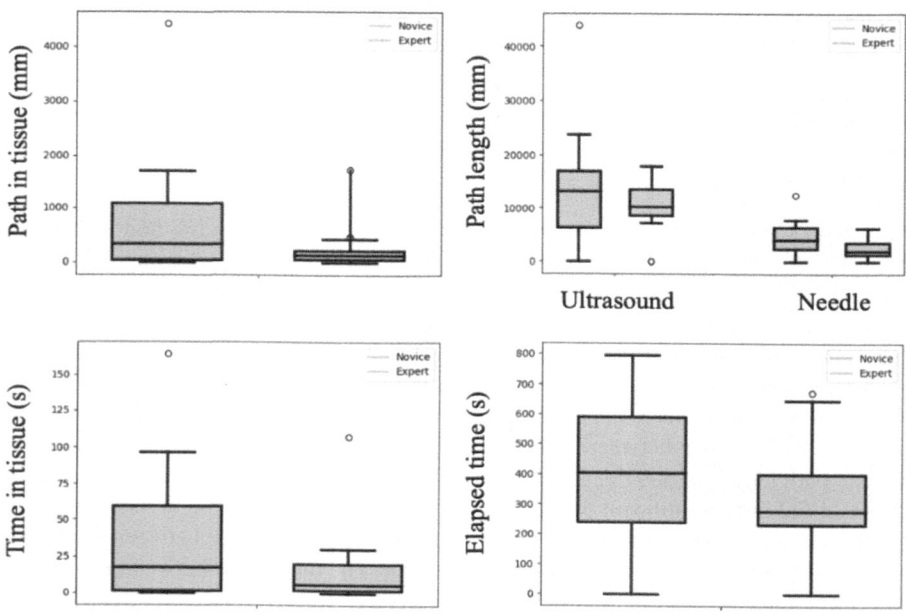

Fig. 2. Value distributions for all metrics in both novice and expert groups

4 Discussion

We were able to successfully set up and record synchronized ultrasound and tracking data for percutaneous nephrostomy in a simulated setting using only open-source software. This is the first time that such research of this nature has been undertaken in Senegal. The Perk Tutor extension was also successfully used to calculate metrics that can be used for skill assessment of ultrasound guided needle insertions. We observed significant differences in three of the five metrics computed. Only the elapsed time and needle path length did not show a significant difference. Despite the difference for these metrics not being significant, the values for novice participants were still substantially larger than the expert participants. This may be the result of the small sample size used for this study. The significant differences between novices and experts affirm the validity of these metrics for percutaneous nephrostomy.

Despite the successful results, this study is subject to limitations. The primary limitation is the small sample size, which included only 12 participants, potentially affecting the generalizability of the results. It is very challenging to collect substantial amounts of this type of data in low-resource settings as there are a very low number of physicians per capita and it is critical that we not take their time away from patients. Additionally, a limitation of the tracking system used in this study is that it requires sensors on instruments, which may interfere with surgeons' usual interaction with the tools. However, we were still able to collect data from the entire pool of urologists and urology residents at Ouakam Military Hospital in Dakar, Senegal. To confirm these findings and ensure broader applicability, future validation in larger samples across multiple centers is essential. We hope to expand this work to additional centers, focusing on refining tracking technologies to reduce costs and improve accessibility, thereby maximizing the impact in medical training.

This research represents a significant advancement for medical computing research in Senegal, marking a major step forward. It has uniquely brought together researchers from both the medical and technology sectors, a collaboration that has not been previously achieved in Senegal. This pioneering work opens numerous future avenues of research, enabling medical researchers to develop curricula around this platform and paving the way for further studies to extend these findings into clinical settings.

5 Conclusion

This study has successfully demonstrated the setup and effective data recording of a simulated procedure in Senegal, leveraging open-source tools to compute validated metrics for skill assessment using tracking. The significant differences observed between novices and experts affirm the reliability of these metrics in evaluating percutaneous nephrostomy skills.

These findings are particularly impactful for medical training in low-resource settings such as Senegal. By utilizing accessible technologies and establishing robust evaluation methods, this research lays a foundation for improving surgical training outcomes in regions where resources are limited. The application of open-source tools not only enhances affordability but also promotes sustainable medical education practices, fostering skill development among healthcare providers in challenging environments.

Acknowledgments. R. Hisey is supported by the NSERC - Canada Graduate Scholarship (Doctoral) and by the FAS International Research Collaborations Fund and Alfred Bader Fellowship in Memory of Jean Royce from Queen's University. G. Fichtinger is supported as an NSERC Canada Research Chair in Computer-Integrated Surgery.

Disclosure of Interests. The authors have no competing interests to declare that are relevant to the content of this article.

References

1. Eugène, K.: Service néphrologie de l'hôpital Aristide LE DANTEC: 20 % des malades hospitalisés souffrent d'insuffisance rénale aiguë (2013). http://www.santetropicale.com/actus. asp?id=15692&action=lire
2. Nor, F.S.M., Draman, C.R., Seman, M.R., Manaf, N.A., Ghani, A.S.A., Hassan, K.A.: Clinical outcomes of acute kidney injury patients treated in a single-center, sub-urban satellite hospital. Saudi J. Kidney Dis. Transplant. **26**(4), 725 (2015). https://doi.org/10.4103/1319-2442.160273
3. Hausegger, K.A., Portugal, H.R.: Percutaneous nephrostomy and antegrade ureteral stenting: technique—indications—complications. Eur. Radiol. **16**, 2016–2030 (2006)
4. Bah, O.R.: percutaneous ultrasound-guided nephrostomy in the urology department of Conakry University Hospital. African J. Urology Androl. **1**(9) (2018)
5. Poursartip, B., LeBel, M.E., Patel, R.V., Naish, M.D., Trejos, A.L: Analysis of energy-based metrics for laparoscopic skills assessment. IEEE Trans. Biomed. Eng. **65**(7), 1532–1542 2018
6. Brown, J.D., et al.: Using contact forces and robot arm accelerations to automatically rate surgeon skill at peg transfer. IEEE Trans. Biomed. Eng. **64**(9), 2263–2275 (2017)
7. Kim, T. S., et al.: Objective assessment of intraoperative technical skill in capsulorhexis using videos of cataract surgery. Int. J. Comput. Assist. Radiol. Surgery. **14**(6), 1097–1105 (2019)
8. Sharon, Y., et al.: Rate of orientation change as a new metric for robot-assisted and open surgical skill evaluation. IEEE Trans. Med. Robot. Bionics. **3**(2), 414–425 (2021)
9. Wang, Y., et al.: Differentiating operator skill during routine fetal ultrasound scanning using probe motion tracking. Medical Ultrasound, and Preterm, Perinatal and Paediatric Image Analysis: First International Workshop, ASMUS 2020, and 5th International Workshop, PIPPI 2020, Held in Conjunction with MICCAI 2020, Lima, Peru, October 4–8, 2020, Proceedings 1. Springer International Publishing (2020)
10. Clinkard, D., et al.: Assessment of lumbar puncture skill in experts and nonexperts using checklists and quantitative tracking of needle trajectories: implications for competency-based medical education. Teach. Learn. Med. **27**(1), 51–56 (2015)
11. Rajan, P., et al.: Design and evaluation of an educational device for ultrasound-guided interventional procedures. In: Proceedings of SPIE 12928, Medical Imaging 2024: Image-Guided Procedures, Robotic Interventions, and Modeling, 129280T (2024)
12. Fedorov, A., et al.: 3D slicer as an image computing platform for the quantitative imaging network. Magnet. Resonance Imaging. **30**(9), 1323–41 (2012)
13. Ungi, T., et al.: Perk tutor: an open-source training platform for ultrasound-guided needle insertions. IEEE Trans Biomed Eng. **59**(12):3475–81 (2012). https://doi.org/10.1109/TBME. 2012.2219307. Epub 2012 Sep 17. PMID: 23008243
14. Keri, Z., et al.: Computerized training system for ultrasound-guided lumbar puncture on abnormal spine models: a randomized controlled trial. Can. J. Anaesth. **62**(7), 777–784 (2015)
15. Holden, M.S., Keri, Z., Ungi, T., Fichtinger, G.: Overall proficiency assessment in point-of-care ultrasound interventions: the stopwatch is not enough. In: BIVPCS 2017 and POCUS 2017, Held in Conjunction with MICCAI 2017, Québec City, QC, Canada, September 14, 2017, Proceedings, pp. 146–153. Springer International Publishing (2017)
16. Lia, H., et al.: Training with Perk Tutor improves ultrasound-guided in-plane needle insertion skill. In: Medical Imaging 2017: Image-Guided Procedures, Robotic Interventions, and Modeling, vol. 10135, pp. 232–239. SPIE, March 2017
17. Lasso, A., Heffter, T., Rankin, A., Pinter, C., Ungi, T., Fichtinger, G.: PLUS: open-source toolkit for ultrasound-guided intervention systems. IEEE Trans Biomed Eng. **61**(10), 2527–37 (2014). https://doi.org/10.1109/TBME.2014.2322864. Epub 2014 May 9. PMID: 24833412; PMCID: PMC4437531

Automatic Segmentation of Medical Images for Ischemic Stroke in CT Scans for the Identification of Sulcal Effacement

Wahabou K. Taba Chabi[1,2,3]([✉]), Sèmèvo Arnaud R. M. Ahouandjinou[1,2,3], Adoté François Xavier Ametepe[1,2,3], and Probus A. F. Kiki[1,3]

[1] Doctoral School of Engineering Sciences (ED-SDI), University of Abomey-Calavi (UAC), Abomey-Calavi, Benin
betiwahab@gmail.com
[2] Laboratory for Research in Computer Science and Applications (LRSIA), Institute for Training and Research in Computer Science (IFRI), University of Abomey-Calavi (UAC), Abomey-Calavi, Benin
[3] Polytechnic School of Abomey-Calavi, LETIA, University of Abomey-Calavi, Abomey-Calavi, Benin

Abstract. Ischemic stroke is a leading cause of morbidity and mortality, requiring rapid and accurate detection for effective treatment. Segmentation of sulcal effacements in CT scan images is crucial for identifying acute strokes. In this paper, we propose a segmentation method that combines the Gray Level Co-occurrence Matrix (GLCM) and the Support Vector Machine (SVM) algorithm to enhance lesion identification efficiency. To achieve this, we developed a two-step approach integrating texturization and classification techniques to segment sulcal effacements in CT scan images. First, CT scan images are normalized and filtered to reduce noise and improve structural clarity, followed by segmentation to identify lesion areas. GLCM is then applied to each region of interest in the CT images to extract textural features such as contrast, correlation, energy, and homogeneity, which describe the properties of brain tissue. These features are used as input for the SVM algorithm, which classifies the effacements by determining the decision boundaries that separate different pixel groups in the feature space. The method is validated on a dataset annotated by experts, with segmentation performance evaluated using accuracy metrics. The results show that integrating GLCM and SVM improves the accuracy of sulcal effacement segmentation by up to 81% compared to traditional methods. The ROC curve of our model indicates good performance, demonstrating more precise detection of ischemic lesions.

Keywords: medical images · image segmentation · CT scanner · ischemic stroke · sulcal effacement

1 Introduction

Each year, more than 15 million people suffer from stroke, resulting in 5 million deaths and 5 million cases of permanent disability, representing a 75% risk rate [1]. Strokes are categorized into ischemic strokes (or cerebral infarctions) and hemorrhagic strokes [2,

3]. Early diagnosis during the acute and subacute phases is crucial for effective treatment. As the saying goes, "time is brain"; fast identification and early treatment can reduce mortality by 30% and mitigate the severity of lesions [4, 5]. CT scans play a vital role in evaluating symptomatic carotid stenosis [6], allowing differentiation between hyperdense hemorrhagic strokes and hypodense ischemic strokes [7]. However, detecting sulcal effacement can be challenging or impossible with this method. Magnetic Resonance Imaging (MRI) is essential for detecting acute ischemic strokes, offering visualization of abnormal tissues and detection of small lesions across all spatial planes [8]. In some countries in sub-Saharan Africa, particularly in Benin, the limited access to advanced medical infrastructure, such as CT scanners and MRI machines, complicates the rapid and accurate diagnosis of strokes. This situation is further exacerbated by the lack of qualified neurology specialists, making the interpretation of exams and the appropriate treatment of patients particularly challenging. To overcome this issue, it is essential to implement an automatic detection system for the diagnosis of acute and subacute ischemic strokes. The goal of this paper is to propose a new method for the segmentation of scanner images and to extract the most relevant textures. During the experiments, a high-performing diagnostic model was developed using specific features to classify brain scanner images. The extracted texture features were integrated into an SVM algorithm to train our model. The optimal parameters of the SVM classifier were adjusted, achieving a maximum performance of 81% in the classification of ischemic strokes.

The remainder of this paper is organized as follows. Beginning with a concise introduction, the Review of Image Segmentation Techniques for Brain Stroke Scans is presented in Sect. 2, Sect. 3 describes the proposed method, and the dataset and experimental results are described in Sect. 4. Section 5 addresses the discussion of the results, and Sect. 6 presents the research conclusion.

2 A Review of Image Segmentation Techniques for Brain Stroke Scans

Studies on acute ischemic stroke segmentation methods based on CT imaging could serve as a reference for developing more effective and efficient stroke detection systems. These systems could assist medical experts in their decision-making, thereby improving the quality and accuracy of **diagnoses. In** recent years, numerous studies have been conducted to identify strokes from CT scan images. Deep learning models based on convolutional neural networks (CNNs) and machine learning methods have been widely used in these works. For example, Ezequiel de la Rosa et al. [9] proposed a model integrating algorithms from different team members of the ISLES'22 challenge. This model achieved a median Dice score of 0.82 and a median lesion F1 score of 0.86, excelling in the extraction of clinical biomarkers such as lesion types and affected vascular territories. Khushboo Verma et al. [9] developed an automatic segmentation algorithm for chronic stroke lesions on T1-weighted MRI images using a deep neural network (DNN), achieving an average Dice similarity coefficient (DSC) of 0.65. Hao Yang et al. [10] proposed IS-Net, an encoder-decoder convolutional neural network for automatic segmentation of ischemic stroke lesions on NCCT images, which compared favorably to state-of-the-art segmentation models with high scores in segmentation criteria and sensitivity. Liangliang Liu et al. [9] proposed a new deep residual attention network (DRANet) for simultaneously segmenting and quantifying ischemic strokes and white matter hyperintensities (WMH) in MRI images. This model, based on the U-net design, applies a novel attention module to extract high-quality features from input images, enhancing DRANet's predictive performance. During the ISLES 2022 challenge, the team of Md Mahfuzur Rahman Siddiquee et al. [10] proposed a method that resamples all images to a common resolution, uses two input MRI modalities (DWI and ADC), and trains a SegResNet semantic segmentation network from MONAI. This solution achieved first place in the challenge in terms of the Dice metric (0.824). Duwei Dai et al. [14] proposed a dual-path U-Net model named I2U-Net. This model encourages the reuse and re-exploration of historical information through rich information interaction between the two paths, allowing deep layers to learn more comprehensive features containing both low-level detail descriptions and high-level semantic abstractions. Extensive experiments on four challenging tasks, including skin lesion, polyp, brain tumor, and multi-organ abdominal segmentation, consistently show that the proposed I2U-Net has superior performance and generalization capabilities compared to other state-of-the-art methods.

This state-of-the-art review provides an overview of current methods and advances in ischemic stroke image segmentation, utilizing deep learning techniques and data from various medical imaging sources. Table 1 presents the top four image segmentation methods along with the performances achieved by the authors.

Table 1. The comparative table of state-of-the-art segmentation methods for ischemic stroke

Author	Model/Approach	Average Dice Score	Other Performances	Strengths
Ezequiel de la Rosa et al. [6]	**Model integrating algorithms from the ISLES'22 team**	0.82 (median)	Lesion F1: 0.86 (median)	Excellent extraction of clinical biomarkers
Khushboo Verma et al. [7]	**Automatic segmentation algorithm using a DNN**	0.65 (average)		Automatic segmentation of chronic lesions
Hao Yang et al. [9]	**IS-Net, an encoder-decoder convolutional neural network**		High scores in segmentation criteria and sensitivity	High performance compared to state-of-the-art models
Md Mahfuzur Rahman Siddiquee et al. [10]	**SegResNet from MONAI**	0.824	1st place in ISLES 2022 challenge	Use of two input MRI modalities, common resampling

3   Proposed Method

This article presents an automatic segmentation model of CT scan images used to detect sulcal effacements. To diagnose acute phase strokes, SVM classifiers were employed. CT scan images were categorized into two main groups: diseased and non-diseased. Using the Gray Level Co-occurrence Matrix (GLCM) texture analysis method [14], essential features such as Contrast, Correlation, Energy, and Homogeneity were extracted and labeled. These features were used to train our models. The performance of the proposed system was evaluated using precision scores obtained from this experimental study.

The described system consists of four stages as shown in Fig. 1: data preprocessing, texture feature extraction, feature selection, model training, and classification.

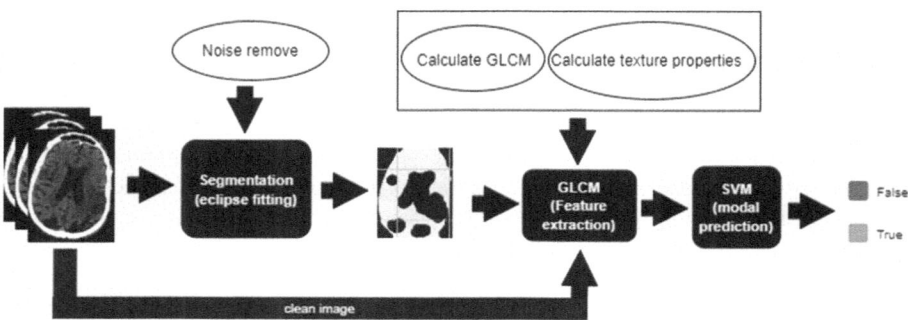

Fig. 1. descriptive diagram

3.1   Pre-processing Phase

Thus, a brain image with a black background and sharper contours, where areas with sulcus effacements are identified. The algorithm for the proposed preprocessing phase is illustrated in Algorithm 1:

Algorithm 1 Proposed ellipse algorithm

Input: Input matrice E with size of h, w
Output: Output matrice E
E[h, w] = 0
c_y = h /2
c_x = w / 2
for i = 0 to h-1 do
    for j = 0 to w-1 do
      if $(((i - c_y) / (h / 2))^2 + ((j - c_x) / (w / 2))^2)$ <= 1 do
        E[i, j] = 1
      endif
    end
end

alg.1: The pseudo-code of the proposed pre-processing.

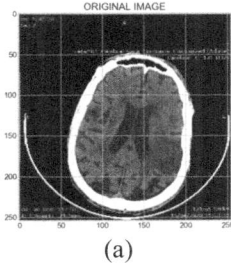
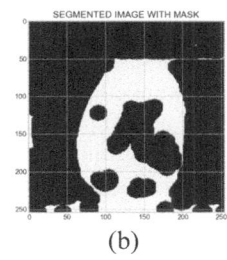
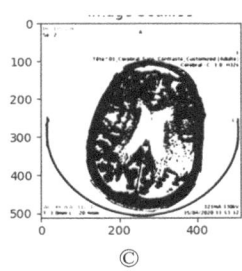

(a)   (b)   ©

The images (b) and (c) are the results of the segmentation of our original image (a), in which we observe the presence of sulcus effacement. The dark black areas in image (b) outside the ventricles represent the sulcus effacement obtained after segmentation. The same areas appear white in image (c) using the thresholding method.

3.2 Feature Extraction Phase

The extraction of textural properties using the GLCM algorithm allowed us to capture relevant and significant information about sulcus effacements present in the images. By combining these features with machine learning techniques such as SVM, we have built an effective model for recognizing sulcus effacements.

3.3 Feature Selection Phase

In the feature selection phase, it is crucial to choose the best subset of specific features related to sulcus effacements from pixels. In this study, the goal is to determine optimal features from the combined deep features obtained during the feature extraction step using the ellipse algorithm. The ellipse algorithm [21] is an image-processing method used to detect and segment regions of interest approximated by ellipses. It typically relies on thresholding and edge detection techniques to identify object contours in an image. It then fits ellipses to these detected contours by minimizing the fitting error, effectively delineating specific areas in an image, such as those exhibiting particular features or specific anatomical structures, as seen in brain CT scans for identifying anomalies or pathologies.

In the classification phase, we employed the SVM classifier to train the selected deep features. It is important to note that SVMs are chosen for their ability to create robust models, handle nonlinear data, and be effective in high-dimensional spaces, while providing good control over model complexity.

4 Results and Discussion

The classification model proposed in this study was developed using the PyCharm software. Additionally, employing a random split approach, 80% of the data was used for training and the remainder for testing. The code for the proposed model is shared at https://github.com/betiwahab/Brain-stroke-ct-scan. Experimental results and the dataset are explained in the following subsections.

4.1 Result

To evaluate our model, we used 254 randomly selected raw images classified as diseased or non-diseased. These images underwent ellipsoidal segmentation to remove irrelevant areas, with a minimum threshold set at 100 pixels. After segmentation, the processed images were subjected to the GLCM algorithm to extract features such as contrast, dissimilarity, homogeneity, energy, correlation, and ASM. The extracted features were used to label our dataset before being input into SVM and KNN models. Our experimental results yielded a confusion matrix showing our method achieved an 81% accuracy, demonstrating the effectiveness of our model in identifying sulcus effacements. ROC curve analysis was used to evaluate the overall performance of the model (Figs. 2 and 3).

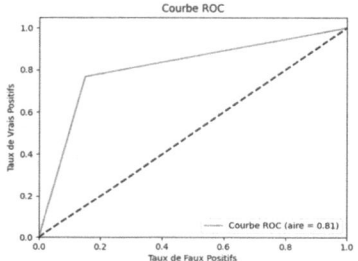

Fig. 2. Receiver Operating Characteristic   **Fig. 3.** Learning curve (ROC) Curve

4.2 Discussion

Our proposed image segmentation model based on SVM to identify sulcus effacements was compared to existing systems in the literature, and the results are presented in Table 2.

Table 2. Positioning table of our model in the article review model

Reference	Feature extraction	Classification	Accuracy
Md Mahfuzur Rahman Siddiquee et al. [10]	SegResNet	SegResNet	82.4%
Ezequiel de la Rosa et al. [6]	CNN	SOFTMAX	82.0%
Proposed model	GLCM	SVM	81%
Khushboo Verma et al. [8]	DNN	DNN	65.0%

Table 2 synthesizes previous research and presents the precision scores obtained. Among these studies, traditional machine learning methods were used by Ezequiel de la Rosa et al. [6], Khushboo Verma et al. [7], and Md Mahfuzur Rahman Siddiquee et al. [10] for classifying brain diseases, employing feature extraction techniques such as CNN, DNN, and SegResNet respectively. Although our model ranks third in the leaderboard ahead of pre-trained methods, it remains an effective method for segmenting sulcus effacement in ischemic stroke images.

In conclusion, our study, based on classical learning methods for ischemic stroke classification, demonstrated comparable performance to studies using pre-trained models, as indicated in Table 3. Additionally, our proposed model's precision score closely matches that obtained by Ezequiel de la Rosa et al. [6], as shown in Table 3. The advantages and limitations of the proposed model are as follows:

Advantages:

- The proposed model identified significant and high-performance features using a selection algorithm after impact zone segmentation.
- Sulcus effacements were classified using the SVM classifier.
- The proposed model achieved precision very close to that of pre-trained deep architectures.

Limitations:

- One limitation of this study could be the optimal search for selected features using the ellipse algorithm.

5 Conclusion

In this study, we proposed a model based on the SVM classifier and a feature selection method using the ellipse algorithm for ischemic stroke classification. To achieve our goal, we utilized a clinical dataset of ischemic stroke consisting of 254 CT scan images. The proposed method is divided into three phases. In the first phase, we preprocessed the images by segmenting impact zones using the ellipse algorithm after a noise removal step. In the second phase, the GLCM method was employed to extract distinctive and significant features, followed by labeling the images. Finally, the quantitative data obtained were used to train the SVM classifier. Experimental results demonstrate that the proposed model achieved an accuracy score of 81% for acute ischemic stroke classification. For future work, we plan to develop a mobile application aimed at aiding healthcare professionals in detecting acute ischemic strokes and other diseases. Additionally, we will explore other models based on Bayesian capsules, machine learning classifiers, and attention modules to assist in the classification of acute and subacute ischemic strokes.

References

1. https://www.emro.who.int/index.html
2. Dede n'dri S., et al.: Diagnostic précoce de l'ischémie cérébrale par IRM de diffusion : premieres experiences en Côte d'Ivoire. J. Africain d'Imagerie Médicale (2023)
3. American Heart Association. https://www.stroke.org/en/help-and-support/resource-library/ lets-talk-about-stroke/carotid-endarterectomy, Accessed 19 June 2024
4. Unité de Chirurgie Carotidienne de l'Institut Mutualiste Montsouris, https://www.carotide. com/scanner-cerebral, Accessed 16 June 2024
5. EL Machkour, M., et al.:Imagerie de l'accident vasculaire cérébral ischémique à la phase aigue. le journal marocain de cardiologie III (2011)

6. Ezequiel de la Rosa et al.: A robust ensemble algorithm for ischemic stroke lesion segmentation: generalizability and clinical utility beyond the ISLES challenge
7. Verma, K., et al.: Automatic Segmentation and Quantitative Assessment of Stroke Lesions on MR Images", Diagnostics 2022; https://doi.org/10.3390/diagnostics12092055;
8. Yang, H., e al.: IS-Net: Automatic ischemic stroke lesion segmentation on CT images. Date of Publication: 20 February 2023 https://doi.org/10.1109/TRPMS.2023.3246496
9. Liu, L., et al.: Attention convolutional neural network for accurate segmentation and quantification of lesions in ischemic stroke disease. Med Image Anal 2020 Oct:65:101791. https://doi.org/10.1016/j.media.2020.101791. Epub 18 July 2020
10. Siddiquee, M.M.R., et al.: Automated ischemic stroke lesion segmentation from 3D MRI ISLES 2022 challenge report (2022)
11. de Langlard, M., et al.: Méthode d'analyse d'image pour la reconnaissance d'objets elliptiques se superposant : application à des images d'écoulement diphasique; Submitted on 28 February 2020 HAL Id: cea-02438728. https://cea.hal.science/cea-02438728

AfriBiobank: Empowering Africa's Medical Imaging Research and Practice Through Data Sharing and Governance

Lukman Enegi Ismaila[1,2,3]([✉]), Houcemeddine Turki[3,4,5], Mohamed Frikha[6], Taliya Weinstein[3], Faith Hunja[3], Chris Fourie[3], and Steve A. Adeshina[7]

[1] Russell H. Morgan Department of Radiology and Radiological Science, Johns Hopkins School of Medicine, Baltimore, MD, USA
lismail1@jhu.edu

[2] F.M. Kirby Research Center for Functional Brain Imaging at Kennedy Krieger Institute, Baltimore, MD, USA

[3] SisonkeBiotik Research Community, Johannesburg, South Africa

[4] Data Engineering and Semantics Research Unit, University of Sfax, Sfax, Tunisia

[5] Faculty of Medicine of Sfax, University of Sfax, Sfax, Tunisia

[6] MIRACL Laboratory, University of Sfax, Sfax, Tunisia

[7] Nile University of Nigeria, Abuja, Nigeria

Abstract. Medical imaging plays a crucial role in healthcare, yet access to sufficient datasets, particularly in Africa, remains limited. This article discusses the challenges and opportunities surrounding medical imaging data sharing and governance in Africa. Drawing on existing research works, we explore the significance of medical imaging data, the limitations in availability, and the factors contributing to this challenge. We discuss the importance of data governance policies, the establishment of medical biobanks, and active steps towards the implementation of federated learning techniques to improve data availability and governance. Furthermore, we propose a medical imaging biobank for Africa (AfriBiobank). To enhance data interoperability, integration, and advanced analysis capabilities, we advocate for the incorporation of a semantic layer utilizing technologies such as semantic web standards and query languages, alongside medical data standards such as *FHIR* and *DICOM*, as well as adherence to linked data principles within the proposed biobank infrastructure. Furthermore, to support the storage and accessibility of data across multiple African countries, we envisage a hardware framework founded on a distributed architecture, leveraging blockchain technology for enhanced cybersecurity measures, alongside the establishment of a robust data center to ensure the scalability of the biobank's data storage capabilities. By addressing these challenges and implementing innovative solutions, Africa can unlock the full potential of scientific advancements in healthcare.

Supplementary Information The online version contains supplementary material available at https://doi.org/10.1007/978-3-031-79103-1_20.

Keywords: medical imaging data · data sharing · data governance · federated learning · Biobank

1   Introduction

Medical imaging data is essential for understanding health outcomes, diagnosing diseases, and informing treatment decisions. However, access to sufficient datasets, especially in Africa, remains a significant challenge. Limited interoperability, insufficient health information infrastructure, and ethical and regulatory compliance issues further exacerbate this challenge [28,29]. While it is not always the front-line concern on the global stage, Africa faces a more delayed version of equity in science. Promoting AI and medical prediction transparency is crucial for bridging the current gap in translating technical innovation and performance gains into tangible improvements in African healthcare service. Sufficient medical data availability to carry out innovative research in Africa remains a key setback. It is without surprise that the Medical Image Computing and Computer-Assisted Interventions (MICCAI) and other related communities implement deliberate steps in enforcing open science models by requesting reviewers to consider whether authors will release code and share links to datasets used as part of the acceptance criteria. While interesting collations of medical data exist around the world, for example, UK Biobank[1], Alzheimer's Disease Neuroimaging Initiative (ADNI)[2], Multimodal Brain Tumor Image Segmentation Benchmark (BraTS) [19], etc., we are unaware of any fully developed medical imaging databases of comparative scale in Africa. Open datasets typically consist of structured digital data available for public use, focusing on data format and accessibility, with fewer ethical restrictions. In contrast, biobanks manage biological samples and associated health data, requiring specialized storage, ethical considerations, and detailed documentation. Access to biobank data is controlled and requires compliance with strict ethical and legal protocols.

The limited availability of medical imaging data is attributed to various issues, including barriers in cross-border transfers of personal data, limited funding, poor data management, lack of computational infrastructure, and cultural and ethical considerations [2,29]. These challenges hinder the transformation of healthcare services. Efforts to improve data-sharing practices and policies have garnered interest from various groups across Africa, advocating for establishing well-articulated data policies, fostering a culture of data sharing, and investing in infrastructure [20]. Additionally, the need for representative datasets of endemic African diseases and the inclusion of rare diseases necessitates the development of the AfriBiobank. African researchers could lead cutting-edge medical research through this initiative by accessing sufficient data resources. Collaboration among regional data silos is crucial to address this challenge and unlock the full potential of medical research in Africa.

This study proposes the following contributions:

---

[1] https://www.ukbiobank.ac.uk/.

[2] https://adni.loni.usc.edu/.

1. An identification of the barriers to availability and open sharing of medical imaging data in Africa.
2. A novel approach of a multistakeholder AfriBiobank for medical imagery data, encompassing technical development and governance factors.
3. Thorough overview of requisite software-level implementation for the AfriBiobank, with a brief discussion around possible hardware suggestions.

While this study provides a holistic description of theoretical steps in achieving the proposed AfriBiobank, the proposal demonstrates an implementation plan based on case reports and an already completed research study. This previous effort guided our work in proposing a multi-stakeholder solution. Our study follows an exploratory approach to describe our observation of the problem with a presentation of a sustainable solution, we envisage a more practical approach in our follow-up study to demonstrate a pilot project and share strategic observation and direction for Biobank implementation in the African region.

2   Related Works

2.1   Previous Availability and Accessibility of Medical Datasets

Efforts towards the implementation of a biobank have been employed to establish comprehensive, reputable, and ethical biobanks in Africa, namely through the Human, Heredity, and Health in Africa (H3Africa)[3] initiative. One notable example of this success has been the Stroke Investigative Research and Educational Network (SIREN) project which has compiled environmental and genetic data accounting for stroke characteristics within African populations across sites in Nigeria and Ghana [3]. This SIREN biobank includes over 3000 brain images. Other African biobanks, primarily located in South Africa, include the Centre for Human Metabolomics (CHM) at the North-West University, which pioneered Africa's inaugural biobank dedicated to rare diseases (RDs) [8], among other public and private biobanks [1], which rely on a variety of biological samples, including blood and pathological samples.

High-income countries benefit from open science in medical imaging inspired by artificial intelligence, thanks to large scale medical image data availability and accessibility of large, public image datasets. However, a gap remains in establishing a comprehensive image biobank to address local needs within Africa [28]. Additionally, initiatives like H3Africa, SIREN, RDs, and others currently implement static medical databases which have resulted in several setbacks regarding scalability, security, and high-volume access [7].

Utilising the technical groundwork established through other African biobanks, which include a sub-imaging base, and a literature review on the current state-of-the-art federated learning and semantic infrastructure, we propose a new medical imaging biobanking system. Our proposal focuses on a more robust implementation of a medical biobank capable of serving as a high-quality

---

[3] https://h3africa.org/.

sustainable data source across the African region with adaptive infrastructure capable of handling large scale data flow and scalability. Our approach envisions the incorporation of advanced technologies such as dynamic data integration, blockchain for enhanced security, and machine learning algorithms to ensure continuous data validation and improvement.

In addition, the ethical, legislative, and regulatory guidelines established from previous African biobank creations have been extensively reviewed. This will ensure our proposed biobank will adhere to the highest ethical standards and regulatory requirements, ensuring the confidentiality and privacy of patient information.

2.2   Ontology Engineering for Biobanking

Ontologies, as a semantic web technology, has been found useful in retrieving data from biobanks. Ontology is used to enhance the interoperability of heterogeneous data. An ontology is a structured vocabulary of general terms (such as "cell", "image", "tissue", or "microscope") [27], representing the types of entities within the domain of reality the ontology aims to cover. These terms are equipped with logical definitions and, thus, support data reasoning.

Ontology framework has been demonstrated as a useful support tool for biobanking implementation [27]. These ontologies are especially relevant because the entities crucial for managing biobanks, including specimens and methods for preparing them like staining, are also relevant to imaging. In the two cases, it is necessary to associate data and the corresponding entities (samples, images) described by this data with demographic and other patient-related data.

Establishing shared infrastructure in biobanking is a major collaborative effort. Some initiatives were taken to create shared infrastructure including the Biobanking and Biomolecular Resources Research Infrastructure [11], which addresses the challenges of data integration stemming from the use of diverse data representations by different groups. This effort has led to the development of a proposed shared terminology for biobanking, as outlined by Fransson et al. (2014) [9], and the establishment of the minimum information about biobank data sharing standard [21], which is part of the minimal information about a biological or biomedical investigation checklist [30]. Brochhausen et al. (2013) [5] have gone a step further by introducing an ontology called Ontologized Minimum Information About Biobank Data Sharing (OMIABIS). This ontology is designed to enable reasoning across the data involved in biobank administration. It provides resources for representing various entities relevant to obtaining information, such as the types of samples stored in a biobank, information about the biobank operators or owners, contact persons, and more. Additionally, OMIABIS incorporates representations of different sample types[4]. Thus, OMIABIS allows Enhanced Interoperability and Data Standardization, Improved Data Quality and Compliance, and Facilitating Advanced Data Analysis and Research. Our

---

[4] The OMIABIS Development Project is available at https://github.com/OMIABIS/omiabis-dev.

proposal uses OMIABIS rather than other ontologies for Biobanking, including OBIB [6], because, while OBIB (Ontology for Biobanking) and other biobanking ontologies provide a useful framework for the management and operation of biobanks, OMIABIS offers a more focused and specialized approach to data sharing and interoperability. Its comprehensive documentation practices and alignment with international standards make it the preferred choice for biobanks aiming to enhance their data-sharing capabilities, ensure ethical compliance, and participate in global research initiatives.

3   Recommendation for Data Sharing

Building large, representative, and comprehensive medical imaging databases is time-consuming and expensive, often resulting in datasets becoming commercially valuable and reducing the incentive to share data, even when they meet safety, privacy, and cultural sensitivity requirements [24]. To address this, we recommend establishing national biobanks for medical imaging data under common ethical oversight. This includes ethics approval from all institutional bodies and broad informed consent forms signed by participants [3], either in their language or with the help of a translator to ensure understanding.

Collaboration between technical experts from academia and industry is essential for developing standards, tools, and infrastructure for responsible data sharing and analysis, including image de-identification. Training programs in data science can empower more researchers to use, collaborate with, and contribute to the AfriBiobank. Strict data protection laws limit cross-border data sharing [24], and limited funding and poor data management practices result in small, fragmented datasets [28].

Federated learning offers a solution to the privacy limitations of healthcare data by transferring only model features, ensuring patient information remains secure while maintaining model performance [8]. Implementing federated learning techniques allows collaborative modeling without institutions relinquishing control of their data. Revising cross-border data transfer policies and fostering a culture of open science are crucial for unlocking Africa's research potential. Regional bodies should draft recommendations to harmonize regulations within economic zones, and standardizing legal frameworks is recommended to ensure uniform governance of biobanks throughout Africa [29]. Lastly, Funding agencies must mandate open data sharing with proper privacy protections. Governance reforms, technology, and infrastructure investments can accelerate Africa's contributions to global health.

4   Proposed System

The AfriBiobank's success will be governed by agreed compliance across its multiple entities or stakeholders (see Supplementary Fig. 1). Institutions retain complete control of their patient database yet require an investment in upscaling their computing infrastructure from healthcare software developers and hardware manufacturers. This initiative promises significant benefits for Government and Non-government agencies, hospitals, research institutions, healthcare product manufacturers, clinicians, and patients, leveraging advanced data sharing and machine learning technologies. Practical engagement will be required to achieve cross-regional understanding as described in Fig. 1 towards the realization of a multisectoral collaboration to provide an enabling environment for high-quality data creation and storage which could improve overall availability in Africa. This could be accomplished through the influence of data-sharing policies, the development of sustainable data policies, and the establishment of our proposed medical imaging AfriBiobanks. The establishment of this AfriBiobank and the adoption of federated learning techniques, ensuring data privacy while enabling institutional collaboration, are proposed as key initiatives to address data availability and governance challenges.

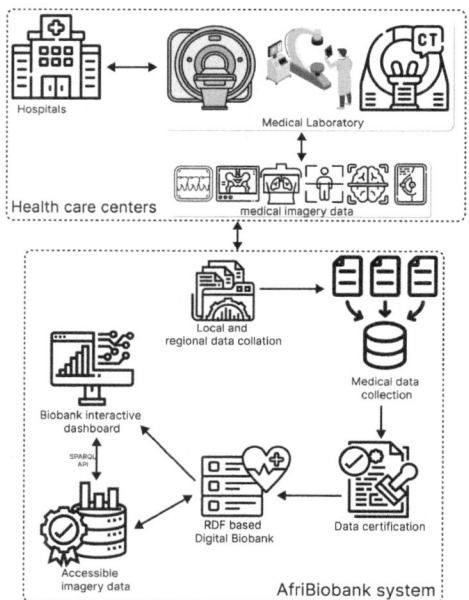

Fig. 1. AfriBiobank inner components showing the flow of data from creation to RDF based biobank and other critical processes.

4.1   Semantic Layer

The semantic layer of the proposed medical imaging AfriBiobank employs various technologies and standards to manage, integrate, and make data accessible. It ensures interoperability by adhering to RDF and Linked Data principles, enabling seamless data integration [13]. OWL ontologies standardize vocabularies across medical sources, facilitating semantic alignment for effective integration and analysis [13]. APIs aid in collecting and aggregating medical data from diverse sources [18]. The semantic layer uses standardized ontologies, such as DICOM and FHIR, to certify and annotate medical data [23,27]. These ontologies, available in OWL format, can be integrated into medical knowledge graphs, enhancing the biobank's utility [23,27]. The system can be internationalized using rdfs:label statements to annotate DICOM and FHIR properties, increasing versatility across regions [32]. By adhering to FAIR principles, the biobank ensures data is discoverable, accessible, and usable [12]. RDF metadata annotates images, making them findable and interoperable, crucial for medical decision support systems [25]. Each image is annotated with RDF triples, capturing essential metadata like patient information and anatomical location [13]. SPARQL queries and APIs allow users to retrieve and visualize data, enabling intuitive exploration as shown in Fig. 2 [32]. Efficient RDF database storage involves choosing a suitable DBMS, like Apache Jena or Virtuoso, and using optimization techniques like partitioning and indexing [17,31].

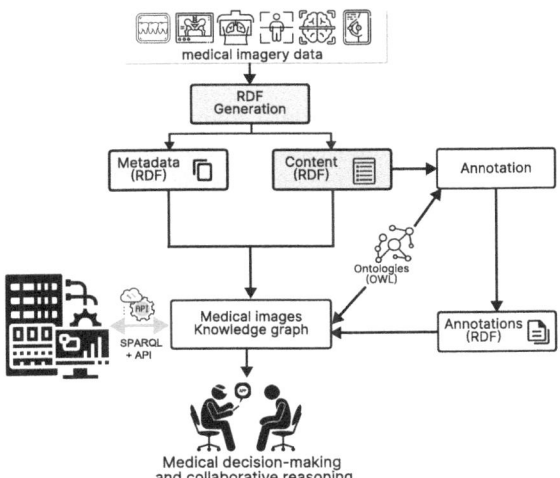

Fig. 2. Representation of the Semantic Layer of the African Medical Imaging Biobank

4.2 Hardware and Security Layer

Robust backup mechanisms, scalability considerations, and security measures are crucial for maintaining data integrity and guarding against potential threats in a medical imaging biobank [14,26]. The hardware infrastructure, including servers, storage systems, network infrastructure, imaging devices, workstations, and backup systems, ensures reliable data storage, efficient transfer, and secure access [14]. Implemented security measures, such as blockchain technology for data integrity, smart contracts for access control, encryption for secure storage, multi-factor authentication for user verification, secure communication protocols for data transmission, regular security audits, penetration testing for vulnerability identification, and data anonymization and pseudonymization for patient privacy, effectively safeguard sensitive patient data [26]. Integrating these security measures bolsters the overall security posture of the biobank, shielding patient data from unauthorized access, tampering, and disclosure, while regular maintenance tasks sustain system efficiency, enabling seamless integration with other data sources and empowering advanced querying and analytical capabilities [14, 26].

4.3   Challenges and Considerations

Establishing the AfriBiobank for biomedical imaging in Sub-Saharan Africa faces challenges, particularly in improving imaging data quality, as 40% of MRI scans are outdated [4]. Addressing these issues requires better training for maintenance staff, funding for new devices, and exploring affordable MRI options like low-field devices [20]. Public participation can be encouraged through incentives and advocacy. Techniques like transfer learning and AI-driven diagnostic tools can mitigate imaging data shortages, enhancing early disease detection and treatment across Africa [22, 28].

Proper staff training in imaging data management and ethics is essential, with educational programs emphasizing consent and privacy [10]. Sustainable financial models, incorporating grants and partnerships, are crucial for long-term operations [10]. Collaborations with other biobanks and research institutions can enhance capabilities. Robust data management systems must ensure data integrity, security, and compliance with regulations like GDPR and HIPAA [10]. Standardizing data collection and ensuring secure, ethical data access are key to maximizing the biobank's impact [10].

Disaster recovery planning requires identifying risks and developing comprehensive plans, including off-site backups [15]. Integrating the biobank with existing healthcare and research infrastructures is vital for streamlined operations and compliance with regulations [16]. Addressing ethical concerns like informed consent and maintaining public trust through transparent communication are essential [16].

5   Conclusion

Unlocking the full potential of scientific advancements in healthcare in Africa requires breaking barriers and fostering collaboration among stakeholders. Data sharing plays a profound role in health research, and initiatives such as medical imaging biobanks and federated learning can facilitate improved data availability and governance. By embracing these initiatives and investing in credible scientific research, Africa's health sector can harness the power of medical data to drive innovation and improve healthcare outcomes.

Acknowledgements. The authors would like to thank *Datasphere Initiative* for their support during Dr. Lukman Fellowship and Dr. Stephen E. Moore for his review of this work.

References

1. Abayomi, A., Christoffels, A., Grewal, R., Karam, L.A., et al.: Challenges of biobanking in South Africa to facilitate indigenous research in an environment burdened with human immunodeficiency virus, tuberculosis, and emerging non-communicable diseases. Biopreservation Biobanking **11**(6), 347–354 (2013)
2. Adebamowo, C.A., Callier, S., Akintola, S., Maduka, O., et al.: The promise of data science for health research in Africa. Nat. Commun. **14**(1) (2023)
3. Akinyemi, R.O., Akinwande, K., Diala, S., Adeleye, O., et al.: Biobanking in a challenging African environment: unique experience from the SIREN project. Biopreservation Biobanking **16**(3), 217–232 (2018)
4. Anazodo, U.C., Ng, J.J., Ehiogu, B., Obungoloch, J., et al.: A framework for advancing sustainable magnetic resonance imaging access in Africa. NMR Biomed. **36**(3), e4846 (2023)
5. Brochhausen, M., Fransson, M.N., Kanaskar, N.V., Eriksson, M., et al.: Developing a semantically rich ontology for the biobank-administration domain. J. Biomed. Semant. **4**(1), 23 (2013)
6. Brochhausen, M., et al.: OBIB-a novel ontology for biobanking. J. Biomed. Semant. **7**(1), 23 (2016)
7. Choudhury, A., Sengupta, D., Aron, S., Ramsay, M.: The H3Africa Consortium: publication outputs of a pan-African genomics collaboration (2013 to 2020), pp. 257–304. BRILL (2022)
8. Conradie, E.H., Malherbe, H., Hendriksz, C.J., Dercksen, M., Vorster, B.C.: An overview of benefits and challenges of rare disease biobanking in Africa, focusing on South Africa. Biopreservation Biobanking **19**(2), 143–150 (2021)
9. Fransson, M.N., Rial-Sebbag, E., Brochhausen, M., Litton, J.E.: Toward a common language for biobanking. Eur. J. Hum. Genet. **23**(1), 22–28 (2014)
10. Hallmans, G., Vaught, J.B.: Best practices for establishing a biobank, pp. 241–260. Humana Press (2010)
11. Harris, J.R., Burton, P., Knoppers, B.M., Lindpaintner, K., et al.: Toward a roadmap in global biobanking for health. Eur. J. Hum. Genet. **20**(11), 1105–1111 (2012)
12. Holub, P., et al.: Enhancing reuse of data and biological material in medical research: from FAIR to FAIR-health. Biopreservation Biobanking **16**(2), 97–105 (2018)
13. Hwang, K.H., Lee, H., Koh, G., Willrett, D., Rubin, D.L.: Building and querying RDF/OWL database of semantically annotated nuclear medicine images. J. Digit. Imaging **30**(1), 4–10 (2017)
14. Im, K., Gui, D., Yong, W.H.: An introduction to hardware, software, and other information technology needs of biomedical biobanks, pp. 17–29. Springer, New York (2018)
15. International Society for Biological and Environmental Repositories: Collection, storage, retrieval and distribution of biological materials for research. Cell Preserv. Technol. **6**(1), 3–58 (2008)

16. Larsson, A.: The need for research infrastructures: a narrative review of large-scale research infrastructures in biobanking. Biopreservation Biobanking **15**(4), 375–383 (2017)
17. Ma, Z., Yan, L., Taniar, D. (eds.): Emerging Technologies and Applications in Data Processing and Management. Advances in Data Mining and Database Management. IGI Global (2019)
18. Medina-Martínez, J.S., Arango-Ossa, J.E., Levine, M.F., Zhou, Y., et al.: Isabl Platform, a digital biobank for processing multimodal patient data. BMC Bioinform. **21**(1), 549 (2020)
19. Menze, B.H., Jakab, A., Bauer, S., Kalpathy-Cramer, J., et al.: The multimodal brain tumor image segmentation benchmark (BRATS). IEEE Trans. Med. Imaging **34**(10), 1993–2024 (2015)
20. Murali, S., et al.: Bringing MRI to low-and middle-income countries: directions, challenges and potential solutions. NMR Biomed. **37**(7), e4992 (2023)
21. Norlin, L., Fransson, M.N., Eriksson, M., Merino-Martinez, R., et al.: A minimum data set for sharing biobank samples, information, and data: MIABIS. Biopreservation Biobanking **10**(4), 343–348 (2012)
22. Pinto-Coelho, L.: How artificial intelligence is shaping medical imaging technology: a survey of innovations and applications. Bioengineering **10**(12), 1435 (2023)
23. Prud'hommeaux, E., Collins, J., Booth, D., Peterson, K.J., et al.: Development of a FHIR RDF data transformation and validation framework and its evaluation. J. Biomed. Inform. **117**, 103755 (2021)
24. Rieke, N., Hancox, J., Li, W., Milletarì, F., et al.: The future of digital health with federated learning. npj Digit. Med. **3**(1), 119 (2020)
25. Satti, F.A., Ali, T., Hussain, J., Khan, W.A., Khattak, A.M., Lee, S.: Ubiquitous Health Profile (UHPr): a big data curation platform for supporting health data interoperability. Computing **102**(11), 2409–2444 (2020). https://doi.org/10.1007/s00607-020-00837-2
26. Shkembi, K., Kochovski, P., Papaioannou, T.G., Barelle, C., Stankovski, V.: Semantic Web and blockchain technologies: convergence, challenges and research trends. J. Web Semant. **79**, 100809 (2023)
27. Smith, B., Arabandi, S., Brochhausen, M., Calhoun, M., et al.: Biomedical imaging ontologies: a survey and proposal for future work. J. Pathol. Inform. **6**(1), 37 (2015)
28. Souza, R., Stanley, E.A., Forkert, N.D.: On the relationship between open science in artificial intelligence for medical imaging and global health equity. In: Workshop on Clinical Image-Based Procedures, pp. 289–300. Springer (2023)
29. Staunton, C., Moodley, K.: Challenges in biobank governance in Sub-Saharan Africa. BMC Med. Ethics **14**(1), 35 (2013)
30. Taylor, C.F., Field, D., Sansone, S.A., Aerts, J., et al.: Promoting coherent minimum reporting guidelines for biological and biomedical investigations: the MIBBI project. Nat. Biotechnol. **26**(8), 889–896 (2008)
31. Tomaszuk, D., Hyland-Wood, D.: RDF 1.1: knowledge representation and data integration language for the web. Symmetry **12**(1), 84 (2020)
32. Turki, H., Hadj Taieb, M.A., Shafee, T., Lubiana, T., et al.: Representing COVID-19 information in collaborative knowledge graphs: the case of Wikidata. Semant. Web **13**(2), 233–264 (2022)

Benchmarking Noise2Void: Superior Denoising of Medical Microscopic Images

Abdourahmane Balde[1]([✉]), Avewe Bassene[2], Sèmèvo Arnaud R. M. Ahouandjinou[3], Ousmane Sall[4], Mamadou Soumboundou[5], Youssou Faye[1], and Lamine Faty[1]

[1] Assane Seck University of Ziguinchor, Ziguinchor, Senegal
`a.b165@zig.univ.sn`, {`yfaye,lamine.faty`}`@univ-zig.sn`
[2] Cheikh Anta Diop University Dakar, Dakar, Senegal
`avewe.bassene@ucad.edu.sn`
[3] University of Abomey-Calavi, Abomey-Calavi, Benin
`arnaud.ahouandjinou@imsp-uac.org`
[4] Cheikh Hamidou KANE Digital University, Dakar, Senegal
`ousmane1.sall@unchk.edu.sn`
[5] Iba Der Thiam University, UMRED, Diamniadio Children's Hospital, Ziguinchor, Senegal
`mamadou.soumboundou@univ-thies.sn`

Abstract. Microscopy is crucial for diagnosing sickle cell disease by enabling the examination of blood samples at the cellular level. This technique reveals abnormalities of red blood cells, aiding in the identification of disease-specific anomalies. However, image quality is frequently degraded by random 'salt and pepper' noise, complicating subsequent processing tasks such as segmentation and feature extraction. This study investigates the Noise2Void deep learning denoising model to enhance the quality of microscopic images of blood smears from sickle cell disease patients. By evaluating the model's performance on clinical data, we demonstrated that N2V outperforms traditional denoising methods as well as the Noise2Noise model, achieving PSNR and SSIM values of 43.98 dB and 0.98, respectively. Our results underscore the potential of N2V to significantly enhance data quality for the development of accurate sickle cell disease detection methods.

Keywords: Denoising · Impulsive Noise · Microscopic Images · Deep Learning · Noise2Void · Sickle Cell Disease

1 Introduction

Microscopy is an essential imaging technique for the diagnosis of sickle cell disease. It enables blood samples to be examined at the cellular level, revealing the abnormalities characteristic of this disease. The quality of microscopic images relies on the signal-to-noise ratio (SNR), which compares signal intensity to background noise [1]. Random 'salt and pepper' noise during acquisition or transmission often disrupts images, degrading their quality and complicating tasks like segmentation, feature extraction, and detection.

U. Anazodo et al. (Eds.): MImA 2024/EMERGE 2024, CCIS 2240, pp. 199–210, 2025.
https://doi.org/10.1007/978-3-031-79103-1_21

Various image processing methods have been proposed to enhance image quality by addressing different types of distortions, which can either affect specific regions or be widespread. In the context of impulse noise [2], this degradation is uniformly distributed across the entire image. Suppression of impulse noise has garnered significant attention and is a thriving area in image analysis [3]. This type of noise is distinguished by abrupt occurrences of very bright or very dark pixels, posing challenges in image interpretation. To ensure the quality of microscopic images and enable reliable processing, reducing or eliminating noise is crucial. Image denoising plays a vital role in diagnosing diseases such as sickle cell disease or cancer.

Image denoising is a computer vision technique aimed at eliminating noise and restoring sharper, more legible images [4]. The effectiveness of the denoising algorithm greatly influences image processing results, presenting a challenge in choosing an approach tailored to the image's noise characteristics. While traditional denoising methods (such as spatial and transform filtering) are widely employed for reducing impulsive noise [5], convolutional neural networks (CNNs) have emerged as powerful solutions, demonstrating good performance in image denoising [6].

This paper conducts a comparative study of various denoising methods proposed in the literature and recommends the most suitable one for computer vision applications in diagnosing disease from image. The main contributions of this paper are following:

- We explore the implementation challenges of applying the Deep Learning Noise2Void (N2V) denoising model to microscopic images of blood smears from sickle cell disease patients
- We assess the N2V model's performance using simulations of clinical data, comparing its outcomes with traditional denoising methods and the Noise2Noise (N2N) model
- We showcase N2V's superior denoising performance, demonstrating a significant advancement in developing sickle cell detection methods compared to other approaches
- Metrics like MSE (Mean Squared Error), PSNR (Peak Signal-to-Noise Ratio), and MSSIM (Mean Structural Similarity Index Measure) are employed to assess the performance of various denoising methods

The remainder of this paper is organized as follows: Sect. 2 presents work related to denoising techniques. Section 3 describes denoising methodology adopted in this study. Section 4 presents the experiments carried out in the study. Section 5 gives the simulation results of our testbed and discuss it in Sect. 6. Finally, Sect. 7 summarizes the paper and outlines future work.

2 Related Works

Numerous studies have focused on noise suppression to enhance image quality and thereby improve the performance of segmentation [7], detection, and classification methods [8].

Xu Weizheng et al. [9] propose a block-averaged filter in a multi-window structure to enhance complex image structures. This approach meets real-time denoising needs but neglects local image statistics, leading to significant edge degradation.

Jona et al. [10] introduce an optimized median filter that corrects pixels affected by salt-and-pepper noise by comparing them with a clear 3×3 image to confirm similarity. The experiments demonstrated strong performance with PSNR of 36.82 dB, MSE of 18.23, and SSIM of 0.99.

Ziad et al. [11] propose a method for median and medium filters. They combine a special mask with special logical index matrices to detect and process only noisy pixels. The results indicate that the enhanced filters effectively manage impulse noise, achieving high PSNRs.

Kumar et al. [12] suggest enhancing the quality of microscopic images by restoring features like texture, edges, contrast, and more. Comparative studies on various filters demonstrate that denoising microscopic images can enhance radiologists' ability to diagnose and predict specific diseases.

Venkatesh et al. [13] conducted a comparative study between the Gaussian filter and the bilateral filter for denoising fluorescence microscopy images. The results showed that bilateral filter was more effective, achieving higher PSNR.

Other studies have explored image correction through transformation methods including the Total Variation filter, the non-local mean filter, Block-matching and 3D filtering (BM3D), among others.

Kazuaki et al. [14] applied the "Total Variation" algorithm to eliminate quantum noise from microscopic images. Their research aimed to optimize the entropy of microscopic images through adjustment of algorithmic hyperparameters, demonstrating effective noise suppression.

Prasad et al. [15] thoroughly examine the efficacy of the NLM (Non-Local Means) filter across various noise levels, investigating the interplay between patch size, window size, and image smoothing parameters. They evaluate performance using MSE, PSNR, and SSIM metrics with MRI and CT images, highlighting the critical role of precise parameter selection in achieving optimal denoising results.

The BM3D method by Dabov et al. [16] effectively denoises images by grouping similar patches, transforming them into wavelets, applying thresholding or Wiener filtering, and then reconstructing the image. However, its performance decreases with higher noise levels.

In recent years, deep learning-based denoising methods have garnered significant attention for their remarkable achievements in image denoising [17]. Tian et al. [18] introduce a comparative study of different deep learning-based denoising methods. Their approach focuses on convolutional neural networks (CNNs) tailored for denoising tasks involving additive white noise, real noise, blind noise, and noisy hybrid images. They propose a comprehensive comparative analysis based on visual image quality.

Chen et al. [19] utilized the ImageNet dataset to train their image-denoising model, leveraging Transformer architectures and contrastive learning to enhance their IPT model's performance across various low-level computer vision benchmarks.

The Noise2Noise (N2N) denoising model proposed by Lehtinen et al. [20] trains on pairs of noisy images to characterize and correct image degradation. However, this approach faces challenges in fields like healthcare [21], where acquiring sufficient training data is difficult.

Addressing these limitations, Krull et al. [22] introduced Noise2Void (N2V), which is tailored for image denoising using a single input image without requiring reference images. This study explores the application of the N2V model to denoise microscopic images of blood smears from patients with sickle cell disease.

3 Methodology

The methodology employed in this study comprises three main steps: initially, RGB microscopic images are converted to grayscale and resized to (256 × 256) following collection. Subsequently, various denoising techniques (including median, mean, Gaussian, bilateral, NLM, TV, BM3D, N2N, and the N2V deep learning method) are applied for impulse noise reduction. Finally, evaluation metrics (MSE, PSNR, SSIM) are used to assess model performances that reveal of the most effective approach.

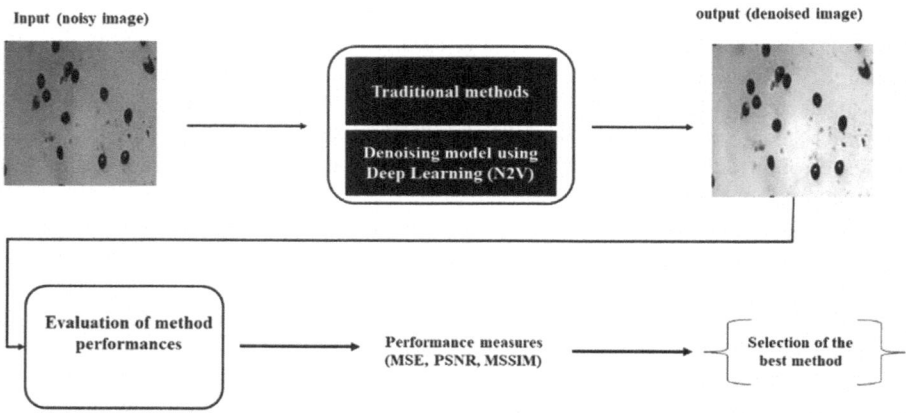

Fig. 1. Methodology for denoising microscopic images of blood smears.

3.1 Data Collection Process

The experimental data used in this study come from patients diagnosed with sickle cell disease. These data take the form of microscopic images acquired using the Emmel test. The experiments were carried out at the UMRED laboratory, Faculty of Health, University of Thiès, and the Diamniadio Children's Hospital, Senegal.

Image quality and visibility were enhanced during the capture process using laboratory methods. The light microscope was set to x100 magnification. The images used were extracted from reading fields and displayed directly on screen. Initially captured in RGB at (3264 × 1856) resolution, they were later resized to (256 × 256).

3.2   Noise Characteristics of Microscopic Images

Blood smear images captured using a light microscope to detect "sickle" shaped red blood cells are prone to various types of noise, including impulse noise and motion artifacts arising from slide preparation and handling.

Impulse noise, also known as "salt and pepper" noise, manifests as random bright or dark pixels due to sensor defects, data transmission errors, or electromagnetic interference during image acquisition. Optical microscopy involves several stages and equipment, such as the microscope sensor, data transfer system, and storage device, each susceptible to environmental or internal electromagnetic disturbances. These disturbances degrade image quality, complicating accurate identification of sickle cells. Therefore, effective denoising methods are crucial for correcting impulse noise and ensuring precise object identification from microscope images. Figure 1 shows an example of a microscope image, in which a set of cells can be seen as the object in the image.

4   Experiments

In this study, we evaluated N2V along with several denoising methods including median, mean, Gaussian, bilateral filters, non-local averaging, total variation, BM3D, and N2N using the aforementioned data. Objective metrics such as MSE, PSNR, and SSIM were employed to assess the performance of these methods. The original image size was reduced to (256 × 256) for our experiments.

4.1   The Noise2Voide Model

The N2V model is a novel approach to image denoising, notable for its capability to learn directly from noisy images without relying on clean reference images. It trains on 64 × 64 pixel patches extracted from noisy images, where during training, certain pixels are randomly masked. The model learns to predict these masked pixels using surrounding unmasked pixels only. This method allows the model to effectively capture local structures and correlations within the data, thereby enhancing its denoising performance [22] (Fig. 2).

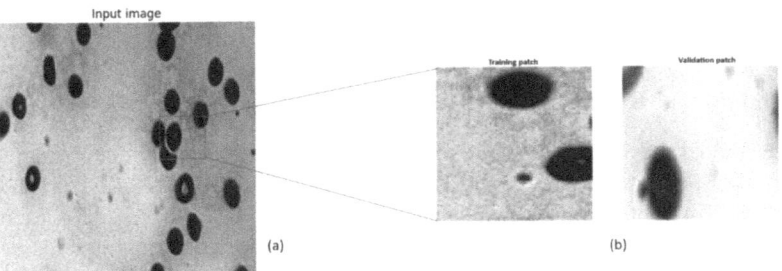

Fig. 2. Image patches used during N2V training. (a) Example image from our da-taset. (b) A patch for training and validation of (a).

The implementation details and training scheme are based on work in [22]. In this work, authors uses the U-Net architecture proposed by Ronneberger et al. [23]. This architecture is of depth 3, with a first convolutional layer (conv) comprising 64 filters of dimension (3 × 3). This number of filters doubles after each subsampling step, which is achieved by (2 × 2) max pooling layers. The last layer is a convolutional layer (conv) with a single filter (3 × 3), padding = 1, stride = 1. The root means square error ('train_loss = 'mse'') is used as the loss function, while batch normalization is activated to stabilize training ('batch_norm = True').

Training is performed on 100 epochs, with an initial learning rate of 0.004, a patch size of 64, and a batch size of 32. To manipulate the masked pixels, we used the uniform with compensation method ('n2v_manipulator = 'uniform_withCP''). A neighborhood radius of 5 pixels ('n2v_neighborhood_radius = 5') is chosen to improve model performance with the help of contextual information about neighborhood pixels.

The images used in this study have a size of (255 × 256) pixels. The dataset consists of 357 images. The N2V model is trained on patches of size (64 × 64) pixels and a total of 45696 patches were generated, with 36556 patches used for training and 9149 patches for validation. This ensures an exhaustive representation of features in the images (Table 1).

Table. 1. Characteristics of the images and patches used for N2V training.

Parameters	Values
Total number of images	357
Image size	(256 × 256)
patches size	(64 × 64)
Total number of patches generated	45696
Number of patches used for training	36556
Patches used for validation	914

The training process for the N2V model is carried out using the Tesla T4 GPU on the Google Cloud. Our test bench includes the Python libraries TensorFlow, Keras, NumPy, and CUDA.

4.2 Traditional Methods

The N2V model was evaluated and compared with other traditional denoising methods (median filter [10], mean filter [9], Gaussian filter [24], bilateral filter [13], NLM [25], BM3D [16], Total Variation [14]) and the IA method (N2N) [20]. The goal is to pinpoint the most suitable denoising method for enhancing image correction tailored to our image structures. These images will be pivotal in future analyses for identifying pertinent features. The subsequent section outlines the metrics employed to evaluate these diverse filtering methods.

4.3 Image Quality Evaluation Metrics

The metrics employed to assess the performance of different image denoising methods include MSE, PSNR, and MSSIM. These metrics offer quantitative and qualitative assessments of error, perceptual quality, and structural similarity, ensuring a comprehensive and precise evaluation of outcomes.

MSE [26] or Mean Square Error: This metric evaluates the similarity between the denoised image I' and the original image I. A lower MSE indicates higher structural similarity, implying that the denoised image closely resembles the original.

$$MSE = \frac{1}{MXN} \sum_{m=1}^{M} \sum_{n=1}^{N} (I(m, n) - I'(m, n))^2 \tag{1}$$

where:

(M and N) represents the image size, I and I' are the original and denoised images respectively.

PSNR [27] or signal-to-noise ratio: This evaluates image quality by comparing the quality of the original signal with the quality of the altered signal. The higher the PSNR, the better the quality of the denoised image compared to the original

$$PSNR = 10 log_{10} \frac{I_{max}}{MSE}^2 \tag{2}$$

MSSIM [28] or Mean Structural Similarity Index: This metric assesses image quality by comparing the structural similarity between two images—the original and the denoised image. MSSIM measures include luminance, contrast, and structure to gauge how closely the denoised image resembles the original.

$$SSIM(x, y) = \frac{(2\mu_x\mu_y + C_1)(2\sigma_{xy} + C_1)}{\left(\mu_x^2 + \mu_y^2 + C_1\right)\left(\sigma_x^2 + \sigma_y^2 + C_2\right)} \tag{3}$$

where,

μ_X is the mean of ithemage X, μ_Y is the mean of the image X, σ_X^2 is the variance of ithe mage X, σ_Y^2 is the variance of imthe age X, σ_{XY} is the covariance between images, et, and C_2 are constants used to stabilize the division in case of low luminance. MSSIM is a unitless measurement with a scale from 0 to 1.

All metrics discussed in this section are evaluated across the various denoising algorithms described earlier, and the results of these evaluations are further elaborated in the following sections.

5 Results

We present experimental results on image denoising using 106 test images from our blood smear microscopy dataset. Figure 3 display the performance metrics, focusing on PSNR and SSIM. The Gaussian filter demonstrated superior performance among classical methods, achieving a PSNR of 40.04 dB and an SSIM of 0.96. Its efficient execution time of 63.75 s makes it a fast and effective denoising technique.

The median filter exhibited moderate performance, achieving a PSNR of 31.97 dB and an MSSIM of 0.86. Conversely, the medium filter, despite its quick execution time of 65.20 s, resulted in a relatively low PSNR of 27.67 dB and an MSSIM of 0.86. On the other hand, the bilateral filter showed a PSNR of 25.73 dB and an MSSIM of 0.88. However, its extended execution time of 482.40 s restricts its practicality for real-time applications.

Among the methods based on transformation and regularization, BM3D offers the best results with a PSNR of 27.91 dB and an MSSIM of 0.84. BM3D's execution time of around 364.21s is relatively long. The Total Variation (TV) method gives a fairly moderate (PSNR, MSSIM) pairing of (26.59 dB, 0.83), although its 47.66s execution time is beneficial for low-latency applications. In addition, the MNL filter showed good performance in terms of image correction.

The N2V achieves an outstanding PSNR of 43.98 dB and an MSSIM of 0.98. In impulse noise processing, N2V outperforms N2N and all other traditional methods. N2V demonstrates significant noise suppression capabilities with superior contrast compared to other methods.

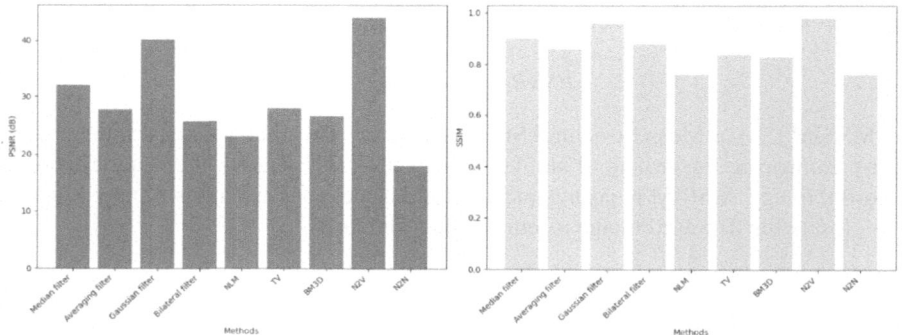

Fig. 3. Comparison of PSNR and SSIM on test data.

Figure 4 depicts microscopic smear images from our dataset. Image (a) serves as a reference image with impulsive noise used in our simulations. Images (b) to (j) demonstrate the visual denoising outcomes for various methods: median filter (b), mean filter (c), Gaussian filter (d), bilateral filter (e), NLM (f), TV (g), BM3D (h), and N2N (j). N2V (i) notably preserves higher intensity values and exhibits better gray-white contrast compared to other methods across all noise levels. In contrast, the Gaussian filter (d) stands out for producing detailed images despite some observed blurring.

The median filter (b) gives very good sharpness with a fast turnaround time. As for BM3D (h), it shows a notable ability to reduce noise, despite a drop in performance proportional to noise intensity. On the other hand, the NLM filter (f) gives the worst results, while TV (g) restores more of the textures of the images studied.

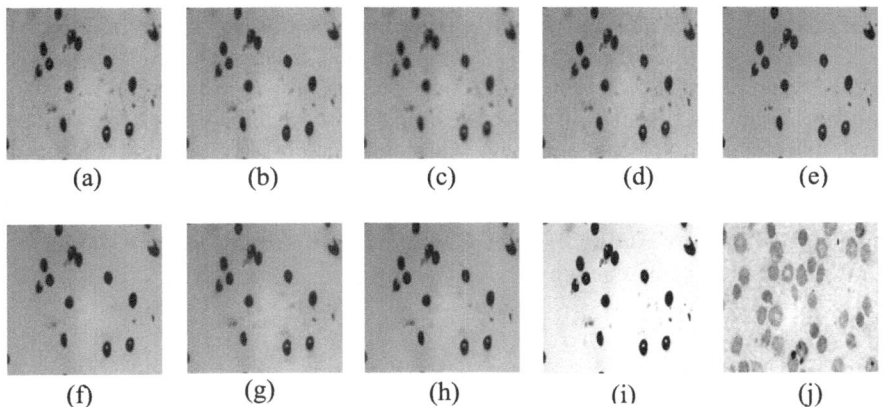

Fig. 4. Visual comparisons of denoising results on a microscopic image of a blood smear from a patient with sickle cell disease, affected by impulse noise during image capture.

6 Evaluation and Discussion

The results presented in Table 2 and Fig. 4 demonstrate that N2V surpasses all other studied denoising methods in terms of PSNR, MSSIM, and MSE for microscopic image processing. Traditional methods, effective in certain scenarios, exhibit significant drawbacks when handling impulse noise, resulting in image blurring (e.g., median and NLM filters) and texture degradation (median filter), which can hinder accurate object identification in image feature studies.

In contrast to conventional filters, N2V preserves fine image details critical for analyzing microscopic images in medical contexts. Its capability to effectively manage diverse noise types across different datasets underscores its strength in image denoising. While traditional methods offer faster execution times, they fall short compared to newer methods like N2V in correcting images intended for rigorous applications such as diagnosing diseases like sickle cell anemia or cancer, where image quality is paramount.

Table. 2. Performance Comparison of image quality.

Metrics	Methods								
	Median filter	Averaging filter	Gaussian filter	Bilateral filter	NLM	TV	BM3D	N2V	N2N
MSE	6.52	16.80	1.26	28.05	40.31	14.68	17.47	1.57	0.016
PSNR (dB)	31.97	27.67	40.04	25.73	23.01	27.91	26.59	43.98	17.98
SSIM	0.90	0.86	0.96	0.88	0.76	0.84	0.83	0.98	0.76
Execution Time (s)	75.47	65.20	63.75	482.40	93.54	364.21	47.66	8850	9426

In this study, our focus was on mitigating impulse noise, which undermines image quality and hinders disease detection, such as sickle cell disease. Our findings highlight N2V as the preferred choice, despite its longer execution time, in settings where image quality is paramount. Its innovative deep-learning approach allows for effective adaptation to diverse noise types and imaging conditions, making N2V highly recommended for precise analysis of medical microscopic images.

This study demonstrates that deep learning-based image restoration offers the potential to effectively correct significant noise and image degradation resulting from acquisition and recording conditions.

7 Conclusion

In this study, we explored various denoising methods suitable for impulse noise suppression in microscopic images of blood smears. Particular attention was paid to the use of the N2V self-supervised learning model, designed to work efficiently with a single noisy image, making it particularly suitable for clinical environments. Our results show that N2V effectively removes impulse noise while preserving critical image structures and outperforming all other evaluated methods in terms of PSNR and SSIM.

This study aimed to enhance the quality of microscopic blood smear images by leveraging deep learning denoising. Given the susceptibility of microscopic images to noise artifacts during acquisition, mitigating these artifacts is vital for developing robust and scalable automated diagnostic methods.

In the future, we plan to use the enhancement capabilities of the N2V model to process images for more needy clinical studies in Africa; proposing an intelligent decision support model for the diagnosis of sickle cell disease. Indeed, sickle cell anemia primarily affects children and adolescents, requiring lifelong treatment from birth. It's shown that over 50% of affected children die before the age of five without clinical follow-up, which can be disastrous in Southern countries.

References

1. Confocal Microscopy - Signal-to-Noise Considerations | Olympus LS. Consulté le: 26 juin 2024. [En ligne]. Disponible sur: https://www.olympus-lifescience.com/en/microscope-res ource/primer/techniques/confocal/signaltonoise/
2. Microscopic image impulse noise filtering of Chinese herbal medicine using pulse coupled neural networks and morphology | IEEE Conference Publication | IEEE Xplore. Consulté le: 26 juin 2024. [En ligne]. Disponible sur: https://ieeexplore.ieee.org/abstract/document/823 0319
3. Wali, A., Naseer, A., Tamoor, M., Gilani, S.A.M., et al.: Recent progress in digital image restoration techniques: a review. Digit. Signal Process. **141**, 104187 (2023). https://doi.org/ 10.1016/j.dsp.2023.104187
4. Denoising of microscopy images: a review of the state-of-the-art, and a new sparsity-based method | IEEE Journals & Magazine | IEEE Xplore ». Consulté le: 26 juin 2024. [En ligne]. Disponible sur: https://ieeexplore.ieee.org/abstract/document/8327626
5. Malinski, L., Radlak, K., Smolka, B., et al.: Is large improvement in efficiency of impulsive noise removal in color images still possible?. Plos One **16**(6), e0253117 (2021). https://doi. org/10.1371/journal.pone.0253117

6. Recent progress in image denoising: A training strategy perspective - Wu - 2023 - IET Image Processing - Wiley Online Library. Consulté le: 26 juin 2024. [En ligne]. Disponible sur: https://ietresearch.onlinelibrary.wiley.com/doi/full/10.1049/ipr2.12748

7. Sen, B., Ganesh, A., Bhan, A., Dixit, S., Goyal, A.: Machine learning based diagnosis and classification of sickle cell anemia in human RBC. In: 2021 Third International Conference on Intelligent Communication Technologies and Virtual Mobile Networks (ICICV), pp. 753–758, févr. 2021, https://doi.org/10.1109/ICICV50876.2021.9388610

8. (PDF) Comparison of deep learning techniques in detection of sickle cell disease. Consulté le: 30 mai 2024. [En ligne]. Disponible sur: https://www.researchgate.net/publication/370 528934_Comparison_of_Deep_Learning_Techniques_in_Detection_of_Sickle_Cell_Dise ase?enrichId=rgreq-ea4fb80fed1111c83debe3aabc4bbc0a-XXX&enrichSource=Y292ZX JQYWdlOzM3MDUyODkzNDtBUzoxMTQzMTI4MTE3OTQ4MDc2NEAxNjkxMjM 2OTkyNDQx&el=1_x_3&_esc=publicationCoverPdf

9. Weizheng, X., Chenqi, X., Zhengru, J., Yueping, H., et al.: Digital image denoising method based on mean filter. In: 2020 International Conference on Computer Engineering and Application (ICCEA), pp. 857–859. mars 2020. https://doi.org/10.1109/ICCEA50009.2020. 00188

10. Jana, B.R., Thotakura, H., Baliyan, A., Sankararao, M., Deshmukh, R.G., Karanam, S.R., et al.: Pixel density based trimmed median filter for removal of noise from surface image. Appl. Nanosci. **13**(2), 1017–1028 (2023). https://doi.org/10.1007/s13204-021-01950-0

11. Al Qadi, Z., Zaini, H.: Improving average and median filters. Int. J. Comput. Sci. Mob. Comput. **12** (2023). https://doi.org/10.47760/ijcsmc.2023.v12i02.001

12. Kumar, A., Srivastava, S., Sarin, R., Irizarry, R.: A comparative study of different denoising and enhancement techniques for blood cell images, pp. 297–303, janv. 2023. https://doi.org/ 10.1049/icp.2023.1506

13. Venkatesh, M., Mohan, K., Seelamantula, C.S.: Directional bilateral filters for smoothing fluorescence microscopy images. AIP Adv. **5**(8), 084805 (2015). https://doi.org/10.1063/1. 4930029

14. Atomic-resolution STEM image denoising by total variation regularization | Microscopy | Oxford Academic. Consulté le: 27 mai 2024. [En ligne]. Disponible sur: https://academic. oup.com/jmicro/article-abstract/71/5/302/6609836?redirectedFrom=fulltext

15. Prasad, P., Anitha, J., Biji, B.: Performance analysis of non-local means denoising on medical images and the impact of filter parameter variation. In: 2024 International Conference on Wireless Communications Signal Processing and Networking (WiSPNET), pp. 1–5, mars 2024. https://doi.org/10.1109/WiSPNET61464.2024.10532872

16. Pakdelazar, O., Gholamali, R.-R.: Improvement of BM3D algorithm and employment to satellite and CFA images denoising, arXiv.org. Consulté le: 28 mai 2024. [En ligne]. Disponible sur: https://arxiv.org/abs/1112.2386v1

17. Wu, W., Chen, M., Xiang, Y., Zhang, Y., Yang, Y.: Recent progress in image denoising: a training strategy perspective. IET Image Process. **17**(6), 1627–1657 (2023). https://doi.org/ 10.1049/ipr2.12748

18. Tian, C., Fei, L., Zheng, W., Xu, Y., Zuo, W., Lin, C.-W.: Deep learning on image denoising: an overview. Neural Netw. **131**, 251–275 (2020). https://doi.org/10.1016/j.neunet.2020.07.025

19. Chen, H., et al.: Pre-Trained image processing transformer. arXiv, 8 novembre 2021. https:// doi.org/10.48550/arXiv.2012.00364

20. Lehtinen, J., et al.: Noise2Noise: learning image restoration without Clean Data, arXiv.org. Consulté le: 1 juin 2024. [En ligne]. Disponible sur: https://arxiv.org/abs/1803.04189v3

21. Buchholz, T.-O., Jordan, M., Pigino, G., Jug, F.:Cryo-CARE: content-aware image restoration for cryo-transmission electron microscopy data, arXiv.org. Consulté le: 1 juin 2024. [En ligne]. Disponible sur: https://arxiv.org/abs/1810.05420v2

22. Krull, A., Buchholz, T.-O., Jug, F.: Noise2Void - learning denoising from single noisy images. arXiv, 5 avril 2019. https://doi.org/10.48550/arXiv.1811.10980
23. Ronneberger, O., Fischer, P., Brox, T.: U-net: convolutional networks for biomedical image segmentation. arXiv, 18 mai 2015. https://doi.org/10.48550/arXiv.1505.04597
24. Krull, A., et al.: Image denoising and the generative accumulation of photons, présenté à Proceedings of the IEEE/CVF Winter Conference on Applications of Computer Vision, 2024, p. 1528-1537. Consulté le: 22 avril 2024. [En ligne]. Disponible sur: https://openaccess.thecvf.com/content/WACV2024/html/Krull_Image_Denoising_and_the_Generative_Accumulation_of_Photons_WACV_2024_paper.html
25. Herbreteau, S., Kervrann, C., et al.: A unified framework of non-local parametric methods for image denoising. arXiv, 21 février 2024. https://doi.org/10.48550/arXiv.2402.13816
26. Évaluation de la qualité des images via FSIM, SSIM, MSE et PSNR : une étude comparative. Consulté le: 25 juin 2024. [En ligne]. Disponible sur: https://www.scirp.org/journal/paperinformation?paperid=90911
27. Signal-to-Noise Ratio - an overview I ScienceDirect Topics. Consulté le: 25 juin 2024. [En ligne]. Disponible sur: https://www.sciencedirect.com/topics/engineering/signal-to-noise-ratio
28. Wang, Z., Bovik, A.C., Sheikh, H.R., Simoncelli, E.P.: Image quality assessment: from error visibility to structural similarity. IEEE Trans. Image Process. **13**(4), 600–612 (2004). https://doi.org/10.1109/TIP.2003.819861

First MICCAI Workshop
on Empowering Medical Information
Computing and Research through
Early-Career Expertise, EMERGE 2024

Self-consistent Deep Approximation of Retinal Traits for Robust and Highly Efficient Vascular Phenotyping of Retinal Colour Fundus Images

Lucas Gago[1]([✉]) [iD], Beatriz Remeseiro[2] [iD], Laura Igual[1] [iD], Amos Storkey[3] [iD], Miguel O. Bernabeu[4] [iD], and Justin Engelmann[4] [iD]

[1] Dept. de Matemàtiques i Informàtica, Universitat de Barcelona, Gran Via de les Corts Catalanes 585, Barcelona, Spain
lgagogag69@alumnes.ub.edu

[2] Dept. of Computer Science, Universidad de Oviedo, Campus de Gijón, Gijón, Spain

[3] School of Informatics, Institute for Adaptive and Neural Computation, The University of Edinburgh, Edinburgh, UK

[4] Centre for Medical Informatics, Usher Institute of Population Health Sciences and Informatics, The University of Edinburgh, Edinburgh, UK

Abstract. Retinal colour fundus images are a fast, low-cost, non-invasive way of imaging the retinal vasculature which could provide information about non-ocular, systemic health. Traditional approaches for retinal vascular phenotyping use handcrafted, multi-step pipelines that are computationally expensive and not robust to common quality issues. Recently, Deep Approximation of Retinal Traits (DART) was proposed which trains a neural network to mimic an existing pipeline in a more efficient and robust way. DART is orders of magnitude faster, more robust and repeatable. However, the original DART was not explicitly trained for repeatability, only provides a single retinal trait, Fractal Dimension (FD), and uses a limited set of augmentations. We propose DARTv2 that increases repeatability with a self-consistency loss, robustness with additional augmentations such as imaging overlays, and utility by adding Vessel Density (VD) as a second retinal trait in addition to FD. DARTv2 shows very high agreement (Pearson 0.9392 for FD and 0.9612 for VD, both $p << 0.05$) with AutoMorph, the pipeline it is based on. DARTv2 is far more robust than AutoMorph and also more robust than the original DART. Finally, DARTv2 is 200 times faster than AutoMorph and 4 times faster than the original DART, while taking up less storage space. DARTv2 will be made available to researchers upon publication.

Keywords: Retinal image analysis · Deep learning · Robustness

1 Introduction

Retinal colour fundus images are pictures of the retina, a layer of tissue at the back of our eyes that allows us to see. These images can be taken non-invasively

---

The original version of the chapter has been revised. The grant information in the acknowledgement section has been added. A correction to this chapter can be found at https://doi.org/10.1007/978-3-031-79103-1_30

U. Anazodo et al. (Eds.): MImA 2024/EMERGE 2024, CCIS 2240, pp. 213–223, 2025.
https://doi.org/10.1007/978-3-031-79103-1_22

in a few seconds with low-cost devices. They are crucial in ophthalmology for retinal disease screening, but also show the retinal vasculature in detail. The retinal blood vessels, in turn, could provide information about general vascular health and serve as a proxy for vascular changes elsewhere in the body, like the heart or the brain [11], a field of study also known as "oculomics" [16]. A common research paradigm is to extract retinal traits that summarise some aspect of the vasculature in a single number, e.g. Fractal Dimension (FD) which captures branching complexity of the blood vessels. Less complex retinal vasculature could indicate poorer vascular health, and indeed lower FD has been associated with cardiovascular [6,15,18] and neurovascular [10,13] disease. Vessel Density (VD), which captures how dense the vasculature is, likewise has shown associations with cardiovascular disease [18].

Retinal traits are traditionally extracted with handcrafted, multi-step pipelines that require high image quality like VAMPIRE [14] or AutoMorph [19]. In practice, a large share of images is excluded due to insufficient quality. In UK Biobank, a dataset collected specifically for research, 25–45% of the images are typically excluded [12,15,18]. These exclusions come at great cost: First, substantially reduced sample sizes and lower statistical power. Second, considerable selection bias as older, non-White, male, and less healthy subjects are more likely to be excluded [3], which exacerbates existing inequalities in healthcare research. Third, using these pipelines in clinical practice is virtually impossible if they fail in a quarter to half of the cases, and doubly so if they systematically fail more often for some subgroups.

Recently, Deep Approximation of Retinal Traits (DART) [4] was proposed to provide a more robust way of computing retinal traits. Follow-up work found DART to be substantially more repeatable than AutoMorph [1], surprisingly at any level of image quality exclusions, including in exclusively high-quality images. Thus, the DART paradigm does not only increase robustness but also repeatability in the absence of quality. A secondary, yet also important benefit is that DART is substantially faster than traditional pipelines, allowing to process images on low-end laptops.

However, the original version of DART had many drawbacks which we address in this work. First, the improved repeatability is only a lucky by-product of the increased robustness. Here, we propose a self-consistency loss to explicitly encourage repeatability. Second, DART used a limited set of data augmentations, and we also extend DART to VD in addition to FD.

2   Methods

2.1   Deep Approximation of Retinal Traits

Briefly, DART approximates an existing pipeline with a neural network, which is trained to give the same output on high-quality images. However, during training, the model receives either original images or augmented versions that have their image quality synthetically degraded. Either way, DART needs to output the same number as the traditional pipeline did for the original, un-degraded

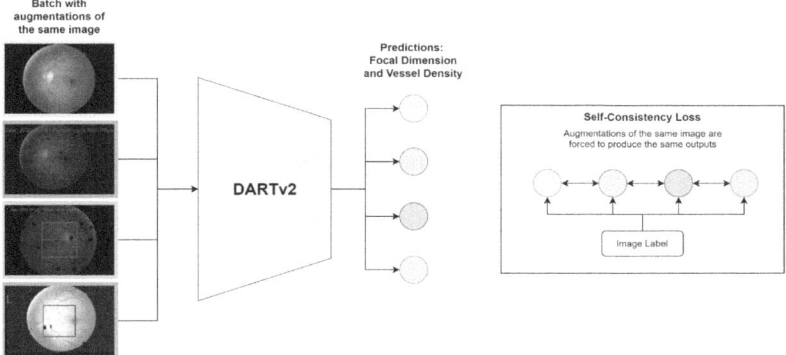

Fig. 1. Overview of DARTv2 training process, where augmented versions of each image are included in each batch to ensure self-consistency.

image. This forces the model to ignore variations in image quality and instead extract all the available information about the retinal trait of interest. Figure 1 shows an overview of how DARTv2 operates. For example, shadows or pathology could obscure parts of the vessels. AutoMorph segments and skeletonises the vasculature, and then computes FD with box counting or VD with averaging, and would give a very low number if part of the vasculature is not segmented. A human clinician, on the other hand, would not be confused by the shadow and instead assess the part of the vasculature that is visible, which is what DART is designed to mimic. The original DART used VAMPIRE [14] to generate ground-truths. In this work, we use AutoMorph [19] which is open-source and fully-automatic.

2.2   Augmentations

We define four levels of augmentation strength as shown in Fig. 2. These include horizontal flips, changing the brightness (lowest level ±5%, highest ±20%), contrast (±5% to ±60%), adding Gaussian blur and simulated imaging noise. We also include simple artefacts that remove multiple small parts of the image to simulate issues like dust or eyelashes, or parts of the images to simulate eyelids and partial shadows. These simulate common imaging issues. However, issues can also occur during the image export. Thus, we additionally simulate text overlays for laterality (left or right), dates and names, and grids indicating where an optical coherence tomography scan is taken. Finally, we simulate images being screenshots rather than proper exports by downsizing to a lower resolution and then back to our target resolution. While theoretically avoidable, in our experience, these are quite common in practice, and thus being robust to them is highly desirable.

2.3   Self-consistency Loss

We explicitly encourage repeatability through a self-consistency loss which penalises poor repeatability across different augmentations. Concretely, during training in each mini-batch, we sample four augmented versions of each image - one of each level of severity - and obtain our model's predictions for each of them. We then use normal mean squared error to penalise deviation from the value to original pipeline provided for the un-augmented image, but additionally also compute the standard deviation across the four versions of each image and add this to our loss. Thus, the model is trained to not only match the original pipeline but to output values that are self-consistent across different levels of image quality which should lead to increased robustness and repeatability in practice.

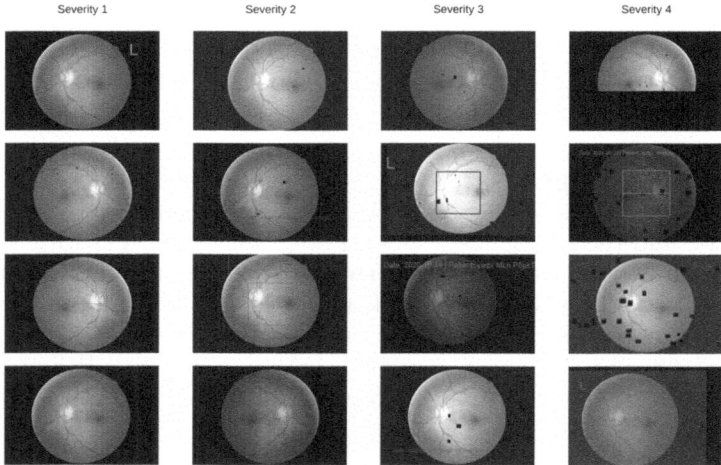

Fig. 2. Randomly sampled augmentations for each of the four levels of severity, using the same original image for illustration purposes.

Given a batch of images $x_i i = 1^B$, where each x_i represents an individual image and B denotes the batch size, for each image in the batch, we apply a set of augmentation functions $S_j j = 1^M$, where each S_j represents a different level of augmentation severity. The self-consistency loss $\mathcal{L}_{\text{self-consistency}}$ is then calculated as:

$$\mathcal{L}_{\text{self-consistency}} = \frac{1}{B} \sum_{i=1}^{B} \left(\frac{1}{M} \sum_{j=1}^{M} \text{Var}\left[f(S_j(x_i))\right] \right) \tag{1}$$

where:

- B is the batch size.
- M is the number of different levels of severity for the augmentation functions.

- f represents the neural network.
- $S_j(x_i)$ represents the augmented version of the image x_i with the j-th level of severity.
- $\text{Var}[\cdot]$ denotes the variance across the network's outputs for the augmented versions of the same image.

2.4 Increasing Robustness Through Data Filtering

For the original version of DART, despite 40% of UK Biobank having already been rejected by VAMPIRE, some poor quality images remained that provided noisy "ground-truths". For DARTv2, we thus aim to avoid these so the model does not replicate undesirable edge cases of the original pipeline. Thus, we filter using QuickQual's "Mega Minified Estimator" [2] which provides a one-dimensional, continuous quality score. QuickQual uses the same EyeQ image quality dataset [5] that AutoMorph's quality algorithm is trained on, but achieves state-of-the-art performance. Recent work found that the repeatability decreases beyond a QuickQual score of 0.8 [1], which indicates a 80% chance of being a bad image. This is about 2.5% of the training data, which we filter out. Note, we do not remove these images from the validation or test sets. Exploration of the training set revealed that there are still some extremely low values, presumably due to failures in the vessel segmentation. We thus clip the lower values of the targets in the training by setting the lowest one-thousandth of values to the 0.1-percentile.

2.5 DARTv2

We use a ConvNeXt [7], specifically the "femto" variant with an overlapping stem from the timm library [17], that was pre-trained on ImageNet. We add a small, perceptron as head with a single hidden layer with 512 hidden units and GELU activations. We normalize images using 0.5 as the mean and standard deviation parameters and resize to 256×256. The model is trained to minimise the sum of the mean squared error with the ground truth targets and consistency loss for 10 epochs with the AdamW [9] optimiser using mini-batches of 256 samples. Prior to the computation of the loss, predictions and targets are normalized to zero mean and unit variance using the training set statistics. We use a cosine learning rate schedule [8] with a linear warmup for the first two epochs and a single cosine cycle, a peak learning rate of 10^{-3}, a weight decay of 10^{-2}, clipping the maximum gradient norm to 0.1. We do not apply weight decay to the biases and initialise the final output layer to use the training set mean targets as output biases and zero weights.

2.6 Data

We use the EyePACS Diabetic Retinopathy dataset on Kaggle, which is openly available and consists of 88,702 colour fundus images acquired with a variety of

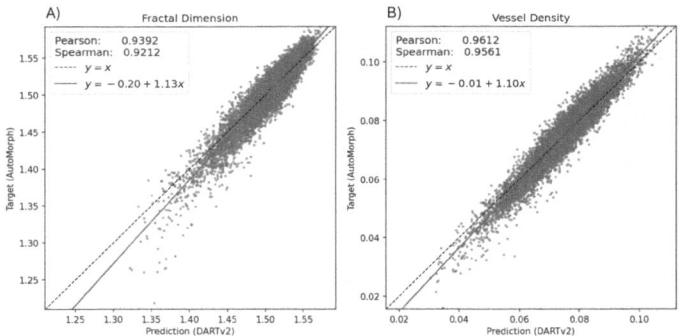

Fig. 3. Agreement between DARTv2 and AutoMorph on the held-out test set for A) Fractal Dimension and B) Vessel Density. The dashed black line indicates the identity line, the red line the best regression fit. (Color figure online)

scanners, and use AutoMorph's FD and VD as ground-truths for training our model. We divided the dataset into train, validation, and test sets, allocating 76.5%, 8.5%, and 15% of the data, respectively. To ensure that each subject appeared only in one of the three sets, we split the data at the subject level. AutoMorph rejected 15.06% of the images due to insufficient image quality, and these images were excluded from further analysis. Thus, our training, validation and test sets contained 56,198, 6,245, and 10,952 images, respectively.

2.7   Evaluation

We quantify the agreement between AutoMorph and DARTv2 using the Pearson and Spearman correlation coefficients. Pearson is the most commonly used correlation and a linear measure. Spearman is a robust measure of correlation and is equivalent to computing the Pearson correlation of the ranks. Furthermore, we also fit a linear regression and report the best regression fit.

To compare the robustness of our model with the original DART and Auto-Morph, we design a synthetic robustness test where images from the test are augmented and we then compare the agreement between each methods output for the original and the augmented image. A robust method should yield very similar values even in the face of augmentations, which would imply both greater robustness and repeatability in practice. While DARTv2 is trained with relatively strong augmentations, including text and OCT region overlays, it would be unfair to consider these as AutoMorph is not expected to be robust to those augmentations. Instead, we consider increasing and decreasing brightness 20% and contrast by 60%. These values were chosen as they visually change the images in a realistic way that is slightly but not overly challenging. In other words, in our opinion, a method for computing retinal traits should be fairly robust in the fact of these changes.

3   Results

3.1   Agreement on Held-Out Test Set

Figure 3 shows the agreement between DARTv2 and AutoMorph on the original images from the held out test set. Generally, agreement is very high, with a Pearson correlation of 0.9392 for FD and 0.9612 for VD (all correlations are $p << 0.05$, as sample sizes are large). Spearman correlations are slightly lower but similar. The best regression fit indicates that the measures are very similar and can be interpreted in the same way. There are some outliers towards the bottom of the plot, where AutoMorph provides an extremely low value, whereas DARTv2 provides a low but not extremely low value. Manual inspection of some of these cases shows that AutoMorph struggles to segment the vasculature in these cases due to poor image quality or the presence of severe retinal pathology. Yet, these images had not been rejected by the AutoMorph quality scoring algorithm. We think that in these cases, AutoMorph outputs erroneously low values and it would be undesirable if DARTv2 replicated this behaviour (Fig. 4).

Fig. 4. Illustration of the augmentations used in our robustness testing using.

Table 1. Pearson correlation between the measurement on the original and augmented images for 1,000 randomly selected test set images. Higher is better, best result in bold. The original version of DART only outputs Fractal Dimension.

	Fractal Dimension				Vessel Density			
	+Brightness	−Brightness	+Contrast	−Contrast	+Brightness	−Brightness	+Contrast	−Contrast
AutoMorph [19]	0.9731	0.6730	0.8348	0.4613	0.9794	0.7390	0.8714	0.5195
DART (original) [4]	0.9777	0.9335	0.9431	**0.8611**				
DARTv2 (ours)	**0.9961**	**0.9407**	**0.9775**	0.8577	**0.9971**	**0.9373**	**0.9844**	**0.8673**
Automorph fail rate	14.10%	1.50%	1.50%	19.00%	13.90%	0.10%	1.10%	13.60%

3.2   Robustness

Table 1 shows the results of the robustness evaluation. For both FD and VD, and for all considered augmentations, DARTv2 was substantially more repeatable than AutoMorph, demonstrating the advantage of a DART-based approach over traditional pipelines. DARTv2 was also more repeatable than the original DART, indicating that our approach improves the robustness of our model.

AutoMorph was unable to process up to 19% of the images depending on the type of augmentation due to numerical issues. The reported Pearson correlations for AutoMorph are excluding these values, which gives a more optimistic estimate of performance for AutoMorph, as difficult cases are the ones where processing fails. No cases, including those difficult cases, were excluded for the original DART or our DARTv2. Thus, it is remarkable that despite this, DART and DARTv2 show substantially higher repeatability than AutoMorph. Indeed, the advantage of DARTv2 is smallest when brightness is increased, but this is after 14.1% and 13.9% of the images failed to be processed by AutoMorph.

Table 2. Inference speed and file sizes.

	AutoMorph	AutoMorph (our optimisation)	DART (original)	DARTv2 (ours)
Images per second	0.36	1.42	77.10	**305.81**
Required disk space	928 MB	928 MB	45 MB	**20 MB**

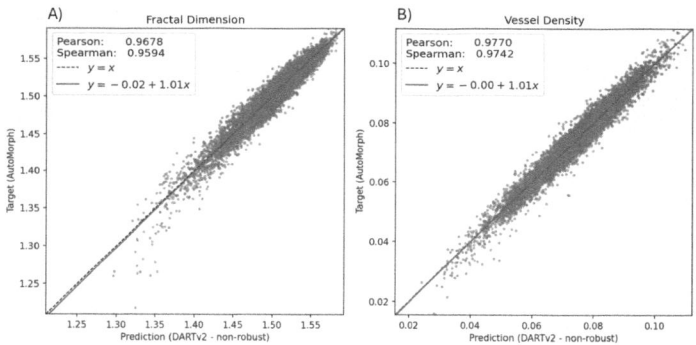

Fig. 5. Agreement between the non-robust version of DARTv2 and AutoMorph on the held-out test set for A) Fractal Dimension and B) Vessel Density. The dashed black line indicates the identity line, the red line the best regression fit. (Color figure online)

3.3 Inference Speed

Inference speed was measured on a desktop workstation with a last-gen high-end gaming GPU (Nvidia RTX 3090) and a four-year-old Intel i9 processor (i9-10900KF). To provide a maximally fair comparison, we measure performance by naively processing images sequentially rather than in batches, as implementing batch processing for AutoMorph is non-trivial while it would be easy to do for DART and DARTv2. Furthermore, we also optimise AutoMorph by removing all processing for retinal traits not considered in this study and by further parallelising non-GPU operations across multiple CPU cores where possible. This allows us to boost the speed of AutoMorph by almost four times.

Table 3. Pearson correlation between the measurement on the original and augmented images for 1,000 randomly selected test set images for our proposed DARTv2 and the non-robust version of DARTv2. Higher is better, best result in bold.

	Fractal Dimension				Vessel Density			
	+Brightness	−Brightness	+Contrast	−Contrast	+Brightness	−Brightness	+Contrast	−Contrast
DARTv2 (ours)	**0.9961**	**0.9407**	**0.9775**	**0.8577**	**0.9971**	**0.9373**	**0.9844**	**0.8673**
DARTv2 - no robustness	0.9750	0.9182	0.9587	0.8251	0.9801	0.9202	0.9660	0.8472

Table 2 shows the results. DARTv2 is more than 800 times faster than Auto-Morph and still 200 times faster than our optimised version. DARTv2 is also 4 times faster than the original DART, primarily due to using a smaller and more efficient model. In terms of filesize, DARTv2 is almost 50 times smaller than AutoMorph and less than half the size of the original DART. While even close to a GB of storage is not unreasonable nowadays, the smaller file sizes also mean faster downloads which will be especially beneficial for researchers without high-speed internet connections.

3.4   Effectiveness of Our Robustness-Enhancing Strategies

To evaluate the effectiveness of our robustness-enhancing strategies, we trained another DARTv2 model in the same way, except for removing the self-consistency loss and our augmentations. Figure 5 shows the agreement of non-robust DARTv2 with AutoMorph on the test set. As expected, agreement is substantially higher when not encouraging robustness as the model is able to learn the behaviour of the original pipeline in edge cases as well, leading to better agreement. However, our goal is not to match the original pipeline perfectly but instead only learn the mimic its consistent behaviour that captures a meaningful aspect of the vasculature. When comparing the robustness of the proposed DARTv2 and the non-robust version (Table 3), we indeed find that DARTv2 is more robust for each of the eight comparisons, indicating the effectiveness of our robustness strategies.

4   Conclusion

We presented DARTv2, an improved model for deep approximation of retinal traits with increased robustness and self-consistency. Our experiments show that DARTv2 not only has very good agreement with AutoMorph on the original images while being substantially more robust, but it is also more robust than the original DART model. Furthermore, DARTv2 is more than 800 times faster than AutoMorph and 4 times faster than the original DART. Our experiments show that our self-consistency loss and augmentation strategies indeed improve robustness. We hope that DARTv2's robustness will allow researchers to exclude fewer images, which would also partially alleviate the selection bias and unfairness introduced by these exclusions. The increased efficiency of DARTv2 could help democratise retinal image analysis.

Future work should expand on the self-consistency loss proposed here and investigate additional strategies for encouraging DART-style models to learn desirable properties. While we expanded on the augmentations used in the original DART, additional augmentations such as simulating the magnification effect due to variations in refractive error should be investigated. Moreover, while Eye-PACS offers significant variability (88,702 images, multiple cameras, diverse ethnicities, healthy and diseased retinas), we agree that evaluating on additional datasets is crucial for establishing generalizability and will include this as a key direction for future work. Finally, in the future additional retinal traits like tortuosity could be added as well as image quality scoring, so researchers can use a single model instead of using DARTv2 and QuickQual separately.

Acknowledgement. This work was partially supported by the MICINN Grant PID2022-136436NB-I00 and AGAUR Grant 2021-SGR-01104.

References

1. Engelmann, J., Moukaddem, D., Gago, L., Strang, N., Bernabeu, M.: Applicability of oculomics for individual risk prediction: repeatability and robustness of retinal fractal dimension using dart and automorph. arXiv preprint (2024)
2. Engelmann, J., Storkey, A., Bernabeu, M.O.: QuickQual: lightweight, convenient retinal image quality scoring with off-the-shelf pretrained models. In: Antony, B., Chen, H., Fang, H., Fu, H., Lee, C.S., Zheng, Y. (eds.) Ophthalmic Medical Image Analysis. LNCS, pp. 32–41. Springer, Cham (2023). https://doi.org/10.1007/978-3-031-44013-7_4
3. Engelmann, J., Storkey, A., Llinares, M.B.: Exclusion of poor quality fundus images biases health research linking retinal traits and systemic health. Invest. Ophthalmol. Vis. Sci. **64**(8), 2922–2922 (2023). iSBN: 1552-5783
4. Engelmann, J., Villaplana-Velasco, A., Storkey, A., Bernabeu, M.O.: Robust and efficient computation of retinal fractal dimension through deep approximation. In: International Workshop on Ophthalmic Medical Image Analysis, pp. 84–93. Springer (2022)
5. Fu, H., et al.: Evaluation of retinal image quality assessment networks in different color-spaces. In: Shen, D., et al. (eds.) MICCAI 2019. LNCS, vol. 11764, pp. 48–56. Springer, Cham (2019). https://doi.org/10.1007/978-3-030-32239-7_6
6. Mordi, I., Trucco, E.: The eyes as a window to the heart: looking beyond the horizon. Br. J. Ophthalmol. **106**(12), 1627 (2022). https://doi.org/10.1136/bjo-2022-322517. http://bjo.bmj.com/content/106/12/1627.abstract
7. Liu, Z., Mao, H., Wu, C.Y., Feichtenhofer, C., Darrell, T., Xie, S.: A convnet for the 2020s. In: Proceedings of the IEEE/CVF Conference on Computer Vision and Pattern Recognition, pp. 11976–11986 (2022)
8. Loshchilov, I., Hutter, F.: SGDR: stochastic gradient descent with warm restarts. arXiv preprint arXiv:1608.03983 (2016)
9. Loshchilov, I., Hutter, F.: Decoupled weight decay regularization. arXiv preprint arXiv:1711.05101 (2017)
10. Luben, R., et al.: Retinal fractal dimension in prevalent dementia: the AlzEye study. Invest. Ophthalmol. Vis. Sci. **63**(7), 4440–F0119 (2022). iSBN: 1552-5783

11. MacGillivray, T.J., Trucco, E., Cameron, J.R., Dhillon, B., Houston, J.G., Van Beek, E.J.R.: Retinal imaging as a source of biomarkers for diagnosis, characterization and prognosis of chronic illness or long-term conditions. Br. J. Radiol. **87**(1040), 20130832 (2014). iSBN: 0007-1285

12. MacGillivray, T.J., et al.: Suitability of UK Biobank retinal images for automatic analysis of morphometric properties of the vasculature. PLoS One **10**(5), e0127914 (2015). iSBN: 1932-6203

13. McGrory, S., et al.: Retinal microvasculature and cerebral small vessel disease in the Lothian Birth Cohort 1936 and Mild Stroke Study. Sci. Rep. **9**(1), 6320 (2019). iSBN: 2045-2322

14. Trucco, E., et al.: Novel VAMPIRE algorithms for quantitative analysis of the retinal vasculature. In: 2013 ISSNIP Biosignals and Biorobotics Conference: Biosignals and Robotics for Better and Safer Living (BRC), pp. 1–4. IEEE (2013)

15. Villaplana-Velasco, A., et al.: Decreased retinal vascular complexity is an early biomarker of MI supported by a shared genetic control. medRxiv (2021). https://doi.org/10.1101/2021.12.16.21267446

16. Wagner, S.K., et al.: Insights into systemic disease through retinal imaging-based oculomics. Transl. Vis. Scie. Technol. **9**(2), 6 (2020). iSBN: 2164-2591

17. Wightman, R.: PyTorch image models (2019). https://github.com/rwightman/pytorch-image-models. https://doi.org/10.5281/zenodo.4414861

18. Zekavat, S.M., et al.: Deep learning of the retina enables phenome-and genome-wide analyses of the microvasculature. Circulation **145**(2), 134–150 (2022). iSBN: 0009-7322

19. Zhou, Y., et al.: AutoMorph: automated retinal vascular morphology quantification via a deep learning pipeline. Transl. Vis. Sci. Technol. **11**(7), 12 (2022). https://doi.org/10.1167/tvst.11.7.12

Non-parametric Neighborhood Test-Time Generalization: Application to Medical Image Classification

Sameer Ambekar[1,2(✉)] , Julia A. Schnabel[1,2,3] , and Daniel M. Lang[1,2]

[1] School of Computation, Information and Technology, Technical University of Munich, Munich, Germany
ambekarsameer@gmail.com
[2] Institute of Machine Learning in Biomedical Imaging, Helmholtz Munich, Munich, Germany
[3] School of Biomedical Engineering and Imaging Sciences, King's College London, London, UK

Abstract. Reliable and stable performance is crucial for the application of computer-aided medical image systems in clinical settings. However, approaches based on deep learning often fail to generalize well under distribution shifts. In medical imaging, such distribution shifts can, for example, be introduced by changes in scanner types or imaging protocols. To counter this, test-time generalization aims to optimize a model that has been trained on single or multiple source domains to an unseen target domain. Common test-time adaptation methods fine-tune model weights utilizing losses with gradient-based optimization, a time-consuming and computationally demanding procedure. In contrast, our approach adopts a non-parametric method that is entirely feedforward and utilizes information from target samples to extract neighborhood information with dynamic voting. By doing so, we avoid fine-tuning or optimization procedures, enabling our method to be more efficient and achieve stable adaptation. We demonstrate the effectiveness of our approach by benchmarking it against different state-of-the-art methods with three backbones on two publicly available medical imaging datasets, consisting of fetal ultrasound and retinal images, and achieve classification accuracy improvements by up to 3.4% and 1.1%, respectively. Moreover, we also demonstrate the utility of our method in practical scenarios, proving efficiency in terms of computational runtime and handling of uncertainty. Our code is publicly available at: https://github.com/compai-lab/2024-miccai-emerge-ambekar.

Keywords: domain adaptation · generalization · unsupervised learning · parameter-free optimization

J. A. Schnabel and D. M. Lang—Shared last-authorship.

1   Introduction

Computer-aided medical imaging systems have achieved significant progress in recent years, with a substantial part of this progress made possible by the advancements of deep learning models [5,7,18,19]. However, a major limitation for their adoption in clinical environments is given by their restricted generalization capacity across unseen data distributions [29]. The reason for this is distribution shifts that can, for example, be caused by variations in scanner types or imaging protocols [20]. To address this, *test-time adaptation and generalization* arose as methods to optimize a trained source model on new incoming target data. Unlike *domain adaptation and generalization* techniques, test-time generalization can consecutively optimize the model on unlabelled data during the test phase without the requirement to access source data, fostering privacy-preserving adaptation to the target domain. Additionally, it allows the model to be optimized continuously without interrupting the inference process, proving especially beneficial in time-sensitive applications where maintaining a flow of real-time decision-making is imperative. Furthermore, the capability of test-time generalization to process data in batches is reflective of real-world scenarios where medical data is also available serially. This aspect enhances its applicability in dynamic clinical workflows.

As shown in Fig. 1(a), test-time generalization [1,8,14,15,17,21,24] methods focus on fine-tuning of the source model based on source model predictions, surrogate models or task predictions. This optimization often involves computation of gradients with norm-based losses followed by finetuning of batch norm layers [23], all the parameters of the model [14] or a linear classification layer [23]. A more recent approach [10] utilizes parameterized ensembles with backpropagation to optimize the last layers of the source model. Even though it is possible to only fine-tune the batch norm parameters [23], gradient-based fine-tuning of model weights, in general, is resource and time-intensive. This increases computational costs and leads to slower adaptation processes, making such methods less practical for real-time applications such as dynamic contrast-enhanced imaging, real-time tumor classification, and rapid stroke identification. Gradient-based finetuning [10,14,17,23] often relies on maximum a posteriori estimation to obtain the target model, which can lead to overfitting when adapting to target data that involves multiple distribution shifts. This reduces the model's generalizability, making it less robust to diverse shifts. Moreover, the potential to converge to local minima and a susceptibility to hyperparameter selections limits these methods efficacy.

By the nature of their feedforward design, non-parametric methods bypass overfitting and negate the need for loss-based gradient finetuning, therefore offering clear advantages, Fig. 1(b). T3A [9] computes class representations or prototypes based on the source model's weights and adjusts the classifier utilizing an entropy threshold. However, by relying only on the entropy of samples, information from the target domain is not fully utilized in the method. A more nuanced approach would be given by application of the source model to directly identify target samples with analogous characteristics to the source features. Such

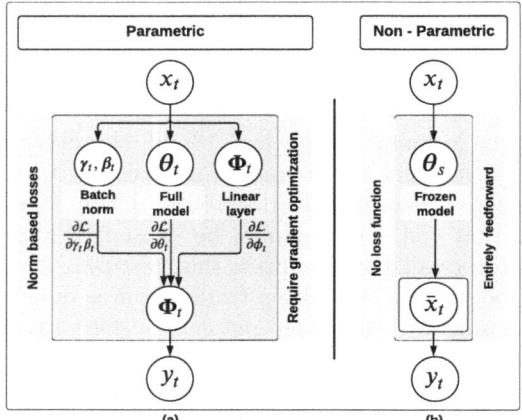

Fig. 1. Data and model interaction scheme. (a) Common test-time generalization techniques utilize norm-based losses for gradient-based finetuning of batch norm layers (β_t and γ_t), full model (θ_t), or linear layers (Φ_t) to obtain target predictions $\mathbf{y}_t$. These methods feature memory and compute constraints and require precise hyperparameter selection with several rounds of backpropagation. (b) Non-parametric approaches such as T3A [9] and ours, obtain $\mathbf{y}_t$ via techniques that operate on frozen source model predictions $\bar{x}_t$ in a feed-forward manner. This neglects the need for additional computational resources and simplifies the generalization process.

a method has the power to increase the utilization of target information and to align closely with the intrinsic data distribution.

Building on these insights, our work introduces a novel, non-parametric method coined Test-time Non-parametric Neighbors (TNN). We leverage neighborhood information between the source prototypes and target data without the need for finetuning. In summary, our contributions are:

- We propose utilizing target neighborhood information with dynamic voting to adjust source-trained classifiers in a non-parametric manner for test-time generalization.
- Our proposed method (TNN) is simple and does not modify the source training process. Yet, it is effective across datasets and requires minimal computation at test time due to its feedforward nature.
- We adopt several state-of-the-art test-time generalization techniques for medical imaging and perform exhaustive comparisons to our approach.

Through comprehensive experiments and ablation studies, we demonstrate the efficacy and potential of TNN in medical imaging contexts, an area where such non-parametric approaches have been underexplored.

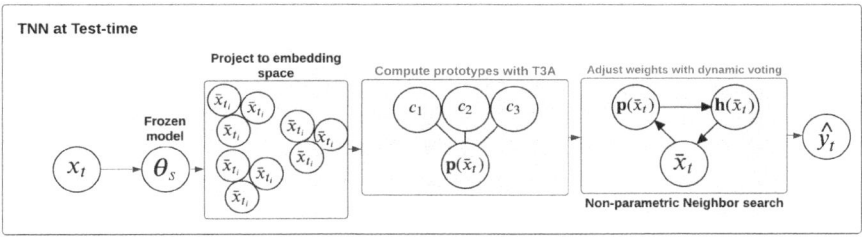

Fig. 2. TNN at test-time. We do not change the source training setup. We initially project the target data to lower dimensional space for class separability with a frozen source model, as the lemma indicates. Further, obtain prototypes for the classes as in [9]. Next, in the embedding space, TNN performs neighborhood search $\mathbf{h}(\bar{x}_t)$ in a non-parametric manner, which is followed by obtaining the classification label $\hat{\mathbf{y}}_t$.

2   Background

Test-time domain generalization [1,9,26] aims to generalize a model $\boldsymbol{\theta}_s$ trained on the source domains $\mathcal{S}$ to an unseen target domain $\mathcal{T}$, with $\mathcal{S}$ usually consisting of several source domains $\{D_s\}_{s=1}^S$. $\mathcal{T}$ which may also consist of several target domains $\{D_t\}_{t=1}^T$. Here, $(\mathbf{x}_s, \mathbf{y}_s)$ and $(\mathbf{x}_t, \mathbf{y}_t)$ denote the image and corresponding label pairs on the source $\{D_s\}_{s=1}^S$ and target domain $\{D_t\}_{t=1}^1$, respectively. The objective of test-time domain generalization is to maximize the log-likelihood of the source model on the target data $p(D_t|\boldsymbol{\theta}_s)$, i.e., $p(\mathbf{y}_t|\mathbf{x}_t, \boldsymbol{\theta}_s)$ (Fig. 2).

Formulation of Parametric Methods. Due to distribution shifts between source and target domains, the source-trained model $\boldsymbol{\theta}_s$ is highly likely to fail on unseen target domains D_t, causing unreliable predictions with high confidence [1,27]. To prevent this, the source model must be generalized to the target domain at test time by transforming $\boldsymbol{\theta}_s$ to $\boldsymbol{\theta}_t$. Most common parametric methods employ fine-tuning based on norm-based losses [10,14,23]. The log-likelihood of the target data is given by:

$$p(\mathbf{y}_t|\mathbf{x}_t, \boldsymbol{\theta}_s) = \int p(\mathbf{y}_t|\mathbf{x}_t, \boldsymbol{\theta}_t) p(\boldsymbol{\theta}_t|\mathbf{x}_t, \boldsymbol{\theta}_s) d\boldsymbol{\theta}_t$$
$$\approx p(\mathbf{y}_t|\mathbf{x}_t, \boldsymbol{\theta}_t^*), \tag{1}$$

with the integration of the distribution $p(\boldsymbol{\theta}_t)$ usually approximated by the maximum a posteriori (MAP) estimation [1]. The final generalized MAP model $\boldsymbol{\theta}_t^*$ is obtained by fine-tuning of the parameters with one or multiple rounds of backpropagation using a norm-based unsupervised loss function, like entropy minimization [23], pseudo labeling [14] or task-specific losses [15,21]. However, fine-tuning the model parameters through gradient optimization with multiple rounds makes parametric methods time-consuming and computationally expensive while also being sensitive to hyperparameter settings.

Non-parametric Methods. To counter the above limitations,recent non-parametric methods such as [9] obtain class representations as prototypes, utilizing the weights of the source-trained linear classifier, i.e., without the need for

MAP approximation or gradient-based optimization. Next, they obtain pseudo labels for the incoming target data based on the distance to those prototypes by applying entropy thresholds. After each incoming batch of target data, significant samples are selected, employing a threshold, and used to update the prototypes via simple adjustments to the classifier.

3   Method

Source Training. Recent studies have shown that utilization of empirical risk minimization (ERM) [6,22] during source training enables models to generalize well under distribution shifts. Other methods, such as [1,25,26,28], included additional objectives to be minimized during source training. However, the requirement to interfere in the training procedure limits the applicability of such approaches. Therefore, we aim to develop a method that does not modify the source training procedure, making it applicable to any pretrained model without any additional requirements. Specifically, as in [6,22], on multiple source domains $\{D_s\}_{s=1}^{S}$, given a source model $\boldsymbol{\theta}_s$, such as ResNet-18, and a loss function $\mathcal{L}$, such as cross-entropy, the total risk is minimized via $\mathbb{E}_{(x_s,y_s) \sim D_s}[\mathcal{L}(\theta_s(x_s), y_s)]$.

Test-Time Generalization via Nearest Neighbors. Our approach can be summarized by: at test-time, in a non-parametric way, we initially compute the source prototypes following [9]. Next, given a batch of target data, we obtain the nearest neighbors for classification and adjust the classifier weights as described below.

Existence of Nearest Neighbors. We propose that for a sufficiently trained model able to separate classes reasonably well in the source domain, cases that are similar in the higher dimensional image domain will lie close to each other in the source learned lower dimensional embedding space. This is ensured by the Johnson-Lindenstrauss (JL) [11] lemma as stated below:

Given a set of points $\{x_i : i = 1, ..., M\}$ in $\mathbb{R}^m$, the JL lemma [11] states that if $n \geq c\epsilon^2 \log M$, with $0 < \epsilon < 1$, then there exists a linear map $A : \mathbb{R}^m \to \mathbb{R}^n$ such that for all $i \neq j$:

$$1 - \epsilon \leq \frac{\|A(x_i) - A(x_j)\|}{\|x_i - x_j\|} \leq 1 + \epsilon.$$

At test-time, utilizing the above lemma, our approach first computes the source model prototypes for each class in lower-dimensional space. To do so, as in [9], we initialize the class-specific prototypes by aggregating the weights of the source-trained linear classifier layer. When receiving new test-time data $\mathbf{x}_t$, we project it into lower-dimensional space using the source-trained model that preserves distances in the embedding space, as ensured by the JL lemma. In this embedding space, we assign $\mathbf{x}_t$ the label of its nearest class prototypes based on a distance measure (see below). Finally, we update the class-specific prototypes with the new sample to reflect new target characteristics. This allows us to classify new samples without the need for extensive computation or any optimization schemes to find the nearest points to the prototypes.

Selecting Neighbors with Dynamic Voting. Since not all of the neighbors provide accurate information about the target data, i.e., some of them are noisy [4], we calculate the distance between the initial prototypes and new classifier weights obtained from incoming neighbor samples at test time for the selection of valid neighbors. We utilized the cosine distance here, while in principle, every other distance metric can be used. Next, we use dynamic voting $h(\bar{\mathbf{x}}_t)$ to obtain the most useful neighbors, i.e., we aggregate each neighbor's prediction and calculate the mean of the new weights obtained to determine the final weights of the classifier. When a new batch of samples arrives, the pseudo labels are predicted based on these classifier weights. This process is repeated iteratively for each new incoming batch of data.

4   Experiments

4.1   Datasets and Implementation Details

Datasets. We validate the effectiveness of TNN on two publicly available datasets. Messidor [3] depicts a retinal image database collected from three independent medical centers, containing 1200 images of diabetic retinopathy, each labeled with a severity score ranging from 0 to 3. Therefore, it consists of 3 domains, 4 classes, and 1200 images in total. Fetal-8 [2] is a maternal-fetal ultrasound dataset that consists of eight classes representing different anatomical planes collected from imaging scanners of two different vendors. Hence, it consists of 2 domains, 8 classes and 12,058 images in total. We consider the problem as a test-time generalization setting, not domain adaptation. Therefore, we train our model on multiple source domains of the Messidor dataset instead of only one. Furthermore, we follow the *leave-one out* evaluation standard from [9,13] and obtain the best source model following the training-domain validation split of [9]. For all the experiments, performance is reported on the target dataset with test-time adaptation [1,9,23] by utilizing accuracy as a metric.

Algorithm 1. TNN for medical images.

Input: $\mathcal{T}$: target domain; learned and frozen θ_s;
Output: θ_t^* with adjusted weights

1: **for** *iter* in N_{iter} **do**
2:    Draw random samples for a batch from $\mathcal{T}$ as $\mathbf{x}_t$
3:    Obtain source prototypes $p(\bar{x}_t)$ from the source-trained model ($\boldsymbol{\theta}_s$) with T3A [9]

4:    Forward pass of $\mathbf{x}_t$ through $\boldsymbol{\theta}_s$ to obtain the points in lower dimensional space
5:    Calculate the distance between source prototypes and the new samples
6:    Obtain subsets of samples and use dynamic voting to obtain classifier
7:    Obtain $\hat{\mathbf{y}}_t$ for the batch of $\mathbf{x}_t$
8: **end for**

Implementational Details. On both datasets, Messidor and Fetal-8, we evaluate the performance of our approach based on three backbones, DenseNet-121, ResNet-18, and ResNet-50. Further, all the in-depth ablation experiments were performed using the ResNet-18 model. As in common test-time methods [6,9], the backbones are pretrained on Imagenet. Baselines and state-of-the-art methods are also implemented for the two datasets and for all of the three backbones, utilizing the Domainbed library [6] with all hyperparameters set to default. The training-validation split strategy of [6,9] is used for the selection of the best source model. At test-time, we perform just forward passes to perform the classification on the online target data. We evaluate the performance of methods following standard test-time generalization evaluation [9]. We utilize a small batch size of 32 for test-time generalization, reflecting real-world practical scenarios.

4.2   Comparisons

We evaluate the performance of our model in reference to different state-of-the-art (SOTA) approaches and a source training strategy employing ERM minimization without adaption to the target domain as the baseline.

State-of-the-Art Comparisons. We compare our approach to existing parametric and non-parametric state-of-the-art methods by re-implementing them on the two datasets for all three backbones. Parametric methods, as shown in Fig. 1, refer to techniques that finetune weights of the source-trained model, utilizing gradient optimization. Non-parametric methods refer to techniques that perform feedforward computation at test-time without any kind of finetuning, optimization, or usage of any external memory bank or an additional model.

Table 1 shows the performance comparisons. TNN achieves the best results on both datasets. Parametric methods achieve comparatively lower performance than the ERM baseline in many cases. One reason for this can be given by the fact that the medical imaging datasets at hand only contain a fraction of the number of samples the parametric methods were designed for. Furthermore, differences between images of distinct classes in the medical domain are way more subtle than in the computer vision domain, with different classes in the retinopathy images of Messidor even depicting a severity score that changes only gradually. Therefore, it is likely that non-parametric methods achieve better performance due to the fact that they do not fine-tune the complete source model weights but rather act upon the source-trained embedding space that should be able to separate classes reasonably well. T3A [9] utilizes the entropy of the samples as a threshold to classify new cases. In contrast, we use detailed neighborhood information for classification. For the Fetal-8 dataset, utilizing the ResNet-18 backbone, the performance of all the generalization methods decreases with reference to the ERM baseline. Reason for this is most likely overfitting due to the small size of the model, but also the small dataset size at test-time. For all the remaining settings, TNN performs better than the other approaches.

Table 1. State-of-the-art comparisons using DenseNet-121, ResNet-18, and ResNet-50 as backbones. We re-implement the methods on the datasets and report the mean accuracy across all the domains. The best results are in **bold**. Our results are averaged over five runs. Our method performs the best consistently with the highest performance improvement.

Methods	Messidor			Fetal-8		
	DenseNet-121	ResNet-18	ResNet-50	DenseNet-121	ResNet-18	ResNet-50
ERM	52.2	53.1	52.1	81.6	84.3	83.6
Parametric						
Tent [23]	44.7 ±0.4	43.9 ±0.5	40.0 ±0.5	73.0 ±0.6	70.3 ±0.4	80.1 ±0.5
SHOT [14]	47.4 ±0.4	51.6 ±0.5	51.9 ±0.6	69.8 ±0.7	70.1 ±0.4	72.4 ±0.7
ShotIM [14]	47.5 ±0.4	51.8 ±0.5	46.0 ±0.6	68.9 ±0.7	69.9 ±0.5	72.0 ±0.7
Non-Parametric						
T3A [9]	51.6 ±0.4	51.8 ±0.3	51.6 ±0.3	82.4 ±0.3	71.8 ±0.4	85.6 ±0.3
TNN (Ours)	**53.1** ±0.2	**53.8** ±0.2	**53.9** ±0.2	**83.3** ±0.2	72.7 ±0.2	**86.1** ±0.2

4.3 Additional Experiments

Addressing Uncertain Scenarios. Ensuring alignment between model output probabilities and the actual likelihood of events is crucial in uncertain scenarios [12]. To quantify this alignment, Table 2 presents the expected calibration error [16] (lower values indicate better calibration) for our approach compared to the Tent model [23] and T3A [9], utilizing a ResNet-18 backbone. We report the ECE error between predicted and ground truth labels, using a public library[1] Consistently, TNN and T3A demonstrate considerably better calibration scores across all domains on the Messidor dataset. Moreover, TNN also takes into account the neighborhood information while addressing uncertainty. By utilizing its feedforward nature for classifier adjustments, TNN achieves a calibrated model.

Computational Cost. In Table 3, we compare the number of parameters required to be trained at test-time and the number of floating point operations per second (FLOPS) consumed by the GPU for Tent [23], T3A [9] and TNN. All of the methods included, including ours, feature the same memory requirements of 1.4 GB for the ResNet-18 model. However, as Tent optimizes the batch normalization layers of the target model at test time, more parameters must be trained than our approach. TNN and T3A are both non-parametric. Thus, they only perform a very limited amount of computational operations but do not need any additional computations to calculate the gradients on the GPU. Therefore, the

---

[1] https://torchmetrics.readthedocs.io/en/v0.8.0/classification/calibration_error.html.

Table 2. Addressing uncertain scenarios. ECE error on the three domains (0-2) of the Messidor dataset. The proposed method consistently reduces the ECE error across all the domains.

	ECE Error ↓			
	0	1	2	Mean ↓
Tent	0.101	0.336	0.130	0.189
T3A	0.001	0.005	0.003	0.003
TNN (Ours)	0.001	0.005	0.003	0.003

Table 3. Computational cost. The number of new parameters to be trained at test-time alongside the TeraFlops consumed on the GPU. TNN and T3A both consume fewer resources and are thus useful for practical scenarios.

	Parameters ↓	Model TFlops ↓
Tent	600000	212992
T3A	0	-
TNN (Ours)	0	-

measurement of TFlops is negligible in this case, considering the vast amount of computations required for weight optimization of parametric approaches. This is especially useful in limited resource settings.

5   Conclusion

We propose using a non-parametric-based neighborhood classification method for medical imaging tasks involving distribution shifts as a novel test-time generalization method. By utilizing target information and neighbors in the embedding space, we sequentially adjust the weights of the classifier, providing an efficient yet powerful generalization technique. A limitation of our method, that is also present for established techniques, is given by the requirement for a shared label space between domains. Moreover, our method relies on prototypes, constructed from the source domain, for target classes separation. In future work, we aim to address classification capabilities for new, unseen categories by application of meta-learning techniques and involvement of additional dataset, for construction of improved prototypes. Furthermore, we provide additional experiments to demonstrate the method's utility in uncertain scenarios and settings that require limited computational resources.

Acknowledgements. This paper is supported by the DAAD programme Konrad Zuse Schools of Excellence in Artificial Intelligence, sponsored by the Federal Ministry of Education and Research. We thank Lina Felsner for her assistance with the diagrams.

References

1. Ambekar, S., Xiao, Z., Shen, J., Zhen, X., Snoek, C.G.: Learning variational neighbor labels for test-time domain generalization. arXiv preprint arXiv:2307.04033 (2023)
2. Burgos-Artizzu, X.P., et al.: Evaluation of deep convolutional neural networks for automatic classification of common maternal fetal ultrasound planes. Sci. Rep. **10**(1), 10200 (2020)
3. Decencière, E., et al.: Feedback on a publicly distributed image database: the messidor database. Image Anal. Stereol. **33**(3), 231–234 (2014)

4. Dubey, A., Ramanathan, V., Pentland, A., Mahajan, D.: Adaptive methods for real-world domain generalization. In: IEEE Conference on Computer Vision and Pattern Recognition, pp. 14340–14349 (2021)

5. Guan, H., Liu, M.: Domain adaptation for medical image analysis: a survey. IEEE Trans. Biomed. Eng. **69**(3), 1173–1185 (2021)

6. Gulrajani, I., Lopez-Paz, D.: In search of lost domain generalization. In: International Conference on Learning Representations (2020)

7. Hannun, A.Y., et al.: Cardiologist-level arrhythmia detection and classification in ambulatory electrocardiograms using a deep neural network. Nat. Med. **25**(1), 65–69 (2019)

8. Huang, Y., et al.: Fourier test-time adaptation with multi-level consistency for robust classification. arXiv preprint arXiv:2306.02544 (2023)

9. Iwasawa, Y., Matsuo, Y.: Test-time classifier adjustment module for model-agnostic domain generalization. In: Advances in Neural Information Processing Systems, vol. 34 (2021)

10. Jang, M., Chung, S.Y., Chung, H.W.: Test-time adaptation via self-training with nearest neighbor information. In: International Conference on Learning Representations (2023)

11. Johnson, W.B.: Extensions of lipshitz mapping into hilbert space. In: Conference Modern Analysis and Probability, pp. 189–206 (1984)

12. Kumar, A., Liang, P.S., Ma, T.: Verified uncertainty calibration. In: Advances in Neural Information Processing Systems, vol. 32 (2019)

13. Li, D., Yang, Y., Song, Y.Z., Hospedales, T.M.: Deeper, broader and artier domain generalization. In: IEEE International Conference on Computer Vision, pp. 5542–5550 (2017)

14. Liang, J., Hu, D., Feng, J.: Do we really need to access the source data? Source hypothesis transfer for unsupervised domain adaptation. In: International Conference on Machine Learning, pp. 6028–6039. PMLR (2020)

15. Liu, Y., Kothari, P., van Delft, B., Bellot-Gurlet, B., Mordan, T., Alahi, A.: TTT++: when does self-supervised test-time training fail or thrive? In: Advances in Neural Information Processing Systems, vol. 34 (2021)

16. Naeini, M.P., Cooper, G., Hauskrecht, M.: Obtaining well calibrated probabilities using Bayesian binning. In: Proceedings of the AAAI Conference on Artificial Intelligence, vol. 29 (2015)

17. Niu, S., et al.: Efficient test-time model adaptation without forgetting. In: International Conference on Machine Learning, pp. 16888–16905. PMLR (2022)

18. Rajpurkar, P., et al.: Deep learning for chest radiograph diagnosis: a retrospective comparison of the chexnext algorithm to practicing radiologists. PLoS Med. **15**(11), e1002686 (2018)

19. Rajpurkar, P., et al.: Chexnet: radiologist-level pneumonia detection on chest x-rays with deep learning. arXiv preprint arXiv:1711.05225 (2017)

20. Roschewitz, M., et al.: Automatic correction of performance drift under acquisition shift in medical image classification. Nat. Commun. **14**(1), 6608 (2023)

21. Sun, Y., Wang, X., Liu, Z., Miller, J., Efros, A., Hardt, M.: Test-time training with self-supervision for generalization under distribution shifts. In: International Conference on Machine Learning, pp. 9229–9248. PMLR (2020)

22. Vapnik, V.: Principles of risk minimization for learning theory. In: Advances in Neural Information Processing Systems, vol. 4 (1991)

23. Wang, D., Shelhamer, E., Liu, S., Olshausen, B., Darrell, T.: Tent: fully test-time adaptation by entropy minimization. In: International Conference on Learning Representations (2021)

24. Wang, M., Deng, W.: Deep visual domain adaptation: a survey. Neurocomputing **312**, 135–153 (2018)
25. Xiao, Z., Zhen, X., Liao, S., Snoek, C.G.M.: Energy-based test sample adaptation for domain generalization. In: International Conference on Learning Representations (2023)
26. Xiao, Z., Zhen, X., Shao, L., Snoek, C.G.M.: Learning to generalize across domains on single test samples. In: International Conference on Learning Representations (2022)
27. Yi, L., Xu, G., Xu, P., Li, J., Pu, R., Ling, C., McLeod, A.I., Wang, B.: When source-free domain adaptation meets learning with noisy labels. In: International Conference on Learning Representations (2023)
28. Zhang, Y., et al.: Adanpc: exploring non-parametric classifier for test-time adaptation. In: International Conference on Machine Learning, pp. 41647–41676. PMLR (2023)
29. Zhou, K., Liu, Z., Qiao, Y., Xiang, T., Loy, C.C.: Domain generalization: a survey. IEEE Trans. Pattern Anal. Mach. Intell. (2022)

Client Security Alone Fails in Federated Learning: 2D and 3D Attack Insights

Santhosh Parampottupadam[1,2]($\boxtimes$) (iD), Ralf Floca[1] (iD), Dimitrios Bounias[1,2] (iD),
Benjamin Hamm[1,2] (iD), Saikat Roy[1] (iD), Sinem Sav[3] (iD), Maximilian Zenk[1,2] (iD),
and Klaus Maier-Hein[1,2,4] (iD)

[1] German Cancer Research Center (DKFZ), Division of Medical Image Computing,
Heidelberg, Germany
santhosh.parampottupadam@dkfz-heidelberg.de
[2] Medical Faculty Heidelberg, Heidelberg University, Heidelberg, Germany
[3] Department of Computer Engineering, Bilkent University, Ankara, Turkey
[4] Pattern Analysis and Learning Group, Department of Radiation Oncology,
Heidelberg University Hospital, 69120 Heidelberg, Germany

Abstract. Federated learning (FL) plays a vital role in boosting both accuracy and privacy in the collaborative medical imaging field. The importance of privacy increases with the diverse security standards across nations and corporations, particularly in healthcare and global FL initiatives. Current research on privacy attacks in federated medical imaging focuses on sophisticated gradient inversion attacks that can reconstruct images from FL communications. These methods demonstrate potential worst-case scenarios, highlighting the need for effective security measures and the adoption of comprehensive zero-trust security frameworks. Our paper introduces a novel method for performing precise reconstruction attacks on the private data of participating clients in FL settings using a malicious server. We conducted experiments on brain tumor MRI and chest CT data sets, implementing existing 2D and novel 3D reconstruction techniques. Our results reveal significant privacy breaches: 35.19% of data reconstructed with 6 clients, 37.21% with 12 clients in 2D, and 62.92% in 3D with 12 clients. This underscores the urgent need for enhanced privacy protections in FL systems. To address these issues, we suggest effective measures to counteract such vulnerabilities by securing gradient, analytic, and linear layers. Our contributions aim to strengthen the security framework of FL in medical imaging, promoting the safe advancement of collaborative healthcare research. The source code is available at: https://github.com/MIC-DKFZ/2D3D-Privacy-Attacks-In-federated-Medical-Imaging.

Keywords: Federated Learning · Privacy Attacks · Medical Imaging · Reconstruction Attacks

1 Introduction

Privacy and regulatory challenges limit the gathering of large medical datasets for deep neural network training [11]. FL [12] addresses these challenges by

U. Anazodo et al. (Eds.): MImA 2024/EMERGE 2024, CCIS 2240, pp. 235–244, 2025.
https://doi.org/10.1007/978-3-031-79103-1_24

enabling collaborative training across hospitals, without centralizing patient data. Hospitals train models locally, share only model updates with a central server, which then aggregates these to improve a global model. Although patient data remains distributed in FL, there is still a significant risk of data leakage through information encoded in model updates, which may be exploited by attackers to reconstruct training data, posing a major privacy concern(e.g. in the case of images by facial reconstruction from MRI data [18]).

With the rise in popularity of FL, its associated software and frameworks, including NVFlare [15], Kaapana [10], MonaiFL [13], and Flower [1], have also gained prominence. Alongside this growth, various privacy attacks have been proposed. This paper focuses on data reconstruction attacks, which aim to recover private training data points as accurately as possible. These attacks utilize methods such as optimization [4,6,7,23,25], analytical techniques [16,22], or exploit vulnerabilities in linear layers [2,5] to compromise FL privacy. Medical FL research has mainly focused on threats through gradient-based optimization [8,11,21,24] and model inversion [20] techniques. These strategies face major challenges, such as high computational resource demands, and typically result in incomplete or blurred reconstructions. This low reconstruction fidelity poses challenges in accurately identifying original records despite being indicative of a privacy breach and is especially relevant in medical imaging, where subtle structural variances are often crucial.

Our research investigates the potential of data reconstruction attacks in FL that can closely replicate the private data of patients, in the scenario of a malicious FL server. We apply an analytical approach [2] to 2D medical imaging and introduce a novel adaptation for 3D medical imaging within the context of FL. This marks the first effort in 3D imaging, achieving superior reconstruction fidelity with minimal architecture modifications from a malicious server. Our study underscores the need for measures to prevent data leakage through malicious server activities, even in the context of a secure client, in terms of software and network security aspects. The used attack highlights the importance of deeply understanding the machine learning architecture involved, when an institution decides to participate in a FL collaboration. In summary, our study presents the subsequent contributions:

- To our knowledge, this is the first demonstration of privacy attacks in FL using 3D medical imaging data with malicious server-shared gradients. Our method reconstructs slices into 3D volumes with minimal modifications, highlighting significant privacy concerns. In simulations with 6 and 12 clients, we achieve high-fidelity 2D and 3D reconstructions without complex optimization, showing the potential for reconstructing identifiable body parts or whole bodies [18].
- We showcase that irrespective of client security, a malicious server can easily attack clients as long as the server or central entity provides the training algorithm, which is common in most FL scenarios.
- We also outline simple but effective techniques to mitigate malicious server-based attacks in cross-consortia medical FL.

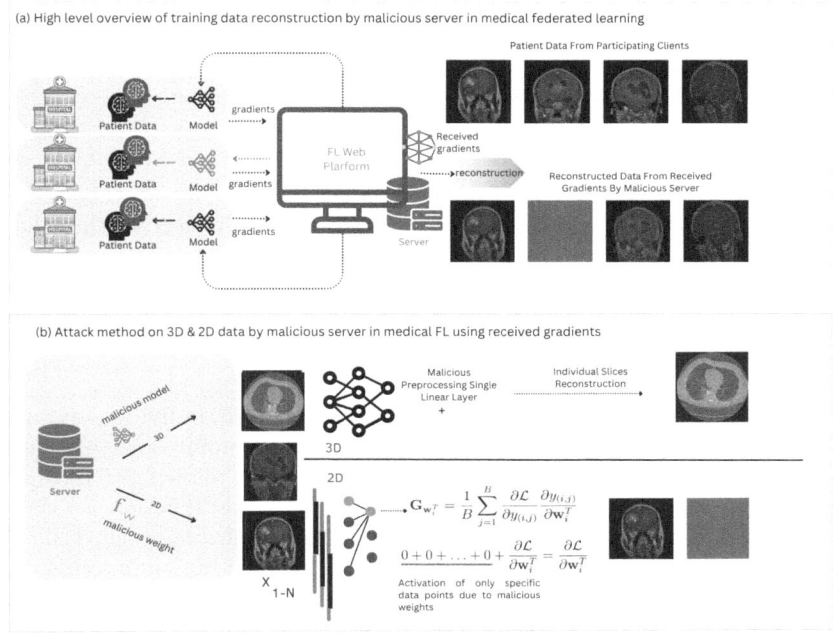

(a) High level overview of training data reconstruction by malicious server in medical federated learning

(b) Attack method on 3D & 2D data by malicious server in medical FL using received gradients

Fig. 1. (a) Illustrates the federated medical imaging setup with hospitals connecting to a central server via a web platform for algorithm exchange and gradient sharing, mimicking real-world practices. (b) Depicts the central server's attack on client 3D and 2D medical imaging data through model preprocessing and malicious weight adjustments, demonstrating successful reconstruction of both data types.

2   Method

In FL with a central server, the server can introduce vulnerabilities when distributing models to clients. This study adapts a 2D attack method from [2] [5] to 3D, wherein a malicious server inserts a linear layer with ReLU activation before the first original network layer to extract sensitive image information from the clients' model updates.

Data Reconstruction Overview: Let $y_i = w_i^T x + b_i$, the output of neuron i in the inserted linear layer. Then the input x can be reconstructed from the gradients of the loss L with respect to both the bias and the weights in the following manner [2,6]:

$$\frac{\partial L}{\partial b_i} = \frac{\partial L}{\partial y_i}\frac{\partial y_i}{\partial b_i} = \frac{\partial L}{\partial y_i}, \quad \frac{\partial y_i}{\partial b_i} = 1, \tag{1}$$

$$\frac{\partial L}{\partial w_i^T} = \frac{\partial L}{\partial y_i}\frac{\partial y_i}{\partial w_i^T} = \frac{\partial L}{\partial b_i}x^T, \tag{2}$$

assuming only a single sample x is passed through the network. If any $\frac{\partial L}{\partial b_i} \neq 0$, perfect reconstruction is achieved by:

$$x^T = \left(\frac{\partial L}{\partial b_i}\right)^{-1} \frac{\partial L}{\partial w_i^T} \tag{3}$$

where ReLU activation is > 0. This shows that the input data can be retrieved from the gradients of the linear layer at the front which receives that data as input. When using mini batches, for the ith neuron, the batch gradient is represented by:

$$G_{w_i}^T = \frac{1}{B} \sum_{j=1}^{B} \frac{\partial L}{\partial y_{(i,j)}} \frac{\partial y_{(i,j)}}{\partial w_i^T} \tag{4}$$

Under the condition that all but one term in Eq. (4) vanish, individual datapoints can be reconstructed from minibatch gradients. As [2] point out, this happens when the ReLU activations are zero for all but one input, allowing to reconstruct samples even when $B > 1$. They further show that reconstruction is even possible when using FedAvg. Initializing the layer with malicious weights allows to increase the chances for the exact reconstruction of individual datapoints. In our experiments, A malicious server added an attack layer, which effectively inserted a linear layer before the actual network. A single federated optimization step was simulated, after which gradients were shared with the server. The server used the gradients from the attack layer to invert the fully connected layer's gradients for data reconstruction.

Adaptation to 3D Data: We introduce a methodology within the FL round where a malicious server employs a custom preprocessor to execute attacks. This preprocessor converts the original 3D volume into 2D slices, facilitating the server's malicious algorithm. This approach extends the 2D attack by injecting specialized preprocessing code, enabling spatial mapping of the extracted data to the corresponding slices to accurately reconstruct the 3D volume. The server then applies the same attack method described in Eq. 3 to reconstruct the 2D slices. These slices are then reassembled into one 3D volume of the target client's data, as we are focusing on a single client's single 3D volume data during a specific epoch. By incorporating trap weights into the initial fully connected (FC) layer, we can increase the attack's accuracy when targeting a specific client's specific batch. For weight row w_i, with N and P indicating negative and positive weights, a neuron activates with ReLU if negative-weighted sums are less than positive-weighted sums. Due to this malicious weight initialisation 5 hold rare inputs, typically only affecting a single data point per mini-batch which can be then extracted using 3.

$$\sum_{n \in N} w_{n,i} x_n < \sum_{p \in P} w_{p,i} x_p. \tag{5}$$

3   Experiments

Brain Tumor MRI Dataset (2D Attack): [3] contains 7023 MRI brain tumor slices with dimensions 512×512. Each slice is labeled with one of the 4 classes Glioma, Meningioma, Pituitary, and No Tumor. For the 6-client experiment, we used a total of 1800 images (300 per client). For the 12-client experiment, we used 2640 images (220 per client). In both setups, the server targeted one client and metrics are based on this target client's dataset. Experiments were conducted on an Nvidia V100 16GB GPU, with training data limited to 1800 and 2640 images due to memory constraints.

MosMedData (3D Attack): [14] includes 1110 anonymized CT lung scans with dimension of $512 \times 512 \times (36\text{--}41)$, comprising 42% male, 56% female participants, with the remaining unknown. The scans were categorized into five classes of COVID-19 severity. For the experiment, we selected volumes with 40 slices.

Technical Setup: We conducted FL simulations with 6 and 12 clients for both 2D and 3D models, with one client also acting as the central server. These settings reflect realistic client numbers typical in medical imaging collaborations. Each client received a local private dataset, generated by a random uniform split of the main datasets. Experiments, focusing on analytic attacks rather than optimization-based ones, were implemented using PyTorch [17].

Reconstruction Attack: 2D vs Proposed 3D: In the 2D Attack, the central server provides algorithms to participating clients with initial weights. For the 4-class classification problem, we employed the ResNet18 [9] architecture, also adapting it for 3D attack. Subsequently, the central server possesses the ability to engage with any client during any epoch round. The 2D attack perform effectively when individual client weight updates are observable. The 2D method fails for 3D data due to its inability to maintain spatial correlations across all three dimensions, leading to suboptimal reconstructions. The original 3D file undergoes conversion into slices before being fed into the model by the model preprocessor, which is initialized by the malicious server. This preprocessing step can be a lightweight snippet initialized by the server. Subsequently, the server initializes malicious weights to target the client data with the malicious model. Once the slices are reconstructed by the server from received gradients, they are converted into a numpy array and patched together to generate the 3D data (Fig. 2).

4   Results and Discussion

For 2D reconstruction metrics, we used Mean Squared Error (MSE) and Structural Similarity Index (SSIM). For 3D reconstruction, we chose SSIM and Peak Signal-to-Noise Ratio (PSNR). PSNR was preferred over MSE for its ability to consider the dynamic range of pixel values in 3D metrics. PSNR assesses image quality by calculating MSE between original and reconstructed images,

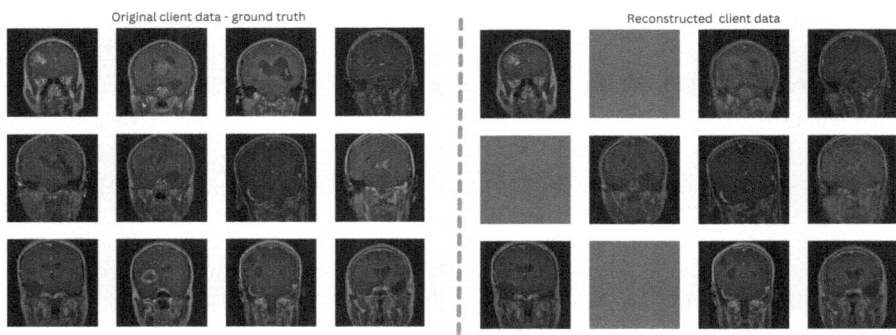

Fig. 2. 2D Brain Tumor MRI Data Reconstruction attack: The left side of the green line depicts the original client data, while the right side displays the corresponding reconstructed images derived from shared gradients of clients reconstructed by a malicious central server. The brown block indicates images not reconstructed, with reconstruction success dependent on the impact of malicious weights on neurons. (Color figure online)

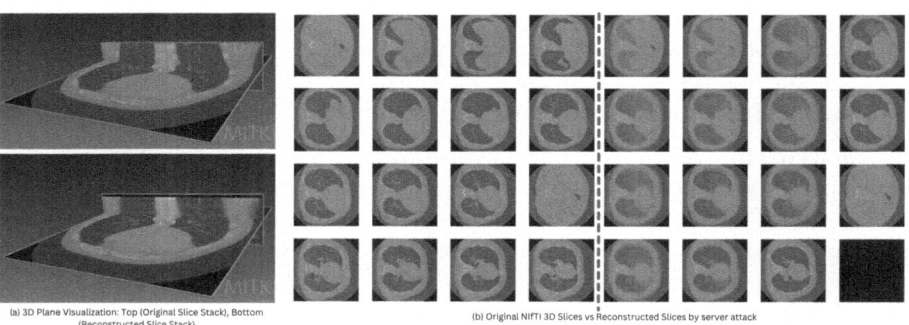

Fig. 3. 3D Attack: (a) depicts the original and reconstructed volume with two image planes cut through each of them, respectively. The axial plane (red frame) shows one slice of the slice stack (completely shown in (b)). The coronal plane (blue frame) shows a cut through all slices of the respective volume, visualized using MITK [19] Figure (b) contrasts original 3D CT scan slices on the left with their reconstructed counterparts on the right, processed by a malicious server from gradients, where black indicates failed reconstruction. (Color figure online)

offering pixel-level accuracy. SSIM evaluates perceptual quality by comparing luminance, contrast, and structure. MSE and PSNR quantify numerical differences, providing accuracy insights, while SSIM measures perceptual similarity, reflecting human visual perception. Together, these metrics offer a comprehensive evaluation of reconstruction quality.

2D Attack. The 2D attack experiment (Table 1) reveals the critical influence of batch size on reconstruction attack efficacy. With six federated clients, increas-

Table 1. Comparative analysis of reconstruction metrics for 2D brain tumor data in FL experiments, focusing on scenarios with 6 and 12 federated clients. The evaluation encompasses varying batch sizes and the number of data points reconstructed from targeted client, utilizing 300 training data points from the targeted client in the 6-client FL scenario and 220 in the 12-client FL scenario.

FL Clients	Batch	Reconstructed images	MSE	SSIM
6	8	114/300	0.0003	0.7171
	12	102/300	0.0007	0.5665
	16	102/300	0.0009	0.5822
	24	99/300	0.0012	0.5212
	32	107/300	0.0025	0.6171
	64	113/300	0.0040	0.4134
	128	102/300	0.0070	0.5139
12	8	89/220	0.0001	0.8666
	12	94/220	0.0001	0.8497
	16	86/220	0.0002	0.8240
	24	86/220	0.0002	0.8489
	32	82/220	0.0002	0.8189
	64	72/220	0.0002	0.8379
	128	64/220	0.0002	0.8070

ing batch size decreases the number of reconstructions and degrades quality, evidenced by higher MSE and lower SSIM. Larger batch sizes lead to fewer affected data points per batch, as malicious weights deactivate other data points, allowing only positive neurons to activate and potentially leak a single data point. This complicates reconstructing gradients of multiple activated neurons, as shown in Fig. 1b of the 2D scenario. Conversely, with twelve federated clients, larger batch sizes improve reconstruction quality, marked by lower MSE and higher SSIM when compared with 6 clients. This suggests that while the quantity of data points susceptible to attack may decrease, the quality of the leaked data points remains high even at larger batch sizes, such as 128.

3D Attack. The results obtained from our investigation into the reconstruction metrics of 3D chest CT scans Fig. 3 present valuable insights into the vulnerability of sensitive data to malicious reconstruction attempts. Table 2 presents the results of one target client's average reconstructed slices from a single 3D volume with a depth of 40 slices, showing the average slices/depth that the server can reconstruct across different batches of a single 3D volume from the target client. Our experiments, conducted with six federated clients, involved varying batch sizes to evaluate the reconstruction process. The results demonstrate a clear trend wherein the quality of reconstruction diminishes as the batch size increases. Specifically, as the number of reconstructed slices per batch rises, both SSIM and PSNR exhibit a gradual decline. For instance, with a batch size of 8, the SSIM is measured at 0.8095 and the PSNR at 65.95, whereas with a batch size of 64, these metrics decrease to 0.6221 and 53.66, respectively. Moreover, the total number of reconstructed slices also plays a crucial role in determining the efficacy of the reconstruction process. We also show that a decrease in the total number of reconstructed slices corresponds to a reduction in both SSIM and PSNR metrics. This suggests that the reconstruction rate is less effective when the batch size increases.

Table 2. Reconstruction metrics for 3D chest CT imaging in FL (reconstructed from the targeted client) were evaluated with 12 federated clients and varied batches, averaged over 3 runs in each case. Images consisted of 40 slices. Among these, some slices are reconstructed, and their metrics are detailed. These slices are then assembled to recreate the original 3D image.

Batch	Average reconstructed slices	SSIM	PSNR
8	34/40	0.8095	65.95
12	32/40	0.7730	69.86
16	30/40	0.7544	65.30
24	22/40	0.7143	65.11
32	19/40	0.7055	51.23
64	14/40	0.6221	53.66

5   Limitations and Mitigation Strategies

In this study, we explore the threat of malicious central servers in FL consortia, even when client security is high. This risk is crucial in FL collaborations across large jurisdictions or competitive environments. Client security alone does not ensure data safety, especially in cross-organizational FL where servers may be malicious. A malicious server can easily attack clients if it provides the training algorithm, which is common in FL. While we tested with ResNet18, this analytic attack can extend to other models. Despite the simplicity of our attack, real-world attackers would likely use more sophisticated methods, necessitating robust countermeasures by institutions.

To address risks associated with malicious server attacks we outline a multifaceted, effective strategy. This includes monitoring server models with a robust scanning architecture to detect unauthorized changes, using encrypted weights, auditing Fully Connected layers and ensuring code integrity through Hashsum Verification. A designated Clearance Officer, or a designated board, can ensure model update and result integrity by vetting changes to preserve trust in the learning process, while a dedicated person/group/third party algorithm monitors FL logs for significant weight changes during iterations. Increasing batch sizes and the number of FL clients in a 3D environment appears to enhance data protection effectively. Although it remains to be seen how common such malicious server attacks are in practice, they represent a significant threat capable of leaking high-fidelity data, particularly in sensitive domains like medical imaging. Due to the complex structure and spatial information, reconstruction is challenging when the algorithm reads the 3D scan as it is. Our future research will focus on reading 3D data directly and reconstructing the entire volume.

In conclusion, reconstructing over 35% of private high-fidelity 2D MRI and 3D CT data reveals significant privacy risks. Hospitals in FL should seek help to ensure patient privacy beyond software and network security.

Acknowledgement. This work was partially supported by the PrivateAIM project, which is funded as part of the Medical Informatics Initiative by the German Federal Ministry of Education and Research (funding code 01ZZ2316A-O).

Disclosure of Interests. The authors have no competing interests to declare that are relevant to the content of this article.

References

1. Beutel, D.J., et al.: Flower: a friendly federated learning research framework. arXiv preprint arXiv:2007.14390 (2020)
2. Boenisch, F., Dziedzic, A., Schuster, R., Shamsabadi, A.S., Shumailov, I., Papernot, N.: When the curious abandon honesty: federated learning is not private (2023)
3. Chaki, J., Wozniak, M.: Brain tumor MRI dataset (2023). https://doi.org/10.21227/1jny-g144
4. Dimitrov, D.I., Balunović, M., Konstantinov, N., Vechev, M.: Data leakage in federated averaging (2022)
5. Fowl, L., Geiping, J., Czaja, W., Goldblum, M., Goldstein, T.: Robbing the fed: directly obtaining private data in federated learning with modified models (2022)
6. Geiping, J., Bauermeister, H., Dröge, H., Moeller, M.: Inverting gradients-how easy is it to break privacy in federated learning? Adv. Neural. Inf. Process. Syst. **33**, 16937–16947 (2020)
7. Haim, N., Vardi, G., Yehudai, G., Shamir, O., Irani, M.: Reconstructing training data from trained neural networks. Adv. Neural. Inf. Process. Syst. **35**, 22911–22924 (2022)
8. Hatamizadeh, A., et al.: Do gradient inversion attacks make federated learning unsafe? IEEE Trans. Med. Imaging (2023)
9. He, K., Zhang, X., Ren, S., Sun, J.: Deep residual learning for image recognition. In: Proceedings of the IEEE Conference on Computer Vision and Pattern Recognition, pp. 770–778 (2016)
10. Kades, K., Scherer, J., Zenk, M., Kempf, M., Maier-Hein, K.: Towards real-world federated learning in medical image analysis using kaapana. In: International Workshop on Distributed, Collaborative, and Federated Learning, pp. 130–140. Springer, Cham (2022)
11. Kaissis, G., et al.: End-to-end privacy preserving deep learning on multi-institutional medical imaging. Nat. Mach. Intell. **3**(6), 473–484 (2021)
12. McMahan, B., Moore, E., Ramage, D., Hampson, S., Arcas, B.A.: Communication-efficient learning of deep networks from decentralized data. In: Artificial Intelligence and Statistics, pp. 1273–1282. PMLR (2017)

13. MonaiFL: Monai Federated Learning. https://docs.monai.io/en/latest/fl.html
14. Morozov, S.P., et al.: MosMedData: data set of 1110 chest CT scans performed during the covid-19 epidemic. Digit. Diagn. **1**(1), 49–59 (2020)
15. NVFlare: NVIDIA FLARE: Federated Learning from Simulation to Real-World. https://developer.nvidia.com/flare
16. Pasquini, D., Francati, D., Ateniese, G.: Eluding secure aggregation in federated learning via model inconsistency (2022)
17. Paszke, A., et al.: Automatic differentiation in pytorch (2017)
18. Schwarz, C.G., et al.: Identification of anonymous MRI research participants with face-recognition software. N. Engl. J. Med. **381**(17), 1684–1686 (2019)
19. MITK (2023). https://github.com/MITK/MITK
20. Teo, Z.L., Zhang, X., Tan, T.F., Ravichandran, N., Yong, L., Ting, D.S.: Federated machine learning in medical imaging and against adversarial attacks: a retrospective multicohort study. Invest. Ophthalmol. Visual Sci. **64**(8), 546–546 (2023)
21. Usynin, D., Rueckert, D., Kaissis, G.: Beyond gradients: exploiting adversarial priors in model inversion attacks. ACM Trans. Priv. Secur. **26**(3), 1–30 (2023)
22. Wen, Y., Geiping, J., Fowl, L., Goldblum, M., Goldstein, T.: Fishing for user data in large-batch federated learning via gradient magnification (2022)
23. Zhao, B., Mopuri, K.R., Bilen, H.: IDLG: improved deep leakage from gradients (2020)
24. Zhou, J., et al.: Personalized and privacy-preserving federated heterogeneous medical image analysis with PPPML-HMI. Comput. Biol. Med. **169**, 107861 (2024)
25. Zhu, L., Liu, Z., Han, S.: Deep leakage from gradients (2019)

Context-Guided Medical Visual Question Answering

Wafa Arsalane[1]([✉]), Philip Chikontwe[3], Miguel Luna[2], Myeongkyun Kang[2], and Sang Hyun Park[1,2,4]

[1] Department of Artificial Intelligence in Interdisciplinary Studies, Daegu Gyeongbuk Institute of Science and Technology (DGIST), Daegu, Korea
`{wafa,shpark13135}@dgist.ac.kr`
[2] Department of Robotics and Mechatronics Engineering, Daegu Gyeongbuk Institute of Science and Technology (DGIST), Daegu, Korea
[3] Department of Biomedical Informatics, Harvard Medical School, Boston, MA, USA
[4] Department of Biostatistics, Epidemiology and Informatics, University of Pennsylvania, Philadelphia, PA, USA

Abstract. Given a medical image and a question in natural language, medical VQA systems are required to predict clinically relevant answers. Integrating information from visual and textual modalities requires complex fusion techniques due to the semantic gap between images and text, as well as the diversity of medical question types. To address this challenge, we propose aligning image and text features in VQA models by using text from medical reports to provide additional context during training. Specifically, we introduce a transformer-based alignment module that learns to align the image with the textual context, thereby incorporating supplementary medical features that can enhance the VQA model's predictive capabilities. During the inference stage, VQA operates robustly without requiring any medical report. Our experiments on the Rad-Restruct dataset demonstrate a significant impact of the proposed strategy and show promising improvements, positioning our approach as competitive with state-of-the-art methods in this task.

Keywords: Medical Visual Question Answering · VQA · Medical Image Interpretation · Radiology

1 Introduction

Medical Visual Question Answering systems can streamline healthcare workflow efficiency by allowing quick retrieval of relevant information from medical images. This can save valuable time for healthcare providers and offer additional insights into diagnostic procedures while assisting in clinical decision-making. Despite holding such potential, the application in the medical field has been minimal due to the small-scale of available datasets, the complex nature of medical images, the diversity of questions and the high level reasoning required to answer them.

© The Author(s), under exclusive license to Springer Nature Switzerland AG 2025
U. Anazodo et al. (Eds.): MImA 2024/EMERGE 2024, CCIS 2240, pp. 245–255, 2025.
https://doi.org/10.1007/978-3-031-79103-1_25

Initial research [5,11] attempted to transfer advances in general VQA to the medical domain. However, Medical VQA faces unique constraints related to the acquisition and processing of data [16]. Usually, constructing medical VQA datasets require costly expert annotation and professional knowledge, *e.g.*, extracting Question-Answer (QA) pairs directly from a medical image needs domain specific expertise. This limitation often restricts the data collection process leading to small size datasets. Unlike general-domain VQA datasets, such as VQA [3], that contains hundreds of thousands of samples, medical VQA datasets are limited to tens of thousand images and QA pairs [4,6,7,14].

Recent research focuses on harnessing the potential of attention-based pre-trained biomedical visual language models such as BioGPT [18] and BLIP-2 [15] in a generative strategy. For instance, Van Sonsbeek *et al.* [24] proposed mapping visual features to a set of tokens that prompt a GPT-XL decoder [23]. For classification-based VQA, capturing high-quality medical features is crucial. For this purpose, Mixture Enhanced Visual Features model (MEVF) [19] uses the Denoising Auto-Encoder (DAE) as a visual extractor to enhance the quality of visual features extracted from medical images. With a typical focus on either visual feature extraction or fusion models, recent works [2,12] have seen a surge in the deployment of vision-language models as image encoders for their significant capability of producing high-quality features. However, such approaches rely heavily on transformers and attention mechanism for feature aggregation and fusion [25]. While attention has proven to be effective in cross-modal settings [2] and text-image fusion [1], this ties the advancement of medical VQA to the development status of attention mechanism and restricts the progress to fusion models. In contrast, this work posits that incorporating additional information to provide textual context during training could enhance both feature extraction and feature fusion.

Inspired by the NLP Question Answering task [20], we propose to use free-text medical reports as additional context to enhance visual feature extraction during training. Specifically, medical reports provide context we use to align image features with the same embedding space as the questions. This alignment improves the quality of image features, enabling accurate responses to questions during the inference stage without requiring additional context input. Our focus is on a closed-ended classification VQA task, where the targets are a predefined set of answers. In our pipeline, we first summarize medical reports using the GPT model [8] to effectively clean and pre-process the raw free-text for better encoding. Then, our model incorporates context features from summarized medical reports through a trainable alignment module that learns meaningful correlations between the textual and visual features. This enhances the image content as a reference for the question and helps identify fine-grained visual details guided by the medical text. To generate the final classification input of the VQA model, a fusion module is trained to combine the aligned image-context features and the encoded question features. Moreover, we use a vision encoder of a pre-trained vision-language model as an image encoder and an LLM-based encoder for text encoding without extra finetuning. By doing so, we significantly reduce the number of trainable parameters and the computational complexity of our model. Experiments on the Rad-Restruct dataset [21] show an increase in performance outperforming the baseline model, with a state-of-the-art accuracy.

In summary, our research contributes in three distinct aspects. First, we propose the first approach to use free-text medical reports as context to guide the prediction of answers in medical VQA. Second, we harness the power of pre-trained models throughout our model while reducing computational complexity by 90%. Finally, we prove through experiments that our strategy outperforms the baseline methods on different metrics with state-of-the-art accuracy.

2   Methodology

Figure 1 presents an overview of our architecture. The primary objective of this method is to utilize free-text reports to enhance the visual reference space, thereby providing a more comprehensive context for answering questions. To achieve this, we first encode the medical report associated with each image to generate textual context, and introduce a module to align this context with image features. Subsequently, we fuse the context-image features with the question vector and feed them into an MLP classifier for final answer prediction. Our framework has 5 main components: 1) a text encoder for encoding the question and context, 2) an image encoder for encoding image features, 3) an alignment module that learns to align and fuse the context and image features, 4) a fusion model that learns to combine the aligned context-image vector with the question features, and 5) a classifier that is trained to predict the answer.

2.1   Report Context Processing

The majority of free-text medical reports are divided into sections *i.e.*, *Impression* and *Findings*. Following prior literature on report generation [13], we focus on the *Impression* section as it provides less general information, and more detailed medical findings. The extracted text is summarized using a prompted GPT model and the output summary is used as context. The summarizing process aims not only to potentially shorten the reports but also to clean and simplify the text descriptions. Herein, we denote the training data as follows: $D = (c_i, q_i, I_i, y_i)_{i=1}^{N}$ where c_i represents the context (text), q_i the question (text), I_i is the image, and y_i the corresponding ground truth label (text).

2.2   Model Architecture

For an input image I_i, we extract the visual features V_I using a pre-trained image encoder E_{image}. Formally,

$$V_I = E_{image}(I_i). \tag{1}$$

Similarly, we obtain feature vectors of the question V_q and context V_c through a text encoder E_{text} as follows:

$$V_q = E_{text}(q_i), \quad V_c = E_{text}(c_i). \tag{2}$$

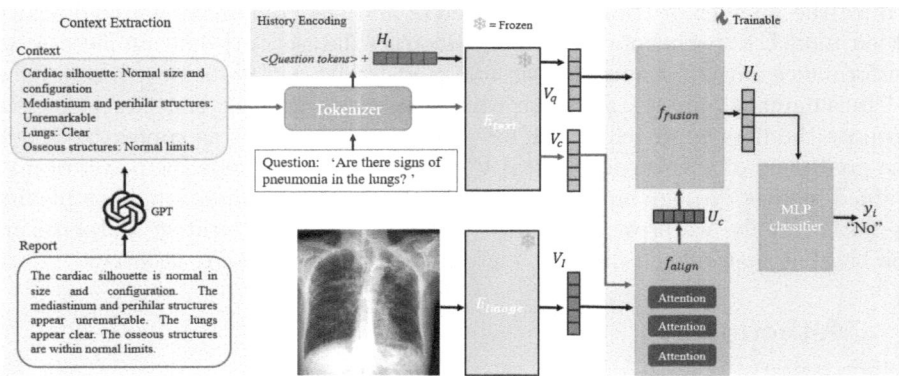

Fig. 1. Model architecture of our proposed context-guided VQA method. E_{text} and E_{image} denote, respectively, an image and a text encoder. f_{align} is an alignment network and f_{fusion} denotes a fusion model. The leftmost section illustrates the process of extracting context from free-text reports using GPT API. The encoders extract V_q, V_c, and V_I, then the context embedding U_c is aligned in f_{align}. Next, f_{fusion} produces a final embedding vector U_i which serves as the input of the MLP classifier that predicts the answer y_i. The history vector H_i contains tokens of all the previous questions with their answers.

To learn correlations between the image and context, we integrate a transformer-based alignment network f_{align} that produces a final context embedding U_c as follows:

$$U_c = f_{align}(V_c, V_I). \tag{3}$$

f_{align} is a stack of three multi-head self-attention layers that compute cross-attention scores between textual and image features. Here, we consider context as query, and image embeddings as key-value pairs. Each layer learns to attend to both image and textual context, producing context-aligned image features. The final aligned representation is obtained via the upper attention layer. The aligned image features implicitly represents image regions that are most relevant to the textual content.

Following the work of Pellegrini *et al.* [21], we adopt an autoregressive strategy and incorporate history information as input. At each step, the history vector H_i contains the previous (higher-level) questions with their answers. For example, the question *"In which part of the body?"* requires prior knowledge from higher level questions such as *"Is there an opacity in the lung?"*. The history vector is constructed by concatenating previous and current question tokens.

Next, a multi-modal transformer-based fusion model f_{fusion} is trained to combine the question and the image-context features to generate a final embedding vector U_i for classification. Formally,

$$U_i = f_{fusion}(U_c, H_i, V_q), \tag{4}$$

Original Report

Summary

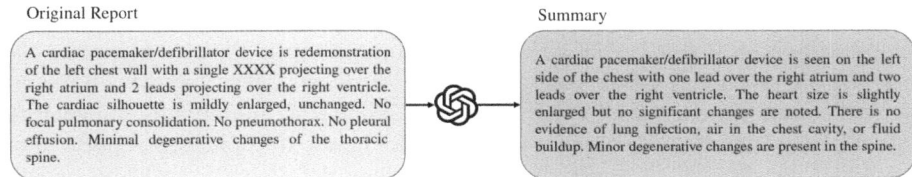

A cardiac pacemaker/defibrillator device is redemonstration of the left chest wall with a single XXXX projecting over the right atrium and 2 leads projecting over the right ventricle. The cardiac silhouette is mildly enlarged, unchanged. No focal pulmonary consolidation. No pneumothorax. No pleural effusion. Minimal degenerative changes of the thoracic spine.

A cardiac pacemaker/defibrillator device is seen on the left side of the chest with one lead over the right atrium and two leads over the right ventricle. The heart size is slightly enlarged but no significant changes are noted. There is no evidence of lung infection, air in the chest cavity, or fluid buildup. Minor degenerative changes are present in the spine.

Fig. 2. An example of a summarized report by the GPT model. The generated summary has been cleaned of special characters, privacy tokens and image-related measurement numbers, and simplified to use more natural language.

where H_i is the history information for the current question q_i. For each image, H_i begins as an empty list. After the first question is answered, the list is updated to include the question and its answer tokens. This process repeats after each subsequent question until the set has been completed. Finally, the MLP classifier predicts the answer y_i given the features of the resulting U_i.

During the training phase, the alignment module is trained using both the context and the image features to produce a vision-context vector. Similar to [21], we train our model with a cross-entropy loss in an autoregressive manner taking in consideration the question history. For inference, our approach assumes an absence of context and takes as input the image and questions. The alignment module creates a context vector using the image through the learned textual representation during the training process.

3   Experiments

Dataset. We use the Rad-Restruct dataset [21], which contains 3720 Xray images, 3597 structured reports, and 180k questions. For each image, a corresponding free-text report is retrieved from the IU-Xray dataset [9]. Images were normalized and cropped to 224×224. We use the original 80-10-10 split for training, validation, and testing, respectively. Rad-Restruct questions are categorized into 3 levels. Level 1, the highest level, contains general-purpose questions such as *"Are there any foreign objects?"*. Level 2 questions ask for more specific findings, like *"Is there pneumonia in the lung?"*, and level 3 questions are related to detailed findings, for example, asking for the degree or a description of a disease, *"What are the attributes?"*. This hierarchical structure allows an autoregressive parsing of the questions, making lower-level questions depend on higher-level ones, which aims to provides more context for difficult questions in level 3 question. In addition, there are 96 classes in total. The questions can be single-choice or multiple-choice, with each having a defined set of possible options. To the best of our knowledge, at the time of the experiment, Rad-Restruct is the only available dataset that enables our experimental design by providing access to all three components, medical images, free-text reports, and QA pairs.

Context Extraction. To extract the context from radiology reports, we use the prompted GPT 3.5 Turbo with the following prompt: *'Please summarize the following X-ray report while keeping the medical terms.'* Fig. 2 shows an example of a summarized report by the GPT API. The GPT model aims to generate natural language text comprehensively. In doing so, it rephrases reports into simplified vocabulary, excluding sensitive medical details that could be beneficial. We address this concern by requesting the retention of medical terms. The length of generated summaries varies based on the original report and content nature. We impose a maximum size of 512 tokens, truncating when needed. These summaries serve as a direct context for training our model.

Training and Evaluation. As an image encoder, we use PubMedClip [10], a variant of the CLIP model [22] designed for medical VQA, it is pre-trained on several medical image modalities. Following the model introduced in [21], we adopt the RadBert model as a text encoder, with the Bert Tokenizer. RadBert is a domain-specific large language model based on Roberta model [17], and was trained on a large corpus of radiology reports. To encode text inputs, we leverage RadBert's pre-trained embeddings, which capture domain-specific semantics and contextual. The encoders are frozen, preserving their pre-trained weights and preventing further parameter modification. As a classifier, we use a 5-layer MLP with batch normalization and a dropout rate of 0.2. The model is trained for 200 epochs on a NVIDIA RTX A6000 GPU. We use Adam Optimizer with a learning rate of 1e-5 and a batch size of 32.

4   Results

Table 1 shows the performance of our Context-VQA method and the state-of-the-art method hi-VQA [21] on the Rad-Restruct dataset. For a fair comparison, we use the same evaluation approach and script as hi-VQA and report our results accordingly using the conventional metrics Accuracy, F1, Precision, and Recall. The main results in Table 1 correspond to the best run scores among multiple runs. As accuracy, we provide the report accuracy, a metric that is specific to the dataset.

Rad-Restruct was built to structure the text reports in the IU-Xray dataset into a form populated by hierarchical questions. Thereby, each image is accompanied by its set of questions referred to as a structured report. The report accuracy metric represents the accuracy of reports that were fully predicted correctly. Thus, a report is considered correct if all questions at all levels for a given image are perfectly answered. Context-VQA outperforms hi-VQA on this specific metric with almost a 10% increase. This translates to an increase of fully predicted reports' questions by 10%. The evaluation was done using the script provided in [21], allowing direct comparison to hi-VQA. Our results significantly outperform the hi-VQA on the other metrics, with precision improved by 29%. We want to clarify that when discussing results, "hi-VQA" refers to the

Table 1. Performance comparison on Rad-Restruct dataset. We compare 5 methods, a visual baseline [21], hi-VQA, both the published results and the released dataset results, re-VQA (our reproduced hi-VQA results) and our method Context-VQA.

	Report Accuracy	F-1	Precision	Recall
Visual baseline [21]	31.3	30.7	65.6	31.2
hi-VQA [21]	32.6	31.7	70.7	32.1
hi-VQA[a]	30.2	32.0	64.6	33.3
re-VQA	29.8	28.7	61.2	29.8
Context-VQA	**39.7**	**32.9**	**90.4**	**33.6**

[a]Released dataset results https://github.com/ChantalMP/Rad-ReStruct/tree/master

Table 2. A comparison of the hi-VQA and Context-VQA for each question level. Context-VQA scores the best accuracy over all levels, while F1 and Recall alternate between the two models on different levels.

	hi-VQA				Context-VQA			
	Accuracy	F-1	Precision	Recall	Accuracy	F-1	Precision	Recall
Level 1	33.6	64.3	**81.0**	**64.5**	**34.7**	**67.2**	80.7	61.2
Level 2 (L2) all	31.0	71.6	85.2	**72.0**	**32.9**	**71.8**	**88.9**	70.8
- L2 diseases	48.1	**73.5**	83.8	71.3	**52.1**	72.8	**89.6**	**72.7**
- L2 signs	71.9	**74.2**	**93.1**	**74.4**	**74.3**	73.7	90.6	73.7
- L2 objects	87.4	67.0	77.1	67.5	**91.4**	**67.2**	**85.0**	**68.6**
- L2 regions	52.4	68.1	82.1	**69.5**	**61.2**	**68.7**	**85.4**	68.3
Level 3	30.2	**4.1**	49.9	**6.2**	**32.5**	3.2	**68.7**	4.2

updated results provided by the authors along with the dataset after the paper was published.

Furthermore, Table 2 shows the performance on each level of questions. Context-VQA steadily achieves state-of-the-art accuracy on all levels. Although, the F-1 score shows a slight decrease for some type of questions in level 2 and level 3, level 1 and level 2 (all) show higher overall performance. In addition, it is noteworthy that our approach, with only 16M trainable parameters was able to perform closely or even better than hi-VQA (164M parameters). We also note that we compare to the released dataset results instead of the scores reported in the paper. Figure 3 illustrates qualitative examples of predictions of Context-VQA compared to the baseline predictions. Questions are arranged sequentially from left to right, level-wise, indicating their hierarchical dependency. In the first instance, hi-VQA predicts a negative response to the initial question, consequently influencing subsequent questions to also receive negative predictions. In examples 2 and 3, we observe that for these cases Context-VQA accurately predicts lower-level questions, which are often challenging and influ-

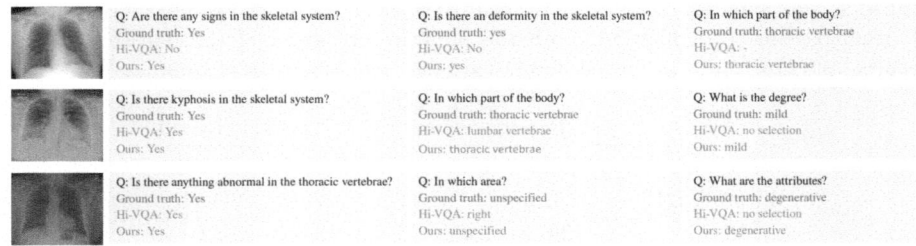

Fig. 3. Examples of the predictions where Context-VQA outperforms the hi-VQA model. The questions are successive from left to right, representing the hierarchical dependency.

Table 3. The computational complexity of our leveraged model against the full architecture and the baseline.

	hi-VQA	Context-VQA	Fully trained Context-VQA
Number of parameters (M)	164	16	232
Training time per epoch	1 h 30 min	35 min	1 h 48 min

enced by preceding questions. This explains the ability of our model to correctly predict sequential questions leading to the improvement in report accuracy, as previously reported.

Computational Complexity. By leveraging the pretrained encoders, we achieve a substantial reduction in the number of trainable parameters, from 232M to 16M, against hi-VQA's 164M parameters. Consequently, the training time is reduced. Experiments on a single A6000 GPU record the training time of 1h30min per epoch for hi-VQA against 35min required by Context-VQA, as demonstrated in Table 3. Notably with fewer parameters and less computational time, Context-VQA still performs well on the Rad-Restruct dataset. The fully trained Context-VQA model refers to our model with all components, including the encoders, trained. With 232 million trainable parameters, this model takes an average of 1 h and 48 min to train for a single epoch. However, due to the risk of overfitting on the relatively small dataset, we have strategically decided against pursuing this approach as it may not yield the most generalizable results.

Ablation Studies. We conducted ablation studies to assess the impact of contextual information on the prediction capabilities of our Context-VQA model. In the first experiment, we integrated the context as an input in the baseline hi-VQA model by tokenizing and feeding it into the text encoder alongside the question. The textual features were then concatenated with the visual features and processed in the fusion module. This experiment aimed to demonstrate the impact of incorporating textual information. The results, labeled as "hi-VQA+context,"

Table 4. A Comparison experiment investigating the impact of the context and the alignment module.

	Report Accuracy	F-1	Precision	Recall
hi-VQA	30.2	32.0	64.6	33.3
hi-VQA + context	32.3	30.4	76.6	30.2
Ours $-f_{align}$	32.6	28.7	80.0	28.8
Context-VQA	**39.7**	**32.9**	**90.4**	**33.6**

are presented in Table 3, showing a performance improvement compared to the baseline model. To further examine the significance of the alignment module, we adopted a similar approach of incorporating contextual information as an additional input in our model without further alignment. The results are reported as "our $-f_{align}$" in Table 4. We note that the recall and F1 scores exhibited a slight decrease, primarily attributed to the lack of encoder updates, which limited improvements in feature extraction.

5   Conclusion

In this work, we proposed a novel approach to enhance medical Visual Question Answering (VQA) systems by leveraging free-text medical reports as contextual information. We use the context by incorporating an adaptive text-image alignment module that learns to align textual and visual features. Through extensive evaluation on Rad-Restruct dataset, we validated the efficacy of integrating context-based information alongside images and questions that provides richer medical features for VQA with significant performance gains. Given the rise of datasets with additional medical reports and EHR, this work paves the way for further exploration and refinement of context-based approaches in advancing the capabilities of medical VQA systems and ultimately improving medical AI systems' outcomes.

Acknowledgement. This work was supported by the National IT Industry Promotion Agency (NIPA), an agency under the MSIT and with the support of the Daegu Digital Innovation Promotion Agency (DIP), the organization under the Daegu Metropolitan Government and Smart Health Care Program funded by the Korean National Police Agency (220222M01).

Disclosure of Interests. The authors have no competing interests to declare that are relevant to the content of this article.

References

1. Amam: An attention-based multimodal alignment model for medical visual question answering. Knowl. Based Syst. **255**, 109763–109763 (2022). https://doi.org/10.1016/j.knosys.2022.109763
2. The multi-modal fusion in visual question answering: a review of attention mechanisms. PeerJ **9**, e1400–e1400 (2023). https://doi.org/10.7717/peerj-cs.1400
3. Antol, S., et al.: VQA: visual question answering. CoRR abs/1505.00468 (2015). http://arxiv.org/abs/1505.00468
4. Ben Abacha, A., Datla, V., Hasan, S., Demner-Fushman, D., Müller, H.: Overview of the VQA-med task at ImageCLEF 2020: visual question answering and generation in the medical domain (2020)
5. Ben Abacha, A., Gayen, S., Lau, J., Rajaraman, S., Demner-Fushman, D.: NLM at ImageCLEF 2018 visual question answering in the medical domain (2018)
6. Ben Abacha, A., Hasan, S., Datla, V., Liu, J., Demner-Fushman, D., Müller, H.: VQA-med: overview of the medical visual question answering task at ImageCLEF 2019. Lecture Notes in Computer Science (2019)
7. Ben Abacha, A., Sarrouti, M., Demner-Fushman, D., Hasan, S., Müller, H.: Overview of the VQA-med task at ImageCLEF 2021: visual question answering and generation in the medical domain (2021)
8. Brown, T.B., et al.: Language models are few-shot learners (2020)
9. Demner-Fushman, D., Antani, S., Simpson, M., Thoma, G.R.: Design and development of a multimodal biomedical information retrieval system. J. Comput. Sci. Eng. **6**(2), 168–177 (2012)
10. Eslami, S., de Melo, G., Meinel, C.: Does clip benefit visual question answering in the medical domain as much as it does in the general domain? arXiv abs/2112.13906 (2021). https://api.semanticscholar.org/CorpusID:245537458
11. Hasan, S., Ling, Y., Farri, D., Liu, J., Müller, H., Lungren, M.: Overview of Image-CLEF 2018 medical domain visual question answering task (2018)
12. Hicks, S., Storås, A.M., Halvorsen, P., de Lange, T., Riegler, M., Thambawita, V.L.: Overview of imageclefmedical 2023 - medical visual question answering for gastrointestinal tract. In: Conference and Labs of the Evaluation Forum (2023). https://api.semanticscholar.org/CorpusID:264441584
13. Jing, B., Xie, P., Xing, E.: On the automatic generation of medical imaging reports. In: Gurevych, I., Miyao, Y. (eds.) Proceedings of the 56th Annual Meeting of the Association for Computational Linguistics (Volume 1: Long Papers), pp. 2577–2586. Association for Computational Linguistics, Melbourne, Australia (2018). https://doi.org/10.18653/v1/P18-1240. https://aclanthology.org/P18-1240
14. Lau, J., Gayen, S., Ben Abacha, A., Demner-Fushman, D.: A dataset of clinically generated visual questions and answers about radiology images. Sci. Data **5**, 180251 (2018). https://doi.org/10.1038/sdata.2018.251
15. Li, J., Li, D., Savarese, S., Hoi, S.: Blip-2: bootstrapping language-image pretraining with frozen image encoders and large language models (2023)
16. Lin, Z., et al.: Medical visual question answering: a survey. CoRR abs/2111.10056 (2021). https://arxiv.org/abs/2111.10056
17. Liu, Y., et al.: Roberta: a robustly optimized BERT pretraining approach. CoRR abs/1907.11692 (2019). http://arxiv.org/abs/1907.11692
18. Luo, R., et al.: Biogpt: generative pre-trained transformer for biomedical text generation and mining. Brief. Bioinform. **23**(6) (2022). https://doi.org/10.1093/bib/bbac409

19. Nguyen, B.D., Do, T.T., Nguyen, B.X., Do, T., Tjiputra, E., Tran, Q.D.: Overcoming data limitation in medical visual question answering (2019)
20. Pandya, H.A., Bhatt, B.S.: Question answering survey: directions, challenges, datasets, evaluation matrices. CoRR abs/2112.03572 (2021). https://arxiv.org/abs/2112.03572
21. Pellegrini, C., Keicher, M., Özsoy, E., Navab, N.: Rad-ReStruct: a novel VQA benchmark and method for structured radiology reporting. In: International Conference on Medical Image Computing and Computer-Assisted Intervention, pp. 409–419. Springer, Cham (2023)
22. Radford, A., et al.: Learning transferable visual models from natural language supervision. CoRR abs/2103.00020 (2021). https://arxiv.org/abs/2103.00020
23. Radford, A., Wu, J., Child, R., Luan, D., Amodei, D., Sutskever, I.: Language models are unsupervised multitask learners (2019). https://api.semanticscholar.org/CorpusID:160025533
24. van Sonsbeek, T., Derakhshani, M.M., Najdenkoska, I., Snoek, C.G.M., Worring, M.: Open-ended medical visual question answering through prefix tuning of language models (2023)
25. Vaswani, A., et al.: Attention is all you need. CoRR abs/1706.03762 (2017). http://arxiv.org/abs/1706.03762

GRAM: Graph Regularizable Assessment Metric

Mariem Touihri[1]([✉])[iD] and Ahmed Nebli[2][iD]

[1] Higher School of Digital Economy (ESEN), University of Manouba,
Ariana, Tunisia
mariem.touihri@esen.tn
[2] Independent Researcher, Jülich, Germany
mr.ahmednebli@gmail.com

Abstract. Here, we propose the Graph Regularizable Assessment Metric $(GRAM)$, a customizable tool for evaluating the quality of generated brain graphs. Current geometric deep learning methods often lack robust quantification techniques for assessing the synthetic brain graphs integrity. $GRAM$ addresses this gap by proportionally combining a set of existing graph metrics to establish a linear correlation between distortions' levels and metric values of ground-truth graphs. To evaluate the performance of our model, we generated a synthetic dataset of structural brain connectomes which was derived from an existing dataset and used to simulate a set of predicted connectomes from a generative model with controlled levels of distortions. Our results show that $GRAM$ outperforms single metrics in quantifying the distortion between generated and original graphs. This approach is a significant step towards establishing a universal graph quality index for graph-based predictive studies.

Keywords: Predicted brain graphs · Quality metrics · Customized metrics

1   Introduction

Brain connectomes are crucial for exploring the connectivity patterns underlying cognitive processes [23]. These connectomes provide a framework for predicting the progression of neurodegenerative diseases by integrating connectomic analyses with established neuroscientific knowledge [4]. For instance, [12,15] introduced geometric deep learning approaches to forecast Alzheimer's disease progression using brain connectome data. Despite their potential, obtaining connectomic data poses significant challenges. One major impediment is the extensive processing required for neuroimages acquired through modalities such as Magnetic Resonance Imaging (MRI). This process is both time-consuming and computationally intensive. Another challenge is the limited availability of sufficient MRI data, which can hinder comprehensive analyses.

A. Nebli—Independant researcher.

U. Anazodo et al. (Eds.): MImA 2024/EMERGE 2024, CCIS 2240, pp. 256–265, 2025.
https://doi.org/10.1007/978-3-031-79103-1_26

One approach to address the challenge of limited MRI data involves the use of generative models to produce synthetic neuroimages. Generative Adversarial Networks (GANs) [8] have shown significant capability in generating realistic brain scans. For instance, [16] proposed a GAN model based on a fully convolutional network and an Auto-Context Model to enhance the realism and accuracy of synthetic images. Similarly, [21] developed a GAN model that produces high-quality, realistic images that simulate the ground-truth brain images- to improve the performance of diagnostic models in medical diagnostics. Despite the potential benefits, using GAN-generated MRI data to study brain connectomes introduces two key challenges. First, the generated MRI data must be indistinguishable from real data both quantitatively and qualitatively. Second, the synthetic data requires additional processing to extract connectivity matrices.

To address the above-mentioned issues, [27] proposed predicting brain connectivity matrices using a graph GAN-based approach. The authors created representative templates from clustered brain graphs to train models that predict the evolution of connectivities for a given brain disease over time. Their novel few-shot learning framework uses minimal training data and employs clustering and Connectional Brain Templates (CBTs) to handle the diversity within brain connectomic data. This ensures robust model training despite limited data. However, unlike images, brain connectomes are virtually impossible to evaluate qualitatively. Instead, quantitative metrics (e.g., centrality measures [7], Average Neighbor Degree [26] and Diversity Index [20]) are, *thus far*, a single way to evaluate the quality of the generated graphs.

In this paper, we highlight the limitations of existing metrics for graph quality assessment and propose a novel universal customizable metric to quantify the quality of generated graphs with an application to a simulated prediction of brain connectivities based on an existing dataset. In particular, we propose Graph Regularizable Assessment Metric ($GRAM$), a customizable framework designed to learn to proportionally combine a set of existing graph metrics in order to evaluate the generated graph's quality. Drawing inspiration from the universal image quality index by [25], $GRAM$ could be considered a first step towards a more universal graph quality index. Our contributions are listed as follows:

1. We propose a new general assumption for quantitatively interpreting the quality of a generated graph based on the linearity between the amount of distortion and the value of the reported metric.
2. We propose a novel general metric based on the weighted combination of existing metrics.
3. Our proposed metric is adjustable depending on the type of graph as well as the chosen metrics to report.

2   Methods

In this section, we present in detail the proposed metric $GRAM$ for quantifying the quality of generated graphs.

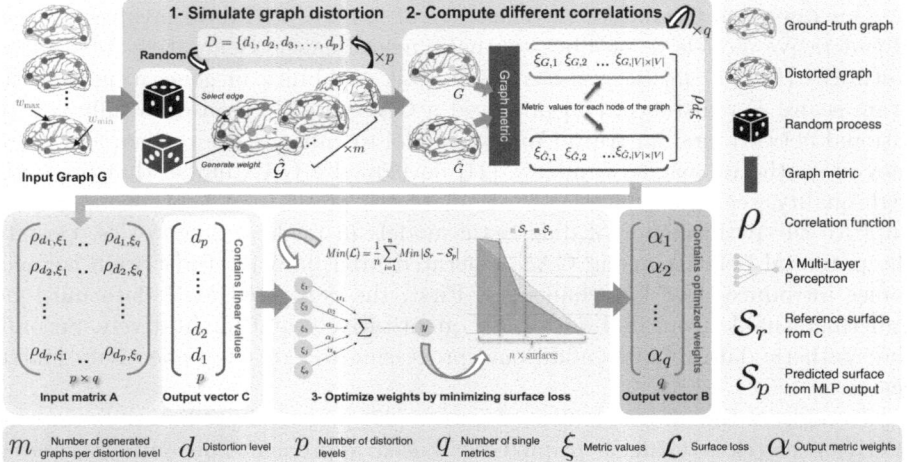

Fig. 1. *Pipeline of the proposed GRAM metric for assessing the quality of directed weighted graphs.* **(1) Simulate graph distortion.** For an input graph G and for a distortion level d we alter weights of randomly selected edges by random generated values in $[w_{\min}, w_{\max}]$ producing m distorted graphs. **(2) Compute different correlations.** First we apply the single metrics to the ground truth graph and the distorted one then calculate the Pearson, Spearman and Kendall's Tau correlations between them. Second for each graph and each correlation coefficient, we generate a matrix A of size $(p \times q)$ containing the correlation values organized by distortion levels vertically and single metrics horizontally. **(3) Optimize weights by minimizing surface loss.** We train $GRAM$ using an MLP to optimize the metrics' weights α_j forming the vector $B(q)$ by minimizing the loss between the predicted surface created by the MLP output $A \times B$ and the reference surface created by the vector $C(p)$ across n surfaces.

2.1   Simulation of Generated Brain Graphs

Let $G = (V, E)$ be a directed weighted graph, V denotes the vertices, and E denotes the weighted edges given by $w : e \to \mathbb{R}$. Let $\hat{G}$ be the simulation of the output of a given generative model F aiming to predict a target brain connectome G such that $\hat{G} \approx G$. The goal of the simulation is to bypass the problem of finding the optimal F to train as well as to control the amount of distortion d between $\hat{G}$ and G, where $d \in]0, 1]$ with s defined as distortion step.

The distortion level between $\hat{G}$ and G is measured by the number of edges $|E|$ with differing weights. Specifically, for an edge $e \in E$ with weight w in G and $\hat{w}$ in $\hat{G}$, the distortion is defined as the proportion of edges for which $w \neq \hat{w}$, regardless of the magnitude of the difference $|w - \hat{w}|$. The objective is to detect any distortion in the generated graph, treating any alteration in edge weights as significant.

As shown in Fig. 1, we define a set of m distorted graphs, each characterized by a distortion level d denoted as $\hat{\mathcal{G}}_d = \{\hat{G}_1, \ldots, \hat{G}_m\}$. The process of generating a suite of distorted graphs across all distortion levels is detailed in Algorithm 1, resulting in the set $\hat{\mathcal{G}} = \{\hat{\mathcal{G}}_s, \ldots, \hat{\mathcal{G}}_1\}$. Initially, we set a predefined distortion step (e,g., 0.1). For each increment of the distortion level d, we randomly select $|\hat{E}|$ edges, where $|\hat{E}| = d \times (|E| - |V|)$. Here, $|E|$ is the total number of edges in the graph, and $|V|$ is the total number of vertices. The term $|E| - |V|$ represents the total number of non-diagonal edges, as diagonal edges (self-loops) are excluded. Therefore, by subtracting $|V|$ from $|E|$, we ensure that we only consider non-diagonal edges. At each selected edge e, we replace its weight with a randomly generated value within the range $[w_{min}, w_{max}]$, where w_{min} and w_{max} are the minimum and maximum weight values in the graph G, respectively. We repeat this process m times to ensure that all the edges are distorted at least once.

Algorithm 1. Generate Distorted Graphs

Require: Directed weighted graph $G = (V, E)$, distortion step s, number of iterations m

Ensure: Set of distorted graphs $\hat{\mathcal{G}}$

 1: Initialize $\hat{\mathcal{G}} \leftarrow \emptyset$
 2: $w_{min} \leftarrow \min\{w(e) \mid e \in E\}$
 3: $w_{max} \leftarrow \max\{w(e) \mid e \in E\}$
 4: **for** d in D with step s **do**
 5:     Initialize $\hat{\mathcal{G}}_d \leftarrow \emptyset$
 6:     $|\hat{E}| \leftarrow d \times (|E| - |V|)$
 7:     **for** i from 1 to m **do**
 8:         $\hat{G} \leftarrow G$
 9:         Select $|\hat{E}|$ random edges from E
10:         **for** each selected edge e **do**
11:             $\hat{w}(e) \leftarrow \text{random}(w_{min}, w_{max})$
12:             Update edge weight in $\hat{G}$ to $\hat{w}(e)$
13:         **end for**
14:         Add $\hat{G}$ to $\hat{\mathcal{G}}_d$
15:     **end for**
16:     Add $\hat{\mathcal{G}}_d$ to $\hat{\mathcal{G}}$
17: **end for**
        return $\hat{\mathcal{G}}$

2.2   Graph Reliability Assessment Metric (GRAM)

Assumption 1: Let $G = (V, E)$ represent a brain graph, where V denotes vertices and E denotes edges, with $w : e \rightarrow \mathbb{R}$ representing the weights of the edges in E. For a metric $\mathcal{M}$ that assesses graph quality, we postulate that the variation in y such that:

$$y = \rho(\mathcal{M}(G), \mathcal{M}(\hat{G})) \tag{1}$$

is **linearly correlated** with the distortion d applied to G. Where $\hat{G}$ is the distorted graph, and ρ is the correlation function. We express $\mathcal{M}$ as follows:

$$\mathcal{M}(\hat{G}) = 1 - k \times d \tag{2}$$

where d is the distortion level expressed as a ratio (e.g., $d = 0.1$ for 10% distortion), and k is a constant scaling factor.

We introduce the $GRAM$: an adjustable and learnable measure for evaluating the quality of generated graphs. Unlike existing metrics, such as centrality measurements that separately assess different graph aspects, $GRAM$ provides a linear approximation of the relationship between distortion evolution and its output value taking into consideration multiple aspects of the graph. For instance, a $GRAM$ value of 0.8 indicates that the graph is 80% similar to the original data. The result of our metric is represented by the y value, which indicates the degree of similarity between the generated and original graphs. We opt for a linear model due to its ease of interpretation and analytical benefits [9]. In graph distortion context, the linear relationship clarifies how changes in edge weights impact overall metrics, enhancing result communication.

To do so, as seen in Fig. 1 for each graph, we define a matrix $A \in \mathbb{R}^{p \times q}$, such that p is the number of distortion levels, q is the number of existing metrics. Within A, each element $A_{i,j}$ represents the correlation between a given metric's output $\xi(G)$ and $\xi(\hat{G})$ applied to G and $\hat{G}$, respectively, at a particular distortion level d. Here i indexes a distinct distortion level, and j refers to a particular metric correlation (e.g., At a distortion increment $s = 0.1$, $A_{1,3}$ corresponds to the third metric correlation at a 10% distortion level). We define C as the reference output of the metric ensuring adherence to Assumption 1. $GRAM$ aims to find the values α_j forming a vector B such that: $A \times B = C$.

We define $GRAM(G)$ as a weighted sum of the metrics' correlations between the ground truth and distorted graphs. Specifically, let $\xi_j(G)$ and $\xi_j(\hat{G})$ denote the j-th metric evaluated on G and the distorted graph $\hat{G}$, respectively. Then:

$$GRAM(G) = \sum_{j=1}^{q} \alpha_j \times \rho(\xi_j(G), \xi_j(\hat{G})) \tag{3}$$

where ρ is the correlation function and α_j are the learnable weights for each metric's correlation.

To solve the vector B, we leverage the universal approximation theorem demonstrated by [10], which establishes that a feedforward neural network featuring a single hidden layer can approximate any continuous function with sufficient neurons and appropriate parameters (weights and biases). Consequently, our approach involves training a Multi-Layer Perceptron (MLP) to determine the parameters within B, taking A as input and C as output.

To train the MLP, we minimize the loss between two surfaces: S_r, formed by the intersection of vector C with the X and Y axes, and S_p, defined by the curve of predicted weights in B, where the vector $A \times B$ intersects the X and Y axes, Fig. 1 (3). The goal is to optimize the predicted weights in B so that the vector $A \times B$ closely approximates vector C, aligning surfaces S_r and S_p.

To do so, we use least squares regression [13] to fit a polynomial function

$$P(x) = a_n x^n + a_{n-1} x^{n-1} + \cdots + a_1 x + a_0$$

to the $A \times B$ output data. This involves finding a polynomial that minimizes the sum of the squared differences between the MLP output data points and the polynomial's predicted values. This method creates a continuous curve that closely follows the data pattern formed by $A \times B$, which we then use to approximate the integral within a specified range.

In our study, the loss is minimized by the following process: first, the total surface created by the MLP values is divided into n distinct parts. Each part is optimized independently to simplify parameter convergence. Finally, we average all the results from the optimizations.

Our proposed surface loss can be defined as follows:

$$\text{Surface Loss} = \frac{1}{n} \sum_{i=1}^{n} \int_{x_{\min}}^{x_{\max}} |f_C(x) - f_{\text{MLP}}(x)| \, dx \qquad (4)$$

where n represents the number of distinct surface parts, $x_{\min}$ and $x_{\max}$ denote the minimum and maximum values of the input range, respectively. The function $f_C(x)$ denotes the line defined by C values, while $f_{\text{MLP}}(x)$ corresponds to the polynomial approximation function of B (the MLP output) multiplied by A.

Training Details. We train our model for 250 epochs, using Google Colab. For optimization, we use Adam optimize [11], with learning rate of 0.01. We used an 80/20 split, resulting in a training set of 70,400 samples and a testing set of 17,600 samples. The training of $GRAM$ took 1 h and 14 min.

3   Results and Discussion

In this section, we evaluate 10 selected individual graph metrics as well as our proposed $GRAM$. Additionally, we discuss each of the findings.

3.1   Dataset

We used a dataset from [22] that contains 88 subjects (48 females, 40 males aged between 18 and 48 years). All subjects are right-handed and healthy. The dataset contains the structural connectomes where each connectome contains 90 brain regions of intrest from the Automated Anatomical Labeling Atlas (AALA) [24]. We emphasize that our simulations resulted in a total of 88,000 graphs.

3.2   Single Metric Evaluation

For this study, we select a step of $s = 0.1$ and a set of ten widely utilized graph metrics in the literature [18]. These metrics are: Betweenness Centrality [6], Closeness Centrality [19], Weighted Degree Centrality [1], Eigen Centrality [2], Pagerank Centrality [17], Katz Centrality [3], Hub-Authority [5], Harmony [14], Average Neighbor Degree [26], Diversity Index [20]. As a baseline for evaluating our proposed $GRAM$, Fig. 2, displays the correlation between the ground truth graphs and the generated ones, for each individual metric across various levels of distortion.

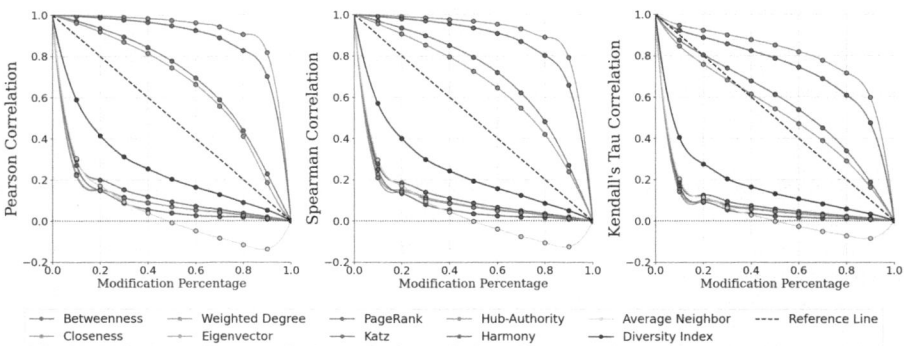

Fig. 2. *Correlations for single metrics.* We plot different correlations (Pearson, Spearman and Kendall's Tau correlations) between the ground truth graphs and the distorted ones, for each individual metric across various levels of distortion.

Figure 2 shows the evolution of the independent metrics differs across the studied correlation coefficients. These evolutions are non-linear and could be visually categorized into two distinct patterns. The first pattern includes metrics such as Betweenness, Closeness, Harmony, and Weighted Degree Centralities which exhibit a moderate progression for values of $d < 0.8$ followed by a rapid decline towards a correlation values of 0. Contrarily, the second pattern shows metrics such as Eigenvector, PageRank, Katz, and Diversity Index. These metrics show a non-linear evolution, characterized by a rapid correlation decline for distortion levels lower than $d = 0.3$. This decline is then followed by a gradual stabilization of the correlations for levels where $d > 0.3$. This observation highlights the insufficiency of relying on a single set of metrics to comprehensively assess generated graph quality. All metrics exhibit non-linear correlations compared to the reference line, thus rendering them unreliable due to their *under-estimation* or *over-estimation* of distortion levels.

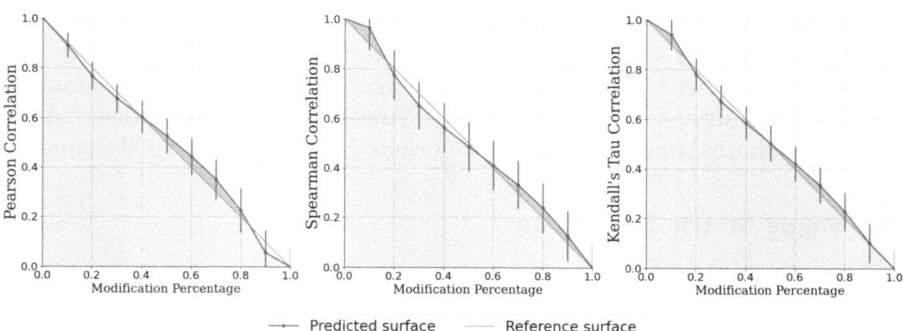

Fig. 3. *GRAM testing results.* The figure illustrates the intersection of $A \times B$ (blue) and reference vector C (orange) as surfaces intersecting the x and y axes, shown for Pearson, Spearman, and Kendall's Tau correlation coefficients. (Color figure online)

3.3   GRAM Evaluation

We generated distorted graphs using a step value of $s = 0.1$ and trained $GRAM$ using the ten previously listed metrics. The training process optimizes two separate surfaces. The loss for the first surface is calculated over the range $[0.1, 0.5]$, while the loss for the second surface is calculated over the range $[0.5, 1]$. Figure 3 shows $GRAM$ testing results based on the previously mentioned correlations (i.e., Pearson correlation coefficient, etc. ...). Visually, the correlational outputs of $GRAM$ seem to closely approximate the target triangular shape created by the reference line and its intersection with x and y axis.

Table 1 shows the weight of each metric as produced by $GRAM$ across various correlation coefficients. The average neighbor degree consistently exhibits high weight values across Pearson, Spearman, and Kendall's Tau correlation coefficients. Yet, almost all the other metrics' weights are close to (0 ± 0.1). This disparity in the correlation's values may be due to the significant overlap in the information captured by the average neighbor degree and closeness centrality with other metrics like betweenness or diversity index. The computation redundancy in some metrics could lead to one of these metrics to be over-represented compared to similar metrics.

Table 1. Optimised α_j values for Pearson ρ_p, Spearman ρ_s, and Kendall's Tau ρ_k correlations

α_j	ρ_p	ρ_s	ρ_k	Single metrics
α_1	0.230	0.250	0.643	Betweenness
α_2	0.473	0.159	0.166	Closeness
α_3	-0.087	0.223	0.106	Weighted Degree
α_4	-0.067	0.056	0.033	Eigenvector
α_5	0.133	0.024	0.129	PageRank
α_6	0.022	0.069	0.006	Katz
α_7	0.001	0.006	-0.059	Hub-Authority
α_8	0.022	-0.159	-0.156	Harmony
α_9	0.309	0.614	0.578	Average Neighbor Degree
α_{10}	0.260	0.501	0.349	Diversity Index

Limitations and Future Directions. This study marks an initial effort to establish a universal metric for assessing the quality of generated brain graphs. One notable limitation is the selection of ten metrics that share similar calculation methods. Another limitation is the evaluation based on a single dataset. Future research should explore a broader range of metrics (e.g., that capture global network properties) and evaluate the model across various graph datasets (e.g., functional vs structural connectomes) for different applications. Additionally, future work should include a comparison between GRAM and other methods for assessing graph quality. Future studies will also incorporate the use of GAN-generated data to further validate our approach.

4 Conclusion

This paper introduced the Graph Regularizable Assessment Metric ($GRAM$) to evaluate the quality of generated brain graphs that could be used as a universal method in reporting the quality of generated graphs in future predictive studies. It combines multiple metrics in a weighted framework, addressing the limitations of existing graph quality metrics. Our proposed method establishes a general assumption for graph quality based on the linearity between distortion and metric values where we used a multi-layer perceptron to optimize metric weights. We test $GRAM$ using a set of simulated structural connectome data on which it demonstrated reasonable reliability in quantifying graph quality. This approach is a significant step towards establishing a universal graph quality index for graph-based predictive studies (e.g., predicting disease progression in Alzheimer's, analyzing brain network development in infants). In future work, we aim to extend $GRAM$ to diverse graph types and datasets.

Code Availability. All codes used for this study are available in: https://github. com/mariemtouihri/GRAM-Metric.

Disclosure of Interests. The authors have no competing interests to declare that are relevant to the content of this article.

References

1. Barrat, A., Barthelemy, M., Pastor-Satorras, R., Vespignani, A.: The architecture of complex weighted networks. Proc. Natl. Acad. Sci. **101**(11), 3747–3752 (2004)
2. Bonacich, P.: Factoring and weighting approaches to status scores and clique identification. J. Math. Sociol. **2**(1), 113–120 (1972)
3. Bonacich, P., Lloyd, P.: Eigenvector-like measures of centrality for asymmetric relations. Soc. Netw. **23**(3), 191–201 (2001)
4. Bullmore, E., Sporns, O.: Complex brain networks: graph theoretical analysis of structural and functional systems. Nat. Rev. Neurosci. **10**(3), 186–198 (2009)
5. Chakrabarti, S., et al.: Mining the web's link structure. Computer **32**(8), 60–67 (1999)
6. Freeman, L.C.: A set of measures of centrality based on betweenness. Sociometry, 35–41 (1977)
7. Freeman, L.C., et al.: Centrality in social networks: conceptual clarification. Soc. Netw. Crit. Concepts Sociol. Londres: Routledge **1**, 238–263 (2002)
8. Goodfellow, I., et al.: Generative adversarial networks. Commun. ACM **63**(11), 139–144 (2020)
9. Hastie, T., Tibshirani, R., Friedman, J.: The Elements of Statistical Learning: Data Mining, Inference, and Prediction, vol. 2. Springer, New York (2009). https://doi. org/10.1007/978-0-387-84858-7

10. Hornik, K., Stinchcombe, M., White, H.: Multilayer feedforward networks are universal approximators. Neural Netw. **2**(5), 359–366 (1989)
11. Kingma, D.P., Ba, J.: Adam: a method for stochastic optimization (2017)
12. Lee, G., Nho, K., Kang, B., Sohn, K.A., Kim, D.: Predicting Alzheimer's disease progression using multi-modal deep learning approach. Sci. Rep. **9**(1), 1952 (2019)
13. Levie, R.D.: Curve fitting with least squares. Crit. Rev. Anal. Chem. **30**(1), 59–74 (2000)
14. Marchiori, M., Latora, V.: Harmony in the small-world. Physica A-Stat. Mech. Appl. **285**, 539–546 (2000)
15. Nebli, A., Kaplan, U.A., Rekik, I.: Deep EvoGraphNet architecture for time-dependent brain graph data synthesis from a single timepoint. In: Rekik, I., Adeli, E., Park, S.H., Valdés Hernández, M.C. (eds.) PRIME 2020. LNCS, vol. 12329, pp. 144–155. Springer, Cham (2020). https://doi.org/10.1007/978-3-030-59354-4_14
16. Nie, D., et al.: Medical image synthesis with deep convolutional adversarial networks. IEEE Trans. Biomed. Eng. **65**(12), 2720–2730 (2018)
17. Page, L., Brin, S., Motwani, R., Winograd, T., et al.: The pagerank citation ranking: bringing order to the web (1999)
18. Rubinov, M., Sporns, O.: Complex network measures of brain connectivity: uses and interpretations. Neuroimage **52**(3), 1059–1069 (2010)
19. Sabidussi, G.: The centrality index of a graph. Psychometrika **31**(4), 581–603 (1966)
20. Shannon, C.E.: A mathematical theory of communication. Bell Syst. Tech. J. **27**(3), 379–423 (1948)
21. Shin, H.-C., et al.: Medical image synthesis for data augmentation and anonymization using generative adversarial networks. In: Gooya, A., Goksel, O., Oguz, I., Burgos, N. (eds.) SASHIMI 2018. LNCS, vol. 11037, pp. 1–11. Springer, Cham (2018). https://doi.org/10.1007/978-3-030-00536-8_1
22. Škoch, A., et al.: Human brain structural connectivity matrices-ready for modelling. Sci. Data **9**(1), 486 (2022)
23. Sporns, O., Zwi, J.D.: The small world of the cerebral cortex. Neuroinformatics **2**, 145–162 (2004)
24. Tzourio-Mazoyer, N., et al.: Automated anatomical labeling of activations in SPM using a macroscopic anatomical parcellation of the MNI MRI single-subject brain. Neuroimage **15**(1), 273–289 (2002)
25. Wang, Z., Bovik, A.: A universal image quality index. IEEE Signal Process. Lett. **9**(3), 81–84 (2002)
26. Yao, D., van der Hoorn, P., Litvak, N.: Average nearest neighbor degrees in scale-free networks. arXiv preprint arXiv:1704.05707 (2017)
27. Özen, G., Nebli, A., Rekik, I.: FLAT-Net: longitudinal brain graph evolution prediction from a few training representative templates. In: Rekik, I., Adeli, E., Park, S.H., Schnabel, J. (eds.) PRIME 2021. LNCS, vol. 12928, pp. 266–278. Springer, Cham (2021). https://doi.org/10.1007/978-3-030-87602-9_25

Unsupervised Analysis of Alzheimer's Disease Signatures Using 3D Deformable Autoencoders

Mehmet Yigit Avci[1], Emily Chan[1,2,4], Veronika Zimmer[1], Daniel Rueckert[1,3,5], Benedikt Wiestler[1,3], Julia A. Schnabel[1,2,4], and Cosmin I. Bercea[1,2(✉)]

[1] Technical University of Munich, Munich, Germany
`cosmin.bercea@tum.de`
[2] Helmholtz Center Munich, Munich, Germany
[3] Klinikum Rechts der Isar, Munich, Germany
[4] King's College London, London, UK
[5] Imperial College London, London, UK

Abstract. With the increasing incidence of neurodegenerative diseases such as Alzheimer's Disease (AD), there is a need for further research that enhances detection and monitoring of the diseases. We present *MORPHADE* (Morphological Autoencoders for Alzheimer's Disease Detection), a novel unsupervised learning approach which uses deformations to allow the analysis of 3D T1-weighted brain images. To the best of our knowledge, this is the first use of deformations with deep unsupervised learning to not only detect, but also localize and assess the severity of structural changes in the brain due to AD. We obtain markedly higher anomaly scores in clinically important areas of the brain in subjects with AD compared to healthy controls, showcasing that our method is able to effectively locate AD-related atrophy. We additionally observe a visual correlation between the severity of atrophy highlighted in our anomaly maps and medial temporal lobe atrophy scores evaluated by a clinical expert. Finally, our method achieves an AUROC of 0.80 in detecting AD, out-performing several supervised and unsupervised baselines. We believe our framework shows promise as a tool towards improved understanding, monitoring and detection of AD. To support further research and application, we have made our code publicly available at github.com/ci-ber/MORPHADE.

Keywords: Unsupervised learning · Registration · Classification

1 Introduction

Due to the increased prevalence of neurodegenerative diseases and their effects on cognitive function, the study of such diseases is a highly active research field. As the leading cause of dementia [1], Alzheimer's disease (AD) is a particular focus

---

M. Y. Avci and E. Chan—These authors contributed equally.

© The Author(s), under exclusive license to Springer Nature Switzerland AG 2025
U. Anazodo et al. (Eds.): MIMA 2024/EMERGE 2024, CCIS 2240, pp. 266–276, 2025.
https://doi.org/10.1007/978-3-031-79103-1_27

of research advancements. However, the complex pathogenesis and progression mechanisms of AD remain only partially understood.

Magnetic resonance imaging (MRI) has shown use in the non-invasive tracking of AD-associated brain changes, such as hippocampal and amygdala atrophy and ventricular dilation [11,16]. Notably, several supervised machine learning methods utilizing MRI have been proposed which yield improvements in AD identification [15,23,24]. However, such methods are restricted by the need for large, annotated data sets. In contrast, unsupervised anomaly detection techniques [2,7,21,25] offer a promising solution by modeling the distribution of healthy brain images to identify and localize anomalies without relying on labeled data.

Nevertheless, unsupervised approaches face challenges in accurately analyzing structural abnormalities, particularly regions of atrophy, which are critical in AD research [5]. Classical techniques using multi atlas-based deformable registration [13] and morphometry methods [3,8] have been proposed to analyze these structural changes. However, such methods allow analysis to be conducted only on a population-level, for instance as deviations from an atlas.

In this work, we propose Morphological Autoencoders for Alzheimer's Disease Detection (*MORPHADE*), a novel unsupervised anomaly detection framework based on deformable autoencoders (AEs) [4] which leverages deformation networks to generate patient-specific anomaly maps from 3D T1-weighted MRI brain scans. These anomaly maps allow not only AD detection, but also crucially reveal the location and degree of atrophy. Our main contributions are as follows:

- We use deformation fields in an unsupervised framework to analyze AD-related changes in the brain. To the best of our knowledge, this is the first use of such an approach using deep learning in the context of AD.
- We extend deformable autoencoders to 3D, utilize adversarial training and propose a dual-deformation strategy to improve reconstruction fidelity and the localization of atrophy.
- We accurately identify AD-affected brain regions, aligning our findings with clinical expectations.
- We assess AD severity by correlating our findings with clinical medial temporal lobe atrophy scores, evaluated by a board-certified clinical expert.
- Through comprehensive validation, we demonstrate superior performance in AD detection compared to unsupervised and even supervised baselines.

2   Background

In unsupervised anomaly detection, reconstruction-based frameworks such as autoencoders (AEs) can be used to learn the distribution of healthy samples and subsequently identify samples that deviate from this norm as anomalous. The encoder E_θ maps an input x to a lower-dimensional latent space and then the decoder D_ϕ learns to reconstruct from this encoded representation. The parameters θ, ϕ of the AE are optimized given healthy input data $\chi = \{x_i, ..., x_n\}$

by minimizing the mean squared error (MSE) between the inputs and their reconstructions:

$$MSE = min_{\theta,\phi} \sum_{i=1}^{N} ||x_i - D_\phi(E_\theta(x_i))||^2. \tag{1}$$

It is then assumed that during inference, the AE will generate a so-called pseudo-healthy reconstruction, in which only in-distribution healthy tissue can be successfully reconstructed and thus any reconstruction errors can be thought of as anomalies. A subject-specific map of anomalies can then be obtained by taking the residual between an input x and its reconstruction $x_{recon} = D_\phi(E_\theta(x))$ as follows:

$$m_{residual} = |x - x_{recon}|. \tag{2}$$

Deformable Autoencoders (AEs) [4] were proposed as a method to alleviate false positives in the anomaly maps due to the limited reconstruction capabilities of traditional AEs. Since the top layers of the AE contain spatial information, deformable AEs use these layers to estimate a dense deformation field Φ that allows local adaptions of the pseudo-healthy reconstruction to the individual anatomy of the subject. The estimation of the deformation field is optimized using local normalized cross correlation (LNCC):

$$\mathcal{L}_{morph} = LNCC(x, x_{morph}) + \beta||\Phi||^2, \tag{3}$$

where β is a weight that is kept relatively high to constrain the deformations to be smooth and local, allowing only small changes to the reconstructions. We therefore refer to this part of the network as the constrained deformer. The improved reconstruction, which we refer to as the morphed reconstruction, x_{morph}, can then be obtained by $x_{morph} = x_{recon} \circ \Phi$.

The authors also propose to use perceptual loss (PL) [12], weighted by the hyperparameter α, in addition to the MSE when optimizing the AE parameters, to promote reconstructions that closely resemble the training distribution:

$$\mathcal{L}_{recon} = \text{MSE}(x, x_{recon}) + \alpha PL(x, x_{recon}). \tag{4}$$

3   Methods and Materials

We propose *MORPHADE*, shown in Fig. 1, which builds upon deformable AEs. Firstly, we employ a 3D convolutional AE to enable the use of 3D images with the framework. Secondly, since PL uses 2D networks pre-trained on ImageNet, we employ an adversarial loss [9] to increase the realness of the reconstructions. We train a discriminator by minimizing this adversarial loss; therefore, the reconstruction loss becomes:

$$\mathcal{L}_{recon} = \text{MSE}(x, x_{recon}) + \gamma \text{Adversarial}(x, x_{recon}), \tag{5}$$

where γ balances the production of realistic reconstructions while maintaining pixel-wise accuracy.

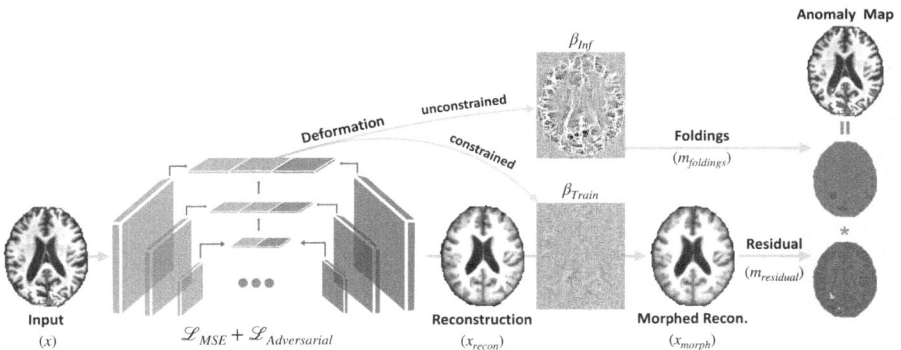

Fig. 1. Our approach, *MORPHADE*, integrates a dual-deformation strategy with a 3D autoencoder and adversarial training. The constrained deformer refines the reconstruction to generate a residual map with reduced false positives, while the unconstrained deformer is used to produce a folding map that highlights anomalies. The residual and folding maps together produce an anomaly map that allows the localization and assessment of the severity of atrophy.

Our major extension to the deformable AEs is the use of a dual-deformation strategy, in which we employ an unconstrained deformer in addition to the constrained deformer, with the aim of improving the localization of atrophic regions. As previously stated, the constrained deformer is trained with a high value of β to improve the generation of the pseudo-healthy reconstructions and thus reduce false positives in the anomaly maps. In contrast, the unconstrained deformer has the goal of reverting the pseudo-healthy reconstruction back to its original anomalous state. The deformer is trained with the same loss as in Eq. 3, but with a low value of β, which allows the creation of unconstrained deformation fields. In such deformation fields, low values of deformation should occur in areas of healthy tissue. Conversely, in regions of atrophy, the deformation field exhibits foldings, or areas in which the mapping of the deformation from the pseudo-healthy reconstruction to the original image is not one-to-one due to the loss of tissue volume. The determinant of the Jacobian of the deformation map, J_{Φ}, can be used to determine local volume changes, with negative values indicating such foldings. Therefore, we highlight the anomalies by using the negative Jacobian values to generate a map of the foldings, $m_{foldings} = \sigma(\max(0, -\det(J_{\Phi})))$, where σ is the Gaussian filtering operation.

We finally multiply these foldings pixel-wise with the residual map from the constrained deformer to generate an anomaly map with reduced false positives and improved atrophy localization:

$$\text{Anomaly Map} = m_{residual} \times m_{foldings}. \tag{6}$$

Implementation. All networks are trained with Adam optimizer. The discriminator is trained with a learning rate of $1.0e^{-4}$, otherwise $5.0e^{-4}$ is used.

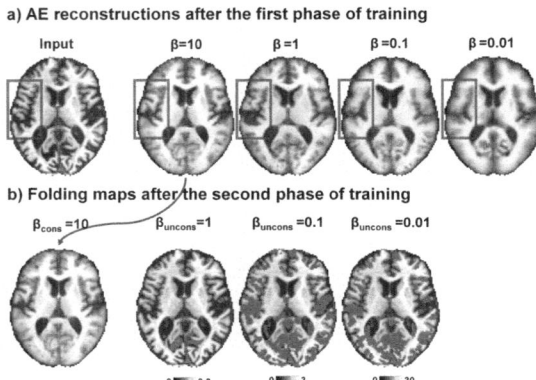

Fig. 2. a) During the first phase of training, a high value of $\beta = 10$ constrains the deformer, promoting the AE to learn to produce less blurry reconstructions. b) During the second phase of training, a lower value of $\beta = 0.01$ allows the deformer to be unconstrained. This unconstrained deformer is then used to generate folding maps (here shown overlaid on the input brain) that enhance the identification of anomalies.

We carry out training in two phases to obtain first the constrained deformer and subsequently the unconstrained deformer. First, the entire framework is trained for 200 epochs with a high value of $\beta = 10$; in this way, we train the constrained deformer, which also encourages the AE to produce sharper reconstructions. We motivate this in Fig. 2a, where we show that using decreasing values of β during training results in blurrier reconstructions. Conversely, a high β value ensures that the AE does not overly rely on the deformations to achieve faithful reconstructions, but is instead forced to learn an accurate representation of the in-distribution data. In the second phase, we train the unconstrained deformer by keeping the weights of the AE frozen while optimizing the deformation parameters with a lower value of $\beta = 0.01$ for 100 epochs. We demonstrate the need for lower β values to produce improved folding maps in Fig. 2b, where it can be seen that using low values accentuates the anomalous regions in the brain. Finally, at inference, we use the constrained deformer to obtain the residual maps and the unconstrained deformer to generate the folding maps.

Dataset and Preprocessing. Data used in the preparation of this article were obtained from the Alzheimer's Disease Neuroimaging Initiative (ADNI) database (adni.loni.usc.edu) [19]. We use skull-stripped T1-weighted MPRAGE images of both male and female patients that are registered to the MNI brain template [17]. Our training set comprises 760 healthy control (HC) samples, with an additional 95 HC samples utilized for validation purposes. For the supervised baseline training, an additional 430 AD samples are used. The test set includes 215 HC samples and 200 samples with AD.

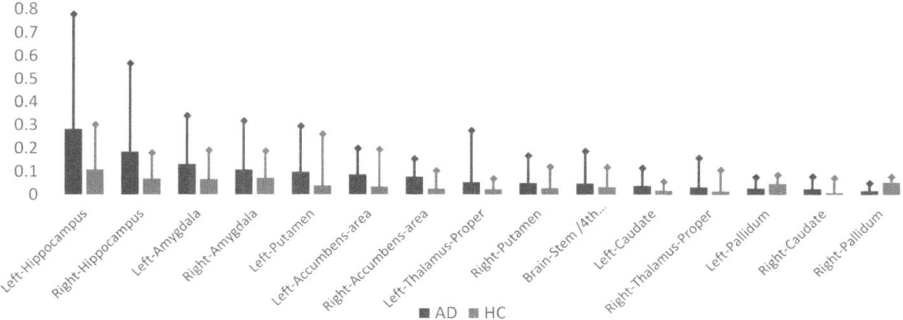

Fig. 3. Anomaly scores for subcortical brain regions for Alzheimer's Disease (AD) and Healthy Control (HC) samples, showcasing markedly higher scores for AD samples in the hippocampus and amygdala, consistent with clinical literature. [6]

4   Experiments and Results

Atrophy Localization. We first validate the effectiveness of our method in identifying atrophy in sub-cortical brain regions affected by AD. To achieve this, we use the FSL FIRST tool [18] to segment these regions and compute mean anomaly scores for each, shown in Fig. 3. Our results indicate that AD patients exhibit notably higher anomaly scores in the hippocampus (left: 0.282 ± 0.495, right: 0.185 ± 0.382) and amygdala (left: 0.132 ± 0.207, right: 0.108 ± 0.208) compared to the hippocampus (left: 0.108 ± 0.193, right: 0.069 ± 0.125) and amygdala (left: 0.066 ± 0.115, right: 0.072 ± 0.111) for the healthy controls. These results are in line with the clinical expectation of these regions being significantly affected by AD pathology [6], indicating that *MORPHADE* is able to identify atrophy in clinically relevant brain regions.

Atrophy Severity. We next evaluate the ability of our method to determine the severity of the localized anomalies by comparing our anomaly maps to medial temporal lobe atrophy (MTA) scores [20] that were assessed by a senior board-certified neuroradiologist. These scores range from 0 to 4 and are assigned based on the degree of structural changes observed in the choroid fissure, the temporal horn of the lateral ventricle, and the hippocampus. Figure 4 shows a visual correlation between the degree of atrophy highlighted in the anomaly map in these key regions and the MTA scores, demonstrating the utility of our method in determining the severity of detected anomalies.

Pathology Detection. In this section, we assess the capability of *MORPHADE* in detecting AD at the patient level. Table 1 shows the Area Under the Receiver Operating Characteristic curve (AUROC) scores obtained when comparing our method to various baselines for identifying subjects with AD compared to healthy control (HC) subjects. Our model achieves an AUROC of 0.80, surpassing even the 3D supervised baselines ResNet [14] and DenseNet [10], with AUROCs of 0.77 and 0.74, respectively.

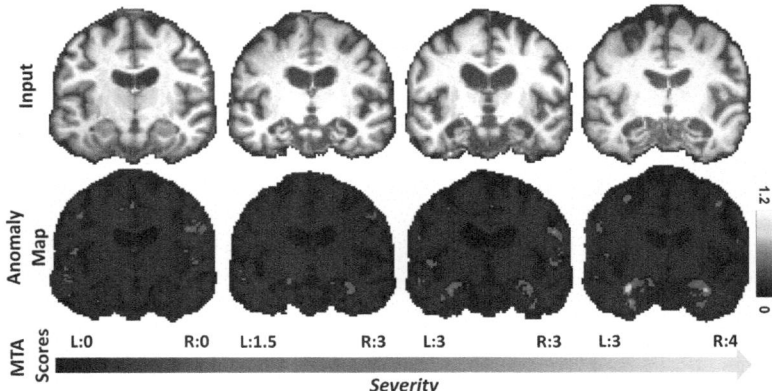

Fig. 4. Anomaly maps for AD patients alongside their corresponding medial tempo-ral lobe atrophy (MTA) scores, demonstrating consistent alignment with AD-related structural changes and clinical MTA assessments.

Furthermore, we obtain improved performance compared to methods pro-posed for unsupervised anomaly detection. These methods are only available in 2D, so were assessed slice-wise with the final anomaly scores obtained by aver-aging over the slices for each patient. f-AnoGAN [21], Ganomaly [2] obtained AUROCs of 0.70 and 0.72, respectively. We also outperform Brainomaly [22] (AUROC 0.78), a method that is not strictly unsupervised since it requires pathological samples during training for improved performance.

We also compare our results to a 3D adversarial AE to illustrate the benefit of utilizing the deformation fields with our method. Figure 5 shows the reconstruc-tions and residual maps obtained for both methods in representative AD and healthy controls (HC) subjects. Our method produces more refined reconstruc-tions compared to the adversarial AE, shown by the improved MAE and SSIM scores. Moreover, the residual maps show fewer false positives for the healthy subject, while accentuating pathological areas for the AD subject. Using these improved residual maps alone for AD detection achieves a superior performance of AUROC 0.77 compared to 0.74 obtained by the adversarial AE.

Finally, we demonstrate the utility of our dual-deformation approach, where AD identification is superior using our method compared to using only the resid-ual maps from the constrained deformer (AUROC 0.77) or the folding maps from the unconstrained deformer (AUROC 0.79). However, it should be noted that using the folding maps alone achieves a high performance; this highlights the effectiveness of using deformations for detecting anomalies, without relying on differences between input and reconstruction.

Table 1. AUROC scores for the classification of AD and Healthy Controls (HC) patients. The best results are shown in **bold**.

Method	AD vs. HC ↑
ResNet (Supervised) [14]	0.77
DenseNet (Supervised) [10]	0.74
Brainomaly [22] (Mixed Supervision)	0.78
f-AnoGAN [21] (Unsupervised)	0.70
Ganomaly [2] (Unsupervised)	0.72
Adversarial AE (Unsupervised)	0.74
MORPHADE (ours) (Unsupervised)	**0.80**
- Only with residual maps ($\beta = 10$)	0.77
- Only with folding maps ($\beta = 0.01$)	**0.79**

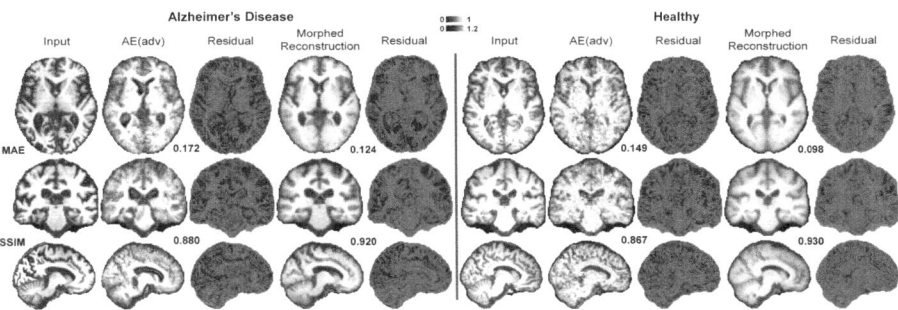

Fig. 5. A comparison of the performance of *MORPHADE* ($\beta = 10$) with adversarial AEs for a subject with AD (left) and a healthy control subject (right). The morphological adjustments facilitated by *MORPHADE* enhance reconstruction fidelity, yielding higher Structure Similarity Index (SSIM) values for our method's morphed reconstructions compared to those of the adversarial AE. The residual maps also demonstrate fewer reconstruction errors for the healthy subject, while highlighting atrophy for the subject with AD.

5   Discussion and Conclusion

In this work, we introduced MORPHADE, a novel framework leveraging 3D deformable AEs for the unsupervised analysis of Alzheimer's Disease using T1-weighted brain MRI. Our approach is unique in employing deformation fields within an unsupervised learning context to analyze, localize, and assess the severity of AD-related atrophy.

Our results demonstrate that MORPHADE can effectively identify and localize atrophy in clinically relevant brain regions, such as the hippocampus and amygdala, which aligns with clinical expectations of AD pathology. Furthermore, the anomaly maps generated by our method show strong visual correspondence with MTA scores, underscoring the potential of our method in clinical assessments. Lastly, MORPHADE achieved an AUROC of 0.80 in detecting

AD, outperforming several supervised and unsupervised baselines. Our method thus identifies AD with high performance without requiring extensive labeled datasets, addressing a significant limitation in current diagnostic approaches.

Future work could explore integrating MORPHADE's deformation metrics with established AD biomarkers, such as tau protein accumulation and amyloid-beta levels, to enhance understanding of disease progression. Additionally, expanding our framework to other neurodegenerative diseases could further validate its versatility and clinical utility.

In conclusion, MORPHADE offers a promising tool for localization and severity assessment of AD-related atrophy, with potential to contribute valuable insights into the progression and diagnosis of neurodegenerative diseases. We believe our approach could enhance the non-invasive monitoring and understanding of AD, paving the way for improved patient outcomes.

Acknowledgments. C.I.B. is funded via the EVUK program ("Next-generation AI for Integrated Diagnostics") of the Free State of Bavaria and partially supported by the Helmholtz Association under the joint research school 'Munich School for Data Science'. This work was also supported by the German Federal Ministry of Health on the basis of a decision by the German Bundestag, under the frame of ERA PerMed.

Disclosure of Interests. The authors have no competing interests to declare that are relevant to the content of this article.

References

1. 2024 Alzheimer's Disease facts and figures. Alzheimer's Dement. **20**(5), 3708–3821 (2024). https://doi.org/10.1002/alz.13809
2. Akcay, S., Atapour-Abarghouei, A., Breckon, T.P.: GANomaly: semi-supervised anomaly detection via adversarial training. In: Jawahar, C.V., Li, H., Mori, G., Schindler, K. (eds.) ACCV 2018. LNCS, vol. 11363, pp. 622–637. Springer, Cham (2019). https://doi.org/10.1007/978-3-030-20893-6_39
3. Ashburner, J., Hutton, C., Frackowiak, R., Johnsrude, I., Price, C., Friston, K.: Identifying global anatomical differences: deformation-based morphometry. Hum. Brain Mapp. **6**(5–6), 348–357 (1998)
4. Bercea, C.I., Rueckert, D., Schnabel, J.A.: What do AEs learn? Challenging common assumptions in unsupervised anomaly detection. In: International Conference on Medical Image Computing and Computer-Assisted Intervention, pp. 304–314. Springer (2023). https://doi.org/10.1007/978-3-031-43904-9_30
5. Bercea, C.I., Wiestler, B., Rueckert, D., Julia, S.: Generalizing unsupervised anomaly detection: towards unbiased pathology screening. In: International Conference on Medical Imaging with Deep Learning (2023)
6. Breijyeh, Z., Karaman, R.: Comprehensive review on Alzheimer's Disease: causes and treatment. Molecules **25**(24), 5789 (2020). https://doi.org/10.3390/molecules25245789
7. Chen, X., Konukoglu, E.: Unsupervised detection of lesions in brain MRI using constrained adversarial auto-encoders. arXiv preprint arXiv:1806.04972 (2018)
8. Chung, M., et al.: A unified statistical approach to deformation-based morphometry. Neuroimage **14**(3), 595–606 (2001). https://doi.org/10.1006/nimg.2001.0862

9. Goodfellow, I.J., et al.: Generative adversarial networks (2014). https://arxiv.org/abs/1406.2661

10. Huang, G., Liu, Z., van der Maaten, L., Weinberger, K.Q.: Densely connected convolutional networks. In: 2017 Proceedings of the IEEE Conference on Computer Vision and Pattern Recognition (2017). https://doi.org/10.1109/CVPR.2017.243

11. Jack, C.R., Jr., et al.: The Alzheimer's Disease Neuroimaging Initiative (ADNI): MRI methods. J. Mag. Reson. Imaging **27**(4), 685–691 (2008). https://doi.org/10.1002/jmri.21049

12. Johnson, J., Alahi, A., Fei-Fei, L.: Perceptual losses for real-time style transfer and super-resolution. In: Leibe, B., Matas, J., Sebe, N., Welling, M. (eds.) ECCV 2016. LNCS, vol. 9906, pp. 694–711. Springer, Cham (2016). https://doi.org/10.1007/978-3-319-46475-6_43

13. Koikkalainen, J., et al.: Multi-template tensor-based morphometry: application to analysis of Alzheimer's Disease. Neuroimage **56**(3), 1134–1144 (2011). https://doi.org/10.1016/j.neuroimage.2011.03.029

14. Korolev, S., Safiullin, A., Belyaev, M., Dodonova, Y.: Residual and plain convolutional neural networks for 3D brain MRI classification (2017). https://doi.org/10.1109/ISBI.2017.7950647

15. Li, H., Shi, X., Zhu, X., Wang, S., Zhang, Z.: FSNet: dual interpretable graph convolutional network for Alzheimer's Disease analysis. IEEE Trans. Emerg. Top. Comput. Intell. **7**(1), 15–25 (2023). https://doi.org/10.1109/TETCI.2022.3183679

16. Liu, M., Zhang, D., Shen, D.: the Alzheimer's Disease Neuroimaging Initiative: hierarchical fusion of features and classifier decisions for Alzheimer's Disease diagnosis. Hum. Brain Mapp. **35**(4), 1305–1319 (2014). https://doi.org/10.1002/hbm.22254

17. Mazziotta, J.C., Toga, A.W., Evans, A., Fox, P., Lancaster, J.: A probabilistic atlas of the human brain: theory and rationale for its development: the International Consortium for Brain Mapping (ICBM). NeuroImage **2**(2, Part A), 89–101 (1995). https://doi.org/10.1006/nimg.1995.1012

18. Patenaude, B., Smith, S., Kennedy, D., Jenkinson, M.: A Bayesian model of shape and appearance for subcortical brain segmentation. Neuroimage **56**(3), 907–922 (2011). https://doi.org/10.1016/j.neuroimage.2011.02.046

19. Petersen, R., et al.: Alzheimer's Disease Neuroimaging Initiative (ADNI). Neurology **74**(3), 201–209 (2010). https://doi.org/10.1212/WNL.0b013e3181cb3e25

20. Scheltens, P., Launer, L., Barkhof, F., Weinstein, H., Van Gool, W.: Visual assessment of medial temporal lobe atrophy on Magnetic Resonance Imaging: interobserver reliability. J. Neurol. **242**, 557–60 (1995). https://doi.org/10.1007/BF00868807

21. Schlegl, T., Seeböck, P., Waldstein, S.M., Langs, G., Schmidt-Erfurth, U.: f-AnoGAN: Fast unsupervised anomaly detection with generative adversarial networks. Med. Image Anal. **54**, 30–44 (2019). https://doi.org/10.1016/j.media.2019.01.010

22. Siddiquee, M., et al.: Brainomaly: unsupervised neurologic disease detection utilizing unannotated T1-weighted brain MR images. In: IEEE/CVF Winter Conference on Applications of Computer Vision, pp. 7558–7567 (2024). https://doi.org/10.1109/wacv57701.2024.00740

23. Wen, J., et al.: Convolutional neural networks for classification of Alzheimer's Disease: overview and reproducible evaluation. Med. Image Anal. **63**, 101694 (2020). https://doi.org/10.1016/j.media.2020.101694

24. Zhang, Y., Teng, Q., Liu, Y., Liu, Y., He, X.: Diagnosis of Alzheimer's Disease based on regional attention with sMRI gray matter slices. J. Neurosci. Methods **365**, 109376 (2022). https://doi.org/10.1016/j.jneumeth.2021.109376

25. Zimmerer, D., Isensee, F., Petersen, J., Kohl, S., Maier-Hein, K.: Unsupervised anomaly localization using variational auto-encoders. In: Shen, D., et al. (eds.) MICCAI 2019. LNCS, vol. 11767, pp. 289–297. Springer, Cham (2019). https://doi.org/10.1007/978-3-030-32251-9_32

Deep Feature Fusion Framework for Alzheimer's Disease Staging Using Neuroimaging Modalities

Aya Gamal[1(✉)] [ID], Mustafa Elattar[1,2] [ID], and Sahar Selim[1,2] [ID]

[1] Medical Imaging and Image Processing Research Group,
Center for Informatics Science, Nile University, Giza, Egypt
`Ay.Gamal@nu.edu.eg`
[2] School of Information Technology and Computer Science, Nile University, Giza,
Egypt

Abstract. Alzheimer's Disease (AD) is a significant neurodegenerative disorder. Detecting AD early is essential for effective management and improving the quality of life for both patients and their families. Recent advancements in medical imaging technology have introduced neuroimaging-based methods for early AD diagnosis. However, the challenges in early AD detection suggest that using a single modality dataset in deep learning (DL) studies, particularly neuroimaging, might not yield precise predictions of AD progression compared to integrating data from multiple imaging modalities. Utilizing information from multi-modal data fusion can enhance the detection of subtle changes and biomarkers, leading to more reliable and accurate diagnosis. In our study, we develop an automated multimodal system that integrates MRI and PET images at an intermediate fusion level, facilitating the early diagnosis of Alzheimer's disease. This fusion approach eliminates the need for extensive preprocessing steps that are typically required in image fusion methods. Our proposed methodology outperforms previous studies in differentiating between individuals with Alzheimer's disease and cognitively normal (CN) individuals, achieving an AUC score of 97.67% and an accuracy (ACC) of 95.24%.

Keywords: Alzheimer's Disease · Neuroimaging Features · 3D Image Classification

1 Introduction

Alzheimer's Disease (AD) is a severe neurodegenerative disease. Early identification of AD is crucial for effective management and enhancement of the quality of life of both patients and their families. Unfortunately, most existing diagnostic techniques rely on subjective assessments of behavioral and cognitive symptoms, leading to potential unreliability and misdiagnosis. In recent years, advances in medical imaging technology have led to the emergence of neuroimaging-based

U. Anazodo et al. (Eds.): MImA 2024/EMERGE 2024, CCIS 2240, pp. 277–288, 2025.
https://doi.org/10.1007/978-3-031-79103-1_28

methods for the early diagnosis of AD. However, these methods often rely on analyzing a single modality, which may fail to capture the full complexity of the disease. Multimodal data fusion has been proposed as a promising approach to address this limitation by combining the information from different modalities.

In a clinical setting, AD is typically diagnosed by systematically examining various aspects of a patient's multiple modalities [23]. These aspects are commonly derived from the diverse information sources of patients, including neuroimaging data, gene sequence data, profile data, and clinical mental state scale data. In contrast to the classification of AD based solely on single-modal neuroimaging, enhanced performance can be attained through the utilization of multimodal classification, involving the integration of diverse information sources. Investigating the synergies among various multimodal neuroimaging modalities significantly contributes to the identification of pathological processes in neurological disorders. This technique has been applied in image classification [25,29] and image registration [10]. The motivation for engaging in multimodal fusion stems from two primary advantages: first, the potential for more robust predictions through the observation of the same phenomenon across multiple modalities [5]; and second, the extraction of complementary information from diverse modalities to enhance the precision of classification results [4].

The multimodal framework comprises essential components that are primarily structured at three key levels. The initial level, known as the integration level, involves defining various modalities of data intended for fusion. Thus, at this stage, a determination is made regarding what should be fused. The subsequent level is the fusion methodology, encompassing the approach employed to combine the identified data guided by the chosen fusion strategy. In the literature, fusion strategies are classified into three groups: early fusion, also known as feature-level fusion, is the process of merging multimodal data by concatenating its features into a vector, which is subsequently inputted into a machine learning model. Intermediate fusion that integrates feature representations gained from one modality at the intermediate layers of a neural network with feature representations learned from other modalities is referred to as joint fusion. Late fusion involves decision-level fusion, in which a distinct model is trained for each modality and the predictions of all models are subsequently integrated to create a final decision. The final level in the framework is the knowledge level, where the final results of the diagnosis are obtained.

Numerous studies have focused on the fusion of diverse modalities for AD diagnosis. Notably, Dwivedi et al. [9], Dong et al. [8], Xu [26], Ning [19], Hao [12], and Zhang [27] have introduced methodologies primarily focused on neuroimaging features, particularly utilizing MRI and PET modalities. Similarly, Khvostikov et al. [14], Kang [13], and Aderghal et al. [2] have directed their attention to the fusion of neuroimaging data, specifically from sMRI and DTI scans. In addition to these, Zuo et al. [30] integrated sMRI, PET, and fMRI data, while Choi and Jin [7] used flurodeoxyglucose and florbetapir PET. Peng et al. [21] combined sMRI, PET, and genetic data, and Lee et al. [17] integrated cognitive performance, demographic information, CSF, and MRI imaging data.

In reviewing these studies, it is evident that the most frequently fused modalities are MRI and PET, indicating their prominent role in multimodal investigations within this research domain. Various approaches to fusing MRI and PET volumes have been explored in the literature. For example, Song et al. [22] introduced a framework for AD diagnosis using a feature-fusion approach to extract semantic information from 3D MRI and PET volumes. They also proposed an image fusion method that outperformed their initial approach by reducing the number of model parameters using a single composite image, although it required multistep preprocessing. Castellano et al. [6] developed a dual branch, multimodal diagnostic model for Alzheimer's Disease using 3D MRI and amyloid PET scans in parallel, demonstrating that these modalities provide complementary insights that enhance predictive accuracy. However, limitations include the selection of only 50 slices from the axial plane, potentially missing comprehensive spatial information, and a loss of temporal resolution in PET scans due to frame averaging.

Kong et al. [16] similarly employed an image fusion technique, while Venugopalan et al. [24] utilized 3D CNNs to extract features from MRI and PET data, demonstrating improved performance over traditional fusion methods despite being limited by dataset sizes. In contrast to CNN-based methods, transformers leverage the self-attention mechanism to capture long-range dependencies within multimodal features. Zhang et al. [28] introduced a model comprising three components: dual 3D CNN encoders for MRI and PET modalities, a Multimodal Transformer Encoder, and a classification head. They employed a feature fusion strategy that utilized a transformer-based cross-attention mechanism to fuse features more effectively. Furthermore, Miao et al. [18] utilized a transformer-based approach for multimodal multiscale fusion networks for the diagnosis of AD by fusing neuroimaging data.

In this study, we developed an automated multimodal system that integrates MRI and PET images at an intermediate fusion level, facilitating the early diagnosis of AD. This fusion method requires minimal preprocessing compared to traditional image-fusion techniques. Our approach surpasses previous studies in distinguishing between AD and CN individuals.

2   Methodology

To preserve modality-specific information for both modalities, we introduced a heuristic intermediate feature fusion framework that can capture complementary information from PET and MRI modalities independently. The components of the proposed feature fusion framework are shown in Fig. 1., which illustrates the intermediate feature fusion approach that preserves modality-specific information while enabling the effective integration of MRI and PET features for classification. The first level of our framework identifies the modalities to be integrated. We then applied preprocessing steps from a streamlined pipeline to MRI and PET scans separately, preparing the data for feature extraction. In the feature extraction step, a 3D pre-trained deep learning model was used as a feature extractor for each modality. Subsequently, we employed an intermediate feature fusion approach by leveraging the feature maps extracted from the previous step and processing them for input into the classification network. Finally, a small and simple 3D CNN network was used as a classification network for the effective classification of the AD stages.

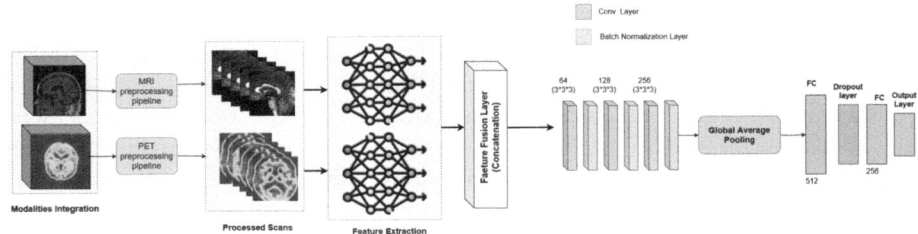

Fig. 1. The proposed intermediate feature fusion framework, highlights the stages of modality integration, feature extraction, and classification.

2.1 Dataset

Our study concentrated on the ADNI dataset (adni.loni.usc.edu), which is widely used to address this problem. We specifically implemented our experiments using structural MRI and 18-fluorodeoxyglucose (FDG)-PET modalities, which are commonly employed noninvasive methods for capturing the characteristics of brain tissue. We collected 3D data from subjects who underwent scans using both of these modalities.

First, we filtered the participants to include only those with data available on both PET and MRI during the same visit and scanning period. In total, 253 subjects participated in this experiment, contributing to a dataset of 822 scans. We aimed to mitigate the risk of data leakage by considering only the first or baseline scans for each subject. This decision ensured an equal number of scans and participants. However, to address the challenge of a small dataset size owing to the constraint of scans from the same time period, each subject could have three to four visits in different years or at least a 6-month gap within the same year. To maintain our principle of avoiding data leakage, we carefully split the data into 80%, 10%, and 10% for the training, validation, and test sets, respectively. Ensuring that a subject's scans do not appear in different sets but all in one place.

The summary of subjects and scans in the dataset is provided in Table 1. gives an overview of the participants included in the study, highlighting the distribution across AD, MCI, and CN groups, which is crucial for understanding the dataset's composition.

Table 1. Summary of participant statistics in the ADNI dataset (MRI and PET).

Class	Subjects	Scans
AD	43	117
MCI	111	433
CN	99	272

Total Number of scans = 822

2.2 Data Preprocessing

Both MRI and FDG-PET images in ADNI underwent various processing stages. Each modality was pre-processed separately. Specifically, the MRI images underwent a series of processing steps, including skull stripping, intensity normalization, uniform resampling to achieve isotropic resolution, 3D cropping to extract only the brain from the black background, resizing all scans to 128×128×128, and the application of histogram equalization to enhance the contrast. The preprocessing pipeline proposed in [11] was applied here, except for the histogram equalization step, which was applied to the scans to enhance the quality and discriminatory power of the images.

Regarding PET scans preprocessing, the initial FDG-PET scans underwent the following processing steps to ensure consistency in PET data across various systems. First, we converted all the PET files into Neuroimaging Informatics Technology Initiative (NIFTI) format files, as all the processed PET image data were in the DICOM format. The dicom2nifti Python package was used to apply the conversion. Similar to MRI, PET images include extensive background regions characterized by zero pixel values beyond the brain tissue. We effectively reduced these non-essential background regions to decrease the volume of the input data via 3D cropping, as in the MRI pipeline. Furthermore, we resized the volume to 128×128×128. Finally, histogram equalization was applied to the PET scans. Figure 2 outlines the PET image processing pipeline, which includes essential steps for standardizing PET images and enhancing their quality for the feature extraction process.

In this study, handling multimodal data posed a significant challenge owing to the limitations of a small sample size. To address this concern, an essential component of the proposed methodology is the augmentation step. We employed various 3D transformations on both MRI and PET data, including 3D random rotation and flipping.

2.3 Networks Architecture

The effectiveness of the 3D CNN models and transfer learning approach in diagnosing AD led us to choose them as the optimal starting point for designing our multimodal framework. The proposed multimodal model architecture is illustrated in Fig. 1. The 3D DenseNet201-based transfer learning model was used as

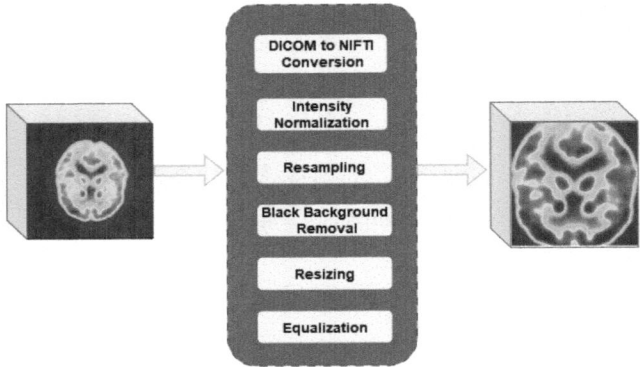

Fig. 2. PET Image Processing Method.

a deep feature extractor for the processed images of both modalities. The feature encoder had four dense blocks, and transition layers were employed between them. After extracting the feature maps from each modality, a concatenation layer was added to the model to fuse the intermediate features and prepare them for the single final network. The last layers in our network form a small and simple 3D CNN. The layer details of the final classification network are shown in Fig. 1.

3   Experiments Setup and Results

In this part of our study, our experiments are organized as follows. Initially, two 3D DenseNet201 models were utilized as feature extractors for both the MRI and PET images. Subsequently, we loaded the weights of each modality independently and incorporated them into the fusion phase. Finally, a straightforward 3D CNN network is applied to the fused features for AD diagnosis. We conducted three classification tasks: AD vs CN, AD vs MCI, and MCI vs CN.

One of the challenges highlighted in the literature is the variability in hyperparameter choices across different studies and experiments. To address this issue, we employ an open-source hyperparameter optimization framework called Optuna [3]. Optuna is compatible with any machine learning or deep learning framework, offering versatility.

We used Optuna's automated hyperparameter optimization algorithms to efficiently explore and evaluate different configurations, facilitating the discovery of optimal model settings. Specifically, we specified the search space for the hyperparameter batch size, learning rate, and input shape by defining their types as categorical, float, and categorical with possible ranges of [5, 8, 16, 32], [0.000001, 0.0001], and [64, 96, 128], respectively. Following the optimization process, Optuna returns the best set of hyperparameters that leads to optimal performance according to the defined objective function.

The optimal configuration obtained was [batch size of 16, an input size of 128, and a learning rate of 3.4885205571560794e-05], achieved in trial 9. All experiments were performed using the TensorFlow deep learning framework [1] in Python. In the training phase of the feature extractors, we employed 200 epochs with a batch size of 16, aligned with the recommendations derived from the Optuna optimization process. Nevertheless, when training the final CNN network, we encountered hardware constraints, compelling us to decrease the batch size to five. Adam optimizer [15] is employed with a learning rate that is recommended from Optuna algorithm and a ReduceLROnPlateau strategy is utilized here to reduce the learning rate when the validation loss has stopped improving. According to the final classification network, all the setups were the same.

In our study, we addressed the challenge of imbalanced classes by implementing oversampling and class-weighting strategies during training of our fusion model. To overcome class imbalance, we applied oversampling to the minority classes using the resample function. This step ensures that each class is adequately represented in the training dataset, thereby preventing the model from being biased towards the majority class. It randomly selects samples with replacements from the provided class indices, effectively duplicating some samples to achieve desired oversampling. This is performed until the size of the minority class matches the size of the majority class, making the class distribution more balanced in the training data. To further mitigate the impact of class imbalance, we computed class weights using the "compute class weight" function from sci-kit-learn [20]. It is used to assign different weights to different classes during model training. In the experiments, the BinaryFocalCrossentropy loss function was employed, which combines the characteristics of both binary cross-entropy (BCE) and focal loss. Binary Cross-Entropy (BCE) serves as the standard loss function for binary classification problems. On the other hand, focal loss is introduced to address class imbalance in binary classification tasks. This is achieved by modulating the cross-entropy loss and downweighting the contribution of well-classified examples where the predicted probability is high. This adjustment allowed the model to prioritize hard-to-classify examples.

We evaluated the performance of our fusion model across three binary classification tasks to recognize the three AD stages, as shown in table 2. The Table presents the performance metrics of our proposed feature fusion method, demonstrating the model's ability to differentiate between AD, MCI, and CN with high accuracy and AUC scores. Notably, the best results were achieved for the AD vs. CN task, with an AUC score of 97.67%, based on a single inference on a hold-out test set. It is important to highlight that the lower performance observed in the AD vs. MCI and MCI vs. CN tasks is expected, as the MCI stage is notoriously challenging to classify due to its overlapping features with both normal aging and early Alzheimer's, which poses a difficulty even for advanced models. Table 3 compares the performance of uni-modal approaches against our proposed feature fusion method, highlighting the substantial improvements in classification accuracy and AUC achieved through multimodal integration.

Table 2. Proposed Feature Fusion Results for 3 classification tasks.

Task	ACC	BA	AUC	F1-score
AD vs CN	95.24	95.71	97.67	93.33
MCI vs CN	80	77.81	86.08	72.86
AD vs MCI	75.0	74.23	80.54	73.4

Table 3. The uni-modal and Proposed Feature Fusion Results for AD vs CN task.

Metric	MRI	PET	Fusion method
ACC	68.75	87.5	95.24
AUC	72.5	94.29	97.67
BA	68.67	87.71	95.71
F1-score	67.15	87.67	93.33

4   Discussion and Conclusion

In this study, our goal was to utilize the power of neuroimaging multimodal data instead of unimodal data. Table 4 benchmarks the performance of our proposed method against other recent studies, showcasing the superiority of our approach in terms of accuracy across multiple classification tasks. Our proposed method outperforms other studies with superior performance for the AD vs. CN task with ACC = 95.24%.

Regarding data subjects used in this paper, instead of utilizing only the baseline scans, we obtained three to four scans for each subject in different years to overcome the small data sizes as much as possible. In addition, we took into consideration the problem of data leakage that could happen through having multiple scans for each subject, so, we split the data very carefully to ensure that the scans of each subject will not appear in different sets. By utilizing the oversampling and class weighting in our experiments, we got a superior performance of the model and we can see this effect clearly through investigating the metrics especially the f1-scores for each class in different tasks.

Integrating 3D augmentation functions significantly improved our experiments and the model's performance. The process followed these steps: first, we applied oversampling to all training data within each classification task, balancing the minor class with the major class after splitting the data into training, validation, and test sets at the subject level. Next, preprocessing was conducted on the oversampled data. Finally, the transformation was applied exclusively to the training data, with an augmentation factor of 5.

As shown in Table 4 there are many studies, some of which follow different approaches for fusing MRI and PET volumes. Song et al. [22] introduced a framework for AD diagnosis with the feature fusion approach (intermediate fusion) to obtain semantic information from the 3D volumes of MRI and PET. In addition, they proposed another fusion method by applying an image fusion process that outperformed the first method. The image fusion approach helps reduce the number of model parameters, as a single composite image is used in the network. However, multistep pre-processing is required to achieve this fusion. Kong et al. [16] presented also an image fusion method which is considered as the early fusion approach where PET and MRI images are fused and fed into the network. In addition, Venugopalan et al. [24] suggested that the deep models for integration also showed improved performance over traditional feature-level and decision-level integrations. However, their study suffers from a limited dataset size. Zhang et al. [28] proposed an adversarial learning approach to enhance the cross-attention mechanism for more effective feature fusion. They focused on subjects with complete T1w and FDG-PET images, utilizing feature fusion with their baseline scans. The effectiveness of their approach was then evaluated on two tasks: AD vs. CN and pMCI vs. sMCI.

Our methodology preserves modality-specific features using intermediate feature fusion, avoiding the extensive preprocessing typically required by image fusion techniques, such as volume registration and alignment. This streamlined approach led to significant improvements, particularly in the AD vs. CN task. By using a simple CNN for feature extraction and classification, our method outperformed others, highlighting the effectiveness of maintaining modality-specific information while minimizing preprocessing complexity

The features extracted from MRI and FDG-PET scans are clinically relevant, capturing critical structural and metabolic information associated with Alzheimer's disease. MRI highlights structural atrophy, while FDG-PET reveals metabolic changes, offering complementary insights for AD diagnosis. However, our current work lacks interpretability methods, essential for translating these features into actionable clinical insights, and faces challenges in multiclass classification due to the complexity of handling 3D data from MRI and PET scans, which requires substantial computational resources. Future work should integrate interpretability techniques to better understand the model's decision-making process and enhance its clinical utility. Additionally, incorporating more rigorous statistical analysis, including reporting central tendencies for cross-validation runs and conducting statistical tests to confirm the significance of observed improvements, is necessary.

In conclusion, our study presented a comprehensive framework for aiding in the early diagnosis of Alzheimer's disease through a focus on neuroimaging features. We specifically chose to focus on fusing neuroimaging features by combining 3D MRI scans with 18-FDG PET scans through the introduction of an intermediate feature fusion method. Our proposed fusion framework demonstrated superior results compared to related studies in the literature.

Table 4. Comparative performance of our classifiers and competitors.

Study	AD vs CN ACC (%)	MCI vs CN ACC (%)	AD vs MCI ACC (%)
Kong et al. (2022) [16]	93.21	86.52	85.63
song et al. (2021) [22] (feature fusion)	93.22	82.37	81.00
song et al. (2021) [22] (image fusion)	94.11	88.48	84.83
Venugopalan et al. (2021) [24]	86	-	-
Castellano et al. (2024) [6]	95.00	-	-
Zhang et al. (2023) [28]	92.9	-	-
Proposed feature fusion method	**95.24**	80	75

Disclosure of Interests. The authors have no competing interests to declare that are relevant to the content of this article.

References

1. Abadi, M., et al.: Tensorflow: large-scale machine learning on heterogeneous systems (2015). https://www.tensorflow.org/. software available from tensorflow.org
2. Aderghal, K., Khvostikov, A., Krylov, A., Benois-Pineau, J., Karim, A., Catheline, G.: Classification of Alzheimer disease on imaging modalities with deep cnns using cross-modal transfer learning, pp. 345–350 (2018). https://doi.org/10.1109/CBMS.2018.00067
3. Akiba, T., Sano, S., Yanase, T., Ohta, T., Koyama, M.: Optuna: a next-generation hyperparameter optimization framework. In: Proceedings of the 25th ACM SIGKDD International Conference on Knowledge Discovery and Data Mining (2019)
4. Bailey, D., et al.: Combined pet/mri: multi-modality multi-parametric imaging is here. In: Summary Report of the 4th International Workshop on PET/MR Imaging; February 23–27, 2015, tüBingen, Germany. Mol. Imag. Biol. **17**, 595–608 (2015)
5. Baltrušaitis, T., Ahuja, C., Morency, L.P.: Multimodal machine learning: a survey and taxonomy. IEEE Trans. Pattern Anal. Mach. Intell. **41**(2), 423–443 (2018)
6. Castellano, G., Esposito, A., Lella, E., Montanaro, G., Vessio, G.: Automated detection of Alzheimer's disease: a multi-modal approach with 3D MRI and amyloid pet. Sci. Rep. **14**(1), 5210 (2024)
7. Choi, H., Jin, K.H.: Predicting cognitive decline with deep learning of brain metabolism and amyloid imaging. CoRR **abs/1704.06033** (2017). http://arxiv.org/abs/1704.06033
8. Dong, A., Zhang, G., Liu, J., Wei, Z.: Latent feature representation learning for Alzheimer's disease classification. Comput. Biol. Med. **150**, 106116 (2022). https://doi.org/10.1016/j.compbiomed.2022.106116
9. Dwivedi, S., Goel, T., Tanveer, M., Murugan, R., Sharma, R.: Multi-modal fusion based deep learning network for effective diagnosis of Alzheimers disease. IEEE MultiMedia (2022)
10. Fan, J., Cao, X., Wang, Q., Yap, P.T., Shen, D.: Adversarial learning for mono-or multi-modal registration. Med. Image Anal. **58**, 101545 (2019)

11. Gamal, A., Elattar, M., Selim, S.: Automatic early diagnosis of Alzheimer's disease using 3D deep ensemble approach. IEEE Access **10**, 115974–115987 (2022). https://doi.org/10.1109/access.2022.3218621
12. Hao, X., Bao, Y., Guo, Y., Ming, Y., Zhang, D., Risacher, S., Saykin, A., Yao, X., Shen, L.: Multi-modal neuroimaging feature selection with consistent metric constraint for diagnosis of alzheimer's disease. Medical Image Analysis **60**, 101625 (12 2019https://doi.org/10.1016/j.media.2019.101625
13. Kang, L., Jiang, J., Jianjun, H., Zhang, T.: Identifying early mild cognitive impairment by multi-modality MRI-based deep learning. Front. Aging Neurosci. **12**, 206 (2020). https://doi.org/10.3389/fnagi.2020.00206
14. Khvostikov, A.V., Aderghal, K., Benois-Pineau, J., Krylov, A.S., Catheline, G.: 3D CNN-based classification using SMRI and MD-DTI images for Alzheimer disease studies. ArXiv **abs/1801.05968** (2018). https://api.semanticscholar.org/CorpusID:12103502
15. Kingma, D., Ba, J.: Adam: a method for stochastic optimization. In: International Conference on Learning Representations (ICLR). San Diega, CA, USA (2015)
16. Kong, Z., et al.: Multi-modal data Alzheimer's disease detection based on 3D convolution. Biomed. Sig. Process. Control **75**, 103565 (2022)
17. Lee, G., Kang, B., Nho, K., Sohn, K.A., Kim, D.: Mildint: deep learning-based multimodal longitudinal data integration framework. Front. Genet. **10** (2019). https://doi.org/10.3389/fgene.2019.00617
18. Miao, S., et al.: MMTFN: multi-modal multi-scale transformer fusion network for Alzheimer's disease diagnosis. Int. J. Imaging Syst. Technol. **34**(1), e22970 (2024)
19. Ning, Z., Xiao, Q., Feng, Q., Chen, W., Zhang, Y.: Relation-induced multi-modal shared representation learning for Alzheimer's disease diagnosis. IEEE Trans. Med. Imaging (2021).https://doi.org/10.1109/TMI.2021.3063150
20. Pedregosa, F., et al.: Scikit-learn: machine learning in python. J. Mach. Learn. Res. **12**(Oct), 2825–2830 (2011)
21. Peng, J., Zhu, X., An, Y., Shen, D.: Structured sparsity regularized multiple kernel learning for Alzheimer's disease diagnosis. Pattern Recogn. **88**, 370–382 (2019). https://doi.org/10.1016/j.patcog.2018.11.027
22. Song, J., Zheng, J., Li, P., Lu, X., Zhu, G., Shen, P.: An effective multimodal image fusion method using MRI and pet for Alzheimer's disease diagnosis. Front. Digital Health **3**, 637386 (2021)
23. Tu, Y., Lin, S., Qiao, J., Zhuang, Y., Zhang, P.: Alzheimer's disease diagnosis via multimodal feature fusion. Comput. Biol. Med. **148**, 105901 (2022)
24. Venugopalan, J., Tong, L., Hassanzadeh, H.R., Wang, M.D.: Multimodal deep learning models for early detection of Alzheimer's disease stage. Sci. Rep. **11**(1), 1–13 (2021)
25. Wang, J., Wang, Q., Wang, S., Shen, D.: Sparse multi-view task-centralized learning for ASD diagnosis. In: Wang, Q., Shi, Y., Suk, H.-I., Suzuki, K. (eds.) MLMI 2017. LNCS, vol. 10541, pp. 159–167. Springer, Cham (2017). https://doi.org/10.1007/978-3-319-67389-9_19
26. Xu, H., Zhong, S., Zhang, Y.: Multi-level fusion network for mild cognitive impairment identification using multi-modal neuroimages. Phys. Med. Biol. **68** (2023). https://doi.org/10.1088/1361-6560/accac8
27. Zhang, Y., Wang, S., Xia, K., Jiang, Y., Qian, P.: Alzheimer's disease multiclass diagnosis via multimodal neuroimaging embedding feature selection and fusion. Inf. Fusion **66**, 170–183 (2021). https://doi.org/10.1016/j.inffus.2020.09.002

28. Zhang, Y., Sun, K., Liu, Y., Shen, D.: Transformer-based multimodal fusion for early diagnosis of Alzheimer's disease using structural MRI and pet. In: 2023 IEEE 20th International Symposium on Biomedical Imaging (ISBI), pp. 1–5. IEEE (2023)
29. Zhou, T., Thung, K.H., Zhu, X., Shen, D.: Effective feature learning and fusion of multimodality data using stage-wise deep neural network for dementia diagnosis. Hum. Brain Mapp. **40**(3), 1001–1016 (2019)
30. Zuo, Q., Lei, B., Shen, Y., Liu, Y., Feng, Z., Wang, S.Q.: Multimodal representations learning and adversarial hypergraph fusion for early Alzheimer's disease prediction, pp. 479–490 (2021). https://doi.org/10.1007/978-3-030-88010-1_40

Explainable Few-Shot Learning for Multiple Sclerosis Detection in Low-Data Regime

Montassar Ben Dhifallah[1,7(✉)] , Dalel Kanzari[1,2] , Salma Naija[3,4] ,
Sana Ben Amor[3,4] , Ahmed Zrig[5,7] , Mezri Maatouk[5,7] ,
Mabrouk Abdelaali[5,7] , Jamel Saad[5,7] , Asma Achour[5,7] ,
Sofiane Gaied Chortane[5,7], Maher Hadhri[6,7] , Ahmed Dahmoul[5,7],
Azza Ben Ali[5,7], Sahar Selim[8] , and Ahmed Nebli[9]

[1] Higher Institute of Applied Sciences and Technology, University of Sousse, Sousse,
Tunisia
montassar.bendhifallah@issatso.u-sousse.tn
[2] Operational Research, Decision and Process Control Laboratory (LARODEC),
41 Liberty Street, Bardo, 2000 Tunis, Tunisia
[3] Department of Neurology, Sahloul Hospital, Sousse, Tunisia
[4] Faculty of Medicine of Sousse, University of Sousse, Sousse, Tunisia
[5] Department of Radiology A, Fattouma Bourguiba Hospital, Monastir, Tunisia
[6] Department of Neurosurgery, Fattouma Bourguiba Hospital, Monastir, Tunisia
[7] Research Unity Interventional Radiology LR18SP08, University of Monastir,
Monastir, Tunisia
[8] School of Information Technology and Computer Science, Nile University, Giza,
Egypt
[9] Independent Researcher, Jülich, Germany

Abstract. Diagnosing multiple sclerosis (MS) accurately is highly challenging due to symptom overlap with other demyelinating diseases. Here, we present DemyeliNeXt, an explainable few-shot learning framework designed to classify MS and other demyelinating diseases from MRI scans. This framework employs a prototypical network with a 3D DenseNet-121 backbone and uses Deep SHAP for feature importance visualization. We train our DemyeliNeXt on a dataset from African populations and we test it for different datasets including MICCAI MSSEG2 public dataset. Our findings demonstrate robust performance across diverse datasets highlighting the model's potential to enhance diagnosis accuracy and generalizability in various clinical settings.

Keywords: Few-Shot Learning · Explainable AI · Multiple Sclerosis · 3D MRI · Deep Learning

A. Nebli—Independent Researcher.

U. Anazodo et al. (Eds.): MImA 2024/EMERGE 2024, CCIS 2240, pp. 289–298, 2025.
https://doi.org/10.1007/978-3-031-79103-1_29

1   Introduction

Multiple sclerosis (MS) is a complex neurological condition that is often misdiagnosed due to its symptom overlap with other conditions such as vasculitis and vascular leukoencephalopathy. Studies indicate that over half of the patients were misdiagnosed for a period exceeding three years [2,12]. Moreover, 70% of these patients had been administered disease-modifying therapies (DMTs), and 31% suffered unnecessary morbidity due to the incorrect diagnosis and treatment [2,12]. This diagnostic challenge results in a prolonged time to achieve a definitive diagnosis, often exceeding several months. Hence, accurate and timely diagnosis is crucial for effective management and treatment planning in MS patients. Advanced imaging techniques and biomarker analyses are increasingly important in differentiating MS from other similar presenting conditions, thereby reducing diagnostic errors and improving patient outcomes. Machine learning provides a robust approach for the analysis of medical images and the diagnosis of MS.

In this context, several studies have employed machine learning models for MS classification. For instance, Wang et al. [15] employed a multi-layer convolutional neural network (CNN) with data augmentation techniques to classify MS. However, the model's explainability remains unexplored. To address this issue, Zhang et al. [17] proposed a classification model for MS subtypes based on VGG19 [10] with global average pooling and utilized Grad-CAM++ [1] for model explanation. While effective in performance and interpretability, this approach did not account for the diversity of MS data, particularly by not comparing it with other similar demyelinating diseases such as vasculitis. To rectify this concern, Huang et al. [3] leveraged a Transformer-based model with a Multiple Instance Learning (MIL) strategy to discriminate between MS and various demyelinating diseases. The authors used Grad-CAM to visualize feature extraction through activation heatmaps. Nevertheless, their study did not incorporate data from low-income countries, such as datasets from the African population. This omission underscores a critical gap, as regional genetic and environmental factors influence disease onset and progression [16]. These factors impact the timeliness and accuracy of MS diagnosis, thereby potentially threatening the patient's life.

Additionally, the collection of MS and other demyelinating diseases data is challenging due to the variability in disease presentation, limited patient availability, and the high cost of medical imaging. Therefore, the application of few-shot learning is essential to leverage limited data effectively. Furthermore, a key finding in MS identification is the presence of white matter lesions in the brain, detectable via Fluid Attenuated Inversion Recovery (FLAIR) sequence of MRI.

This study focuses on distinguishing MS from other demyelinating diseases. We introduce DemyeliNeXt, an explainable few-shot learning framework for the classification of MS and other demyelinating diseases. Our approach employs a prototypical network with a 3D DenseNet-121 backbone, which integrates spatial information from FLAIR MR (Magnetic Resonance) images to classify them as MS vs other demyelinating diseases (NON-MS). Additionally, the framework provides model interpretability through the Deep SHAP model for visualizing the most important features leading to the classification of the input MRI. The primary contributions of our work are as follows:

1. Application of Few-Shot Learning: We apply few-shot learning for the detection of multiple sclerosis (MS).
2. Emphasis on Explainability: Our method integrates explainability mechanisms to enhance interpretability, making it more suitable for clinical settings.
3. Utilization of African 3D MRI Data: We trained our model using 3D MRI data from African populations, which are often underrepresented in medical datasets. By benchmarking our model against MICCAI MS public dataset, we demonstrated its robust performance, thereby validating its generalizability across diverse populations.

2 Proposed Method

In this section, we explain the key building blocks of our proposed DemyeliNeXt architecture for explainable MS identification from other demyelinating diseases.

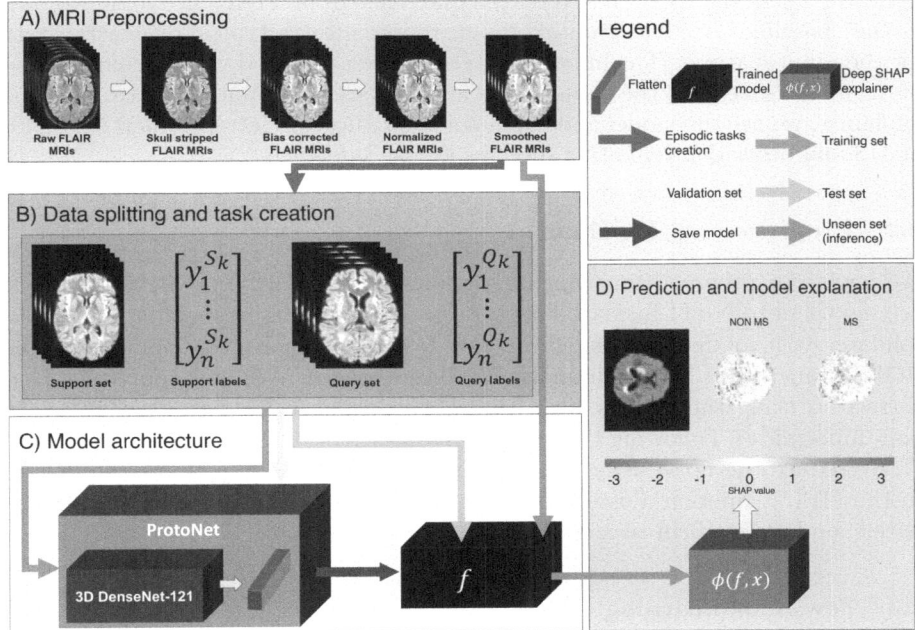

Fig. 1. *DemyeliNeXt Pipeline.* (A) Preprocessing MRI scans: includes skull stripping, bias correction normalization, and FLAIR MRI smoothing. (B) Data splitting into support and query sets. (C) Training a prototypical network with 3D DenseNet-121 backbone. (D) Model testing on unseen MRIs with explanations provided using Deep SHAP.

2.1 Architecture Overview

In this study, we introduce DemyeliNeXt, a four-stage pipeline designed for the classification of multiple sclerosis (MS) and other demyelinating diseases from MRI scans, while also providing model interpretability. Figure 1 illustrates the first stage (Sect. 2.2), which involves a preprocessing pipeline for FLAIR MRI scans. Here, raw FLAIR images are normalized, while noise and artifacts are reduced. In the second stage, the MRI scans are divided into training, validation, and testing sets. Each set contains a support set (S) with labeled examples to update model parameters and a query set (Q) with unlabeled examples for performance evaluation.

The third stage (Sect. 2.3) involves training a 3D DenseNet-based (DenseNet-121) [4] prototypical network to classify the preprocessed MRIs. The training process utilizes N^{tr} training tasks, each comprising N_{shots} support examples for model weight updates and N_{query} query examples for performance assessment. In the final stage, we employ Deep SHAP [7] to approximate the model for interpretability. Deep SHAP, inspired by DeepLIFT [9], assigns importance scores to each input feature by propagating neuron contributions backward through the network. These scores are based on the difference from a reference input, known as the "baseline" or "background" input, representing a typical or neutral state for the input features. The importance scores are computed via the combination of the model's weights, the actual input and the baseline input. After training the explainer, we use the model and explainer to predict and interpret new examples of MS and other demyelinating diseases during inference.

2.2 Preprocessing Pipeline

We begin our preprocessing pipeline by anonymizing DICOM MRI scans, converting them to NIfTI format. This process removes patient metadata and consolidates each volume into a single file. Next, we perform skull stripping using the ROBEX algorithm [5] to eliminate non-brain tissues. We then apply bias field correction using the N4ITK algorithm [14] to remove low-frequency intensity non-uniformities. Following this, we normalize MRI intensities to a range of 0 to 1. We reduce the noise using a Gaussian filter. Finally, we reorient the images to the "IPL" (Inferior, Posterior, Left) orientation, resample them to isotropic voxels, and resize them to a standard format.

2.3 Few Shot Learning

Prototypical Network. Prototypical Networks (ProtoNet) [11] seek to find a metric space in which samples from the same class are close to one another. This approach makes the model particularly useful in settings with limited labeled data. Based on the prototype concept [11], the model depicts each class using the mean of its embedded support set S. Prototypical Networks then determine query samples Q based on their proximity to these prototypes. To generate the image embeddings, we use a 3D DenseNet-121 [4] as a backbone. We employed

Euclidian distance for our ProtoNet to calculate the distance between the support samples and query samples. We create dataset episodes using a sampler that follows uniform distribution to load data from the dataset for each label.

Loss Function. We use binary cross-entropy loss:

$$\mathcal{L} = -\left[y \log(p) + (1-y) \log(1-p) \right] \tag{1}$$

where y and p are the MS label and the predicted probability of MS from the model respectively. We use ADAM [6] as an optimizer with step LR scheduler to decay the learning rate.

2.4 Explainability with Deep SHAP

Deep SHAP [7] approximates explanations for deep neural network models using SHAP (SHapley Additive exPlanations) values to quantify feature importance. This method integrates concepts from a deep learning explanation technique called DeepLIFT [9] that uses Shapley values [8]. We apply Deep SHAP to interpret our trained 3D DenseNet-based ProtoNet model using preprocessed MRI scans from the testing dataset. This approach creates a simplified explanation model, assessing the importance of each voxel in our testing MRIs, visualized through feature importance plots.

2.5 Model Inference and Explanation

After training and evaluating the model, we perform inference on unseen examples where we pass them to the explainer to check the used feature importance of the model on the classification of the new examples.

3 Results and Discussion

In this section, we provide a quantitative evaluation of our model on three distinct datasets and we display the findings of the used Deep SHAP.

3.1 Employed Datasets

In this work, we utilized three labeled datasets, summarized in Table 1. We trained, validated, and tested using a set that comprises 182 FLAIR MRI scans from 121 patients with multiple sclerosis (MS) and other demyelinating diseases (NON-MS). The dataset was split randomly and patient-wise into three different sets as follows: 70% for training, 15% for validation and 15% for testing. This dataset is sourced from the radiology department at CHU Fattouma Bourguiba Monastir (FBM), Tunisia. It includes 3D and axial scans: 91 scans from 52 MS patients and 91 scans from 69 patients with other demyelinating diseases such as vasculitis and vascular leukopathy.

We tested our model on a set containing 91 FLAIR MRI scans from 36 MS patients, obtained from the MRI center of CHU Sahloul Sousse (SS), Tunisia. Additionally, we used 80 3D FLAIR MRI scans from 40 patients in the MIC-CAI 2021 MS Segmentation Challenge (MSSEG-2) as a benchmark dataset. We randomly sampled data from each set to create episodes consisting of a support set and a query set. Prior to training, gamma correction was applied to all scans using $\gamma = 2.5$. No further data augmentation was performed.

Table 1. Datasets statistics

Source	Number of patients	Number of scans	Age	Gender
CHU FBM, Tunisia	MS: 52 NON-MS: 69	MS: 91 NON-MS: 91	21–63	MS: 22M/30F NON-MS: 19M/50F
CHU SS, Tunisia	36 MS	91	NA	4M/32F
MSSEG-2	40 MS	80	NA	NA

3.2 Experimental Settings

Parameter Settings. For model training, we used an ADAM optimizer [6] with a learning rate of 0.001. We applied learning rate decay for every single step by 0.1 using a step scheduler. We employed dropout with 20% rate. As for Deep SHAP explainer training, we adopted 90 background examples. We trained our model and our explainer on the Nvidia RTX 3090 GPU.

Hyperparameter Settings. We conducted three distinct training experiments using 2-way ($K = 2$) classification. Validation was performed with 100 episodes ($N^{val} = 100$) every 500 training episodes. Testing was also conducted with 100 episodes. Each training lasted for 1000 episodes. Detailed hyperparameters for each experiment are listed below:

- **Experiment A**: Trained with 5 examples in both support and query sets ($N_{shots} = 5$, $N_{query} = 5$).
- **Experiment B**: Trained with 3 examples in both support and query sets ($N_{shots} = 3$, $N_{query} = 3$).
- **Experiment C**: Trained with 1 example in both support and query sets ($N_{shots} = 1$, $N_{query} = 1$).
- **Test 1**: We used the saved model from Experiment A to test on 91 scans from CHU SS MS dataset and on 13 scans from CHU FBM NON-MS test set.
- **Test 2**: We used the saved model from Experiment A to test on 80 scans from MSSEG-2 and on 13 scans CHU FBM NON-MS test set.

3.3 DemyeliNeXt Evaluation

Table 2 shows the classification accuracy, precision, recall, specificity, and F1 scores for the different experiments detailed in Sect. 3.2. Across all experiments, Test 2, which involved training on an African dataset and testing on a combination of African and European datasets, achieved the highest classification accuracy. This result may indicate that our model has the ability to generalize well across different populations despite the differences in socio-economic conditions between the subjects in each of the datasets.

In contrast, Experiment C and B, which utilized one, and three shots and queries, respectively, demonstrated the lowest performance. This indicates that reducing the number of shots below a certain threshold adversely affects model accuracy. These findings suggest that while reducing shots can decrease computational demands, maintaining an adequate number of shots is critical for reliable performance (see experiment A). In particular, one could generally recommend using the model trained in Experiment A as a guide for practitioners in balancing computational efficiency with diagnosis accuracy for MS.

Table 2. Experiments results

Experiments/Tests	Accuracy	MS specific Accuracy	NON-MS specific Accuracy	Precision	Recall	Specificity	F1 score
A: 5 shots 5 queries (Dataset: CHU FBM)	78.8%	–	–	0.75	0.87	0.71	0.8
B: 3 shots 3 queries (Dataset: CHU FBM)	63.83%	–	–	0.62	0.72	0.56	0.67
C: 1 shot 1 query (Dataset: CHU FBM)	65.0%	–	–	0.64	0.68	0.62	0.66
Test 1 5 shots 5 queries (Dataset: CHU SS MS + CHU FBM NON-MS)	75.5%	68.6%	82.4%	0.8	0.69	0.82	0.74
Test 2: 5 shots 5 queries (Dataset: MSSEG-2 + CHU FBM NON MS)	**87.8%**	**85%**	**90.6%**	**0.9**	**0.85**	**0.91**	**0.87**

Figure 2 illustrates the explanation of our model backbone on unseen MS and NON-MS examples with lesion annotation. The plot highlights the features utilized by our trained ProtoNet model for classification that are explained by the Deep SHAP method. We evaluated the explainer results using the key diagnostic features outlined in the McDonald criteria [13], which include lesion size, number of lesions, lesion location, lesion contrast, and lesion shape. The Deep SHAP explainer seems to identify some of the key features for classification, specially the lesions in MS example (Fig. 2 B). However, one should note that there is a risk that the included features in the explanation could be deemed irrelevant to clinicians.

Limitations and Future Studies. Despite the promising results, DemyeliNeXt has a few limitations that warrant further investigation. For instance, our approach currently utilizes only FLAIR MRI scans; incorporating other imaging modalities like T1-weighted and T2-weighted MRIs could potentially enhance diagnostic accuracy. While Deep SHAP provides some level of explainability, the clinical relevance of the highlighted features remains uncertain, indicating a need for further refinement. In future studies, we aim to benchmark against state-of-the-art methods. We will also focus on expanding the dataset to include

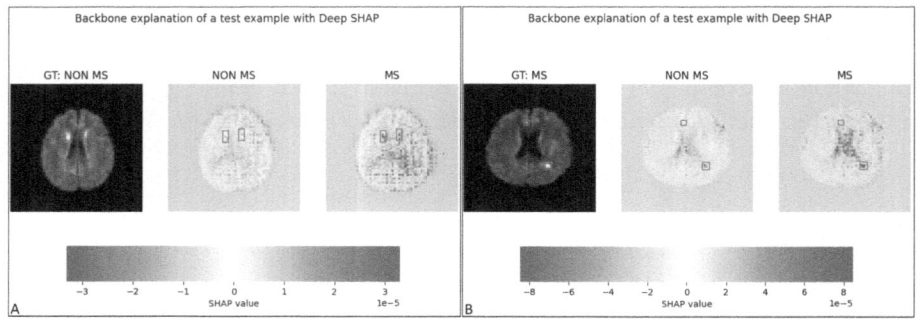

Fig. 2. *Deep SHAP Explanation of MS and NON-MS Examples.* **A:** Explanation of NON-MS example. **B:** Explanation of MS example. For each of the subfigures (A and B), the left panel displays an annotated MRI section of a patient with a NON-MS demyelinating disease (A) and a patient with MS disease (B). The center panel highlights the features identified by our model for classifying the case as NON-MS using Deep SHAP. The right panel shows the features identified for classification as MS using Deep SHAP. Lesions' locations are highlighted with orange rectangles across all panels. For the two right hand side panels, blue indicates the features excluded by the model, while red shows the important features for each class

diverse minority populations, integrating multimodal imaging techniques, as well as developing more clinically relevant explainability methods with their evaluation.

4 Conclusion

In this study, we introduced DemyeliNeXt, an explainable few-shot learning framework designed for the classification of multiple sclerosis (MS) and other demyelinating diseases in an African population. By incorporating the Deep SHAP model, we provided visual explanations for the model's decisions, enhancing its interpretability. Our findings, derived from MRI data of underrepresented African populations, demonstrate that this approach can generalize effectively to non-African datasets. Although the classification accuracy decreases with fewer shots, the method remains computationally efficient and can aid practitioners in improving diagnostic accuracy. In future work, we aim to extend our framework by including more minority populations and integrating additional neuroimaging modalities, thereby enhancing the generalizability and robustness of our model.

Acknowledgments. The data collection for this study was conducted under the agreement of the head of radiology department of CHU Fattouma Bourguiba Monastir, Tunisia, the head of neurology department of CHU Sahloul Sousse, Tunisia and the director of MRI center of Sahloul, Sousse, Tunisia.

Code Availability. We provide the code repository of our method on GitHub at this link: https://github.com/Montassar-bdh/DemyeliNeXt.

Disclosure of Interests. The authors have no competing interests to declare that are relevant to the content of this article.

References

1. Chattopadhay, A., Sarkar, A., Howlader, P., Balasubramanian, V.N.: Grad-CAM++: generalized gradient-based visual explanations for deep convolutional networks, pp. 839–847 (2018). https://doi.org/10.1109/WACV.2018.00097
2. Gaitán, M.I., Correale, J.: Multiple sclerosis misdiagnosis: a persistent problem to solve. Front. Neurol. **10**, 466 (2019)
3. Huang, C., et al.: Transformer-based deep-learning algorithm for discriminating demyelinating diseases of the central nervous system with neuroimaging. Front. Immunol. **13**, 897959 (2022). https://doi.org/10.3389/fimmu.2022.897959
4. Huang, G., Liu, Z., van der Maaten, L., Weinberger, K.Q.: Densely connected convolutional networks (2017)
5. Iglesias, J.E., Liu, C.Y., Thompson, P.M., Tu, Z.: Robust brain extraction across datasets and comparison with publicly available methods **30**(9), 1617–163. https://doi.org/10.1109/TMI.2011.2138152
6. Kingma, D.P., Ba, J.: Adam: a method for stochastic optimization. arXiv preprint arXiv:1412.6980 (2014)
7. Lundberg, S.M., Lee, S.I.: A unified approach to interpreting model predictions, pp. 4765–4774 (2017). http://papers.nips.cc/paper/7062-a-unified-approach-to-interpreting-model-predictions.pdf
8. Shapley, L.S.: A value for n-person games, pp. 307–317 (1953). https://doi.org/10.1515/9781400881970-018
9. Shrikumar, A., Greenside, P., Kundaje, A.: Learning important features through propagating activation differences, pp. 3145–3153 (2017)
10. Simonyan, K., Zisserman, A.: Very deep convolutional networks for large-scale image recognition. CoRR abs/1409.1556 (2014). https://api.semanticscholar.org/CorpusID:14124313
11. Snell, J., Swersky, K., Zemel, R.: Prototypical networks for few-shot learning **30** (2017). https://proceedings.neurips.cc/paper_files/paper/2017/file/cb8da6767461f2812ae4290eac7cbc42-Paper.pdf
12. Solomon, A.J., et al.: The contemporary spectrum of multiple sclerosis misdiagnosis: a multicenter study. Neurology **87**(13), 1393–1399 (2016)

13. Thompson, A.J., et al.: Diagnosis of multiple sclerosis: 2017 revisions of the McDonald criteria **17**(2), 162–173. https://doi.org/10.1016/S1474-4422(17)30470-2. https://linkinghub.elsevier.com/retrieve/pii/S1474442217304702

14. Tustison, N.J., et al.: N4ITK: improved N3 bias correction. IEEE Trans. Med. Imaging **29**(6), 1310–1320 (2010). https://doi.org/10.1109/tmi.2010.2046908

15. Wang, S.H., et al.: Multiple sclerosis identification by 14-layer convolutional neural network with batch normalization, dropout, and stochastic pooling. Front. Neurosci. **12**, 818 (2018). https://doi.org/10.3389/fnins.2018.00818

16. Waubant, E., et al.: Environmental and genetic risk factors for MS: an integrated review. Ann. Clin. Transl. Neurol. **6**(9), 1905–1922 (2019)

17. Zhang, Y., Hong, D., McClement, D., Oladosu, O., Pridham, G., Slaney, G.: Grad-CAM helps interpret the deep learning models trained to classify multiple sclerosis types using clinical brain magnetic resonance imaging. J. Neurosci. Methods **353**, 109098 (2021). https://doi.org/10.1016/j.jneumeth.2021.109098

Correction to: Self-consistent Deep Approximation of Retinal Traits for Robust and Highly Efficient Vascular Phenotyping of Retinal Colour Fundus Images

Lucas Gago ⓘ, Beatriz Remeseiro ⓘ, Laura Igual ⓘ, Amos Storkey ⓘ, Miguel O. Bernabeu ⓘ, and Justin Engelmann ⓘ

Correction to:
Chapter 22 in: U. Anazodo et al. (Eds.):
Medical Information Computing, **CCIS 2240,**
https://doi.org/10.1007/978-3-031-79103-1_22

The original version of the chapter was inadvertently published without grant information in the acknowledgement section. It has been corrected.

The updated version of this chapter can be found at
https://doi.org/10.1007/978-3-031-79103-1_22

Author Index

U. Anazodo et al. (Eds.): MImA 2024/EMERGE 2024, CCIS 2240, pp. 299–301, 2025.
https://doi.org/10.1007/978-3-031-79103-1

The manufacturer's authorised representative in the EU is Springer
Nature Customer Service Centre GmbH, Europaplatz 3, 69115 Heidelberg,
Germany. If you have any concerns regarding our products, please
contact ProductSafety@springernature.com

Printed and bound by CPI Group (UK) Ltd, Croydon, CR0 4YY

29/04/2026

02099551-0001